Communications
in Computer and Information Science   67

Joaquim Filipe   Ana Fred
Bernadette Sharp (Eds.)

Agents and Artificial Intelligence

International Conference, ICAART 2009
Porto, Portugal, January 19-21, 2009
Revised Selected Papers

 Springer

Volume Editors

Joaquim Filipe
Departament of Systems and Informatics
Polytechnic Institute of Setúbal
Estefanilha, Setúbal, Portugal
E-mail: j.filipe@est.ips.pt

Ana Fred
IST - Technical University of Lisbon
Lisbon, Portugal
E-mail: afred@lx.it.pt

Bernadette Sharp
Staffordshire University
School of Computing
Beaconside, Stafford, UK
E-mail: b.sharp@staffs.ac.uk

Library of Congress Control Number: 2010921985

CR Subject Classification (1998): I.2, I.2.3, I.2.6, H.5.2, H.2.8, I.2.9

ISSN   1865-0929
ISBN-10   3-642-11818-6 Springer Berlin Heidelberg New York
ISBN-13   978-3-642-11818-0 Springer Berlin Heidelberg New York

Typesetting: Camera-ready by author, data conversion by Scientific Publishing Services, Chennai, India
Printed on acid-free paper   06/3180   5 4 3 2 1 0

Preface

The present book includes a set of selected papers from the First International Conference on Agents and Artificial Intelligence (ICAART 2009), held in Porto, Portugal, during January 19–21, 2009. The conference was organized in two simultaneous tracks: "Artificial Intelligence and Agents." The book is based on the same structure.

ICAART 2009 received 161 paper submissions, from more than 37 different countries in all continents. After a blind review process, only 26 where accepted as full papers, of which 21 were selected for inclusion in this book, based on the classifications provided by the Program Committee. The selected papers reflect the interdisciplinary nature of the conference. The diversity of topics is an important feature of this conference, enabling an overall perception of several important scientific and technological trends. These high-quality standards will be maintained and reinforced at ICAART 2010, to be held in Valencia, Spain, and in future editions of this conference.

Furthermore, ICAART 2009 included five plenary keynote lectures given by Juan Carlos Augusto (University of Ulster), Marco Dorigo (IRIDIA, Free University of Brussels), Timo Honkela (Helsinki University of Technology), Edward H. Shortliffe (Arizona State University) and Paulo Urbano (University of Lisbon). We would like to express our appreciation to all of them and in particular to those who took the time to contribute with a paper to this book.

On behalf of the conference Organizing Committee, we would like to thank all participants. First of all to the authors, whose quality work is the essence of the conference, and to the members of the Program Committee, who helped us with their expertise and diligence in reviewing the papers. As we all know, producing a conference requires the effort of many individuals. We wish to thank also all the members of our Organizing Committee, whose work and commitment were invaluable.

December 2009

Joaquim Filipe
Ana Fred
Bernadette Sharp

Organization

Conference Co-chairs

Joaquim Filipe   Polytechnic Institute of Setúbal / INSTICC, Portugal
Ana Fred   Technical University of Lisbon/IT, Portugal

Program Chair

Bernadette Sharp   Staffordshire University, UK

Organizing Committee

Sérgio Brissos   INSTICC, Portugal
Marina Carvalho   INSTICC, Portugal
Helder Coelhas   INSTICC, Portugal
Vera Coelho   INSTICC, Portugal
Andreia Costa   INSTICC, Portugal
Bruno Encarnação   INSTICC, Portugal
Bárbara Lima   INSTICC, Portugal
Raquel Martins   INSTICC, Portugal
Elton Mendes   INSTICC, Portugal
Carla Mota   INSTICC, Portugal
Vitor Pedrosa   INSTICC, Portugal
Vera Rosário   INSTICC, Portugal
José Varela   INSTICC, Portugal

Program Committee

Scott A. DeLoach, USA
Rodrigo Agerri, UK
Thomas Ågotnes, Norway
H. Levent Akin, Turkey
Klaus-Dieter Althoff, Germany
Andreas Andreou, Cyprus
Antonio Bahamonde, Spain
Matteo Baldoni, Italy
Steve Barker, UK
Mike Barley, New Zealand
Punam Bedi, India
Nabil Belacel, Canada
Orlando Belo, Portugal

Carlos Bento, Portugal
Carole Bernon, France
Aurélie Beynier, France
Guido Boella, Italy
Sander Bohte, The Netherlands
Tibor Bosse, The Netherlands
Djamel Bouchaffra, USA
Jean-Louis Boulanger, France
Danielle Boulanger, France
Paolo Bresciani, Belgium
Jan Broersen, The Netherlands
Silvia Calegari, Italy
Valérie Camps, France

Luigi Ceccaroni, Spain
Vincent Chevrier, France
Alan Colman, Australia
Dan Corkill, USA
Paulo Cortez, Portugal
Stephen Cranefield, New Zealand
John Debenham, Australia
Giovanna Di Marzo Serugendo, UK
Nicola Di Mauro, Italy
Gaël Dias, Portugal
Juergen Dix, Germany
Julie Dugdale, France
Amal El Fallah Seghrouchni, France
Yagil Engel, USA
Yaniv Eytani, USA
Stefano Ferilli, Italy
Antonio Fernández-Caballero, Spain
Edilson Ferneda, Brazil
Roberto Flores, USA
Naoki Fukuta, Japan
Wai-Keung Fung, Canada
Alessandro Garcia, UK
Peter Geczy, Japan
Joseph Giampapa, USA
Paolo Giorgini, Italy
Andrea Giovannucci, Spain
Marie-Pierre Gleizes, France
Jorge Gomez Sanz, Spain
Dominic Greenwood, Switzerland
Sven Groppe, Germany
Rafael H. Bordini, UK
Wladyslaw Homenda, Poland
Wei-Chiang Hong, Taiwan
Mark Hoogendoorn, The Netherlands
Marc-Philippe Huget, France
Fuyuki Ishikawa, Japan
François Jacquenet, France
Wojtek Jamroga, Germany
Gion K. Svedberg, Sweden
Nikos Karacapilidis, Greece
Graham Kendall, UK
Yves Kodratoff, France
Joanna Kolodziej, Poland
Boris Kovalerchuk, USA

Stan Kurkovsky, USA
Candace L. Sidner, USA
Nuno Lau, Portugal
Stephane Loiseau, France
Bernd Ludwig, Germany
Xudong Luo, UK
Laurent Magnin, France
Pierre Maret, France
Goreti Marreiros, Portugal
Tokuro Matsuo, Japan
Rui Mendes, Portugal
Tomoharu Nakashima, Japan
Paulo Novais, Portugal
Luis Nunes, Portugal
Andrea Omicini, Italy
Krzysztof Patan, Poland
Wojciech Penczek, Poland
Loris Penserini, Italy
Gabriel Pereira Lopes, Portugal
Dana Petcu, Romania
Eric Platon, Japan
Agostino Poggi, Italy
Carlos Ramos, Portugal
Marek Reformat, Canada
Alessandro Ricci, Italy
Fátima Rodrigues, Portugal
Juha Röning, Finland
Rosaldo Rossetti, Portugal
Elie Sanchez, France
Viviane Silva, Brazil
Iryna Skrypnyk, Finland
Adam Slowik, Poland
Chun-Yi Su, Canada
Ramayah T., Malaysia
Ryszard Tadeusiewicz, Poland
Luís Torgo, Portugal
Paola Turci, Italy
Leon van der Torre, Luxembourg
Katja Verbeeck, Belgium
Jose Vidal, USA
Danny Weyns, Belgium
Cees Witteveen, The Netherlands
Bozena Wozna, Poland

Auxiliary Reviewers

Tristan Behrens, Germany
Marenglen Biba, Italy
Nils Bulling, Germany
Yann-Michael De hauwere, Belgium
Susana Irene Díaz Rodríguez, Spain
F. Jorge F. Duarte, Portugal
Roberto Ghizzioli, Switzerland

Michael Koester, Germany
Sara Montagna, Italy
Peter Novak, Germany
Mariusz Nowostawski, New Zealand
Martin Purvis, New Zealand
Bjoern Raupach, Belgium
Peter Vrancx, Belgium

Invited Speakers

Edward H. Shortliffe   Arizona State University, USA
Marco Dorigo   IRIDIA Université Libre de Bruxelles, Belgium
Timo Honkela   Helsinki University of Technology, Finland
Juan Carlos Augusto   University of Ulster, UK
Paulo Urbano   Faculdade de Ciências da Universidade de Lisboa,
                          Portugal

Table of Contents

Part II: Agents

Invited Speakers

Past, Present and Future of Ambient Intelligence and Smart Environments

Juan Carlos Augusto

University of Ulster, Jordanstown, U.K.
jc.augusto@ulster.ac.uk

Abstract. We are gradually making a transition to a new era where computers become truly intertwined with our daily lives. Up to not so long ago, we were able to know clearly where computers were and in which way they affected our lives. This has been gradually blurred and now computing devices of various types are all around us, embedded in different objects we interact with and in that way they influence our lives. There are indications that this trend is irreversible and that computing and society will now interact with each other in far richer ways than before, to the point that computing will become transparent to humans and still intrinsically involved in our daily living. This paper provides a brief overview of the evolution of these fields, describes some of the current developments, and points at some of the immediate challenges that researchers in these area face.

1 The Past

For centuries humans have witnessed scientific and technological leaps that changed the lives of their generation, and those to come, forever. We are no exception. In fact so much of those advances are occurring now, in a more or less unperceived way. Slowly and silently technology is becoming interwoven in our lives in the form of a variety of devices which are starting to be used by people of all ages.

The technological advances in miniaturization of microprocessors (Figure 1) have made possible a significant development for Ambient Intelligence. Computing power is now embedded in many different objects like home appliances (e.g., programmable washing machines, microwave ovens, robotic hovering machines, and robotic mowers), they travel with us outside the home (e.g., mobile phones and PDAs), and they help guide us to and from our home (e.g., car suspension and fuel consumption and GPS navigation). Computers that require reduced power and that are tailored to accomplish very specific tasks are gradually spreading through almost every level of our society.

This widespread availability of resources forms the technological layer for the realization of Ambient Intelligence. Having the necessary technology is not enough for an area of science to flourish. Previous experiences of people with computers over recent decades have created an interesting context where people's expectations of these systems are growing and their fear of using them has decreased. Concomitantly with this difference in the way society perceives technology there is also a change in the way services are handled. An important example of this is the decentralization of health care and development of health and social care assistive technologies. For various reasons governments and health professionals are departing away from the hospital-centric

J. Filipe, A. Fred, and B. Sharp (Eds.): ICAART 2009, CCIS 67, pp. 3–15, 2010.

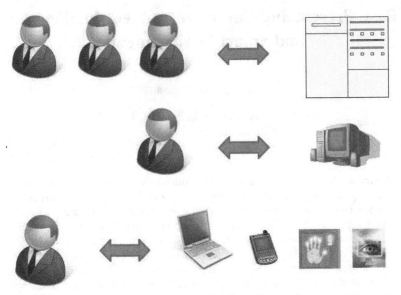

Fig. 1. Historical evolution and shift on availability of computing power per person

health care system enabling this shift of care from the secondary care environment to primary care. Subsequently, there is an effort to move away from the traditional concept of patients being admitted into hospitals rather to enable a more flexible system whereby people are cared for closer to home, within their communities. Smart homes are one such example of a technological development which facilitates this trend of bringing the health and social care system to the patient as opposed to bringing the patient into the health system.

For example, the South Eastern Health & Social Care Trust of Northern Ireland has established a Connected Health project intended to support to approximately 1,000 by 2011 in the community. In the arena of telecare the programme includes the fitting of sensors into private and social housing. The programme is intended to help maintain elderly and vulnerable people stay as safe as possible in their home. In addition, it is intended to increase their level of autonomy, independence and health status particularly if they have a long-term chronic condition, which can be significantly detrimental to their lifestyle.

Developments, competencies and drivers are converging at the same time in history and all of the necessary components are in place; that is the need to distribute technology around us, the will to change the way our society interacts with technology, the available technological knowledge and all the elements to satisfy the demand are converging.

2   The Present

The areas of Ambient Intelligence and Smart Environments are being defined naturally as work in the area progresses and on demand by everyday life problems and real applications. Although Ambient Intelligence [1,2,3] and Smart Environments [4] are strongly related, we can distinguish them by going back to the old "mind/brain" metaphor used in AI. The first one is more concerned with the specific techniques to make an environment behave intelligently whilst the second one is more related with the intelligent interconnection of resources and their collective behavior. Both overlap hugely and share many common objectives and it is difficult to tell apart one from the other. These areas gradually evolved in the last decades, motivated by seminal work conducted at Xerox Labs under the paradigm of the *disappearing computer* [5]:

"The most profound technologies are those that disappear. They weave themselves into the fabric of everyday life until they are indistinguishable from it."

This concept indicated the possibility for some technology to become fully integrated in everyday life and at the same time emphasized the degree of transparency of a technology as a measure of the success for that technology on being fully adopted by society.

This developments evolved into the areas of ubiquitous and pervasive computing which in turn were complemented by other pre-existing areas of computing (for example artificial intelligence, HCI, etc.) to create areas with consistent goals which emphasize different aspects of the resulting systems.

2.1   Definition

Ambient Intelligence can be defined as follows:

"A digital environment that supports people in their daily lives by assisting them in a sensible way." [6]

In order to be sensible, a system has to be intelligent. That is how a trained assistant, e.g. a nurse, typically behaves. It will help when needed but will restrain to intervene unless is necessary. Being sensible demands recognizing the user, learning or knowing her/his preferences and the capability to exhibit empathy with the user's mood and current overall situation.

2.2   A Multi-disciplinary Area

Ambient Intelligence and Smart Environments systems nourish from many well established areas of computing and engineering. It also mixes with many other professions through their many application domains, e.g., health and social care. Figure 2 highlights some of those technological and scientific pillars.

2.3   Basic Architecture

Systems for Ambient Intelligence can be organized in various ways but some features are to be found in all those architectures. Little effort has been directed towards identifying what all AmI systems have in common and on studying these systems as a new category of artificial entities. This section aims at rectifying that by providing both a view of a system from the point of view of a basic system architecture and then a

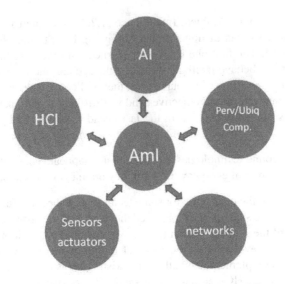

Fig. 2. Interaction in between AmI and other disciplines

complementary view from the point of view of information flow. A complementary explanation can be found in [6].

What are the essential components of an AmI system? Basically an AmI system has a real environment and occupants that interact with that environment in some typical way for that combination environment/occupants. Hence we can define an AmI system as follows:

$$AmISystem = \langle E, IC, I \rangle$$

such that:

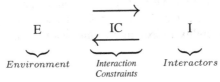

such that:

E: is the Environment. For example, a house, a hospital, a factory, a street, a city, an airplane, an airport, a train, or a bus station.

IC: is a set of Interaction Constraints. It specify the possible ways in which elements of E and I can interact with each other. Some elements that can be typically further specified here are $\langle S, A, C, IR \rangle$ where:

S is a set of sensors, A is a set of actuators, C is a set of contexts of interest and IR is a set of Interaction Rules. Sensors capture information from the environment. Actuators allow the system to act upon and influence the environment. The set of contexts of interest distinguish those situations where we expect the system to act. The set of interaction rules establishes the protocol on how the system put all the previous elements together to make decisions and trigger actions.

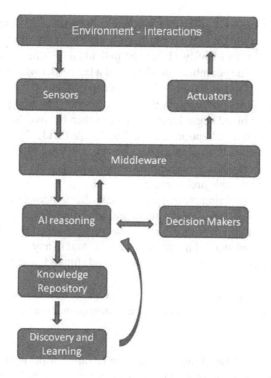

Fig. 3. Generic Architecture for AmI-SmE systems

I: is a set of Interactors (usually beneficiaries, it can be people, pets or robots). They can interact with the system in various ways, IR should capture the ways this interaction is conducted.

The definition above tell us of the essential elements, they can be further refined to any arbitrary level of detail and instantiated to the specific details of different application domains. Still it is more declarative in nature stating what is important rather than how it all works. Figure 3 highlights the flow of information and how the different components of an AmI system interact with each other gathering information from the real world, understanding it, taking decisions and using those decisions to interact with the real world again.

Sensors, Actuators and Middleware. A distinctive feature of the systems we are addressing in this article is that they are immersed in the real, physical, world. As such they have to interact directly with an environment. These systems have to gather information of that environment in real-time through sensing devices and after some reasoning they usually have to act.

Given the importance of sensing/actuating devices this area for research and development is very actively pursuing the production of new sensing devices or the expansion of the capabilities of current devices. There are nowadays sensors that can detect wide range of situations and measure a variety of substances.

The most widely known is probably infrared sensors that can detect movement as it has become fairly common to have anti-burglar alarms which are based in that technology. The possibility to identify objects or individuals is one of the most popular sensing options [7,8]. They combine an ID tag and a tag reader which can detect the ID tag based on proximity. Other sensors allow can detect weight, the presence chemicals, gases, humidity, brightness or temperature. Other devices can read physiological data like blood pressure or blood sugar levels which can then be used for healthcare. More details on these technological options can be found in [9,10,11,12].

Sensors can be physically connected to a network or wireless [13], each option with advantages and disadvantages, for example, the first ones are more reliable but the second ones offer a more flexible architecture.

Sensors and actuators bring their own problems for system implementation. First there is a cost associated. Then, all of them are, to different degrees, unreliable [14]. There are problems of compatibility between sensors produced by different manufacturer and they require substantial maintenance effort. And in any case they can generate vast amounts of data that has to be somehow stored, filtered, merged and interpreted [15,16].

Important European projects have been devoted to the development of efficient middleware that can provide a viable architecture to interconnect sensors (see for example, Amigo and Persona).

Artificial Intelligence. One of the most exciting technical aspects of a system exhibiting Ambient Intelligence is the capability to act autonomously in the benefit of humans. This implies both a hard challenge and a tremendous responsibility. We will focus on the former and will address the latter in another section.

There are several aspects that will have a strong influence in the intelligence a system can exhibit:

- Learning and Activity Recognition: it means the system is capable to analyze the vast amount of data produced by sensor triggering and out of that it and can make sense of the events that happen in a particular environment. It means the system should be able to group together events as recorded by the sensors into conceptual clusters. For example, from the movement and RFID sensors installed in the kitchen and in other objects like cups, kettles, cupboard, water taps, etc. the system identifies that a person is preparing coffee, which in term is part of making breakfast, etc. [17,18].
- Context-awareness: all Ambient Intelligence systems take place in an environment. What we do in this area is to smarten up the environment deploying hardware and software that links the environment with a computing system which is supposed to operate in the interest of a human or group of humans. To operate successfully such systems must understand the context [19] and the evolution of that context, i.e. its dynamics [20].
- Reasoning: cognitive inference is essential for the system to infer whether it has to act or not and what action(s) should be taken. A variety of methods exist here, ranging from systems which are more rule-based [20] to those based in biologically inspired models [21].

- Multiagents: have an important role in providing a flexible paradigm to model the different levels of autonomy and dependency that each component can have in a Smart Environment [22]. One problem so far which is preventing full exploitation of the multiagent technology is that what has been used so far is merely forcing the diverse needs of AmI-SmE systems to pass through the sieve of traditional agents. More effort has to be put on developing the type of multiagent architectures that are needed to develop AmI-SmE sytems (for an attempt in that direction see [23]).
- Robots: provide a valuable tool both as an interface and as an actuator within a smart environment. Robots can provide an element of socialization [24]. They can also be disguised in the way of a tool that users can benefit from like an intelligent wheelchair which can help navigate a house to users with mobility challenges [25].

Human-Computer Interaction. Weiser's initial vision was very emphatic on the requirement that technology only will be successful if it becomes adopted to the extent of not being noticed, very much the way we use a fridge or a washing machine nowadays. Humans should be able to use devices in a way that does not demand vast amounts of training and specialization. Needles to say most of what it is on offer today in the areas of AmI and SmE fall short in this aspect. It is also fair to say that there is a significant part of the community which is doing interesting progress and is working extremely hard to achieve this aim.

Gesture recognition [26], gaze tracking [27], facial expression recognition [28], emotion recognition [29], and spoken dialogue [30], either isolated or combined to form multi-modal interfaces [31], are some of a range of options becoming available to facilitate communication between humans and the system in a natural way [32].

Images also help assess a situation where safety can be compromised. The Wireless Sensor Networks Lab at Stanford University uses a network of video cameras to infer a sequence of body postures and hence detect possible hazards like a fall [33].

2.4 Applications

The range of possible applications for Ambient Intelligence and Smart Environments is vast and we can look at the future of the area with expectation and hope that it will bring to everyday life a range of available solutions. Here we list some emerging applications driven by the demand of users, companies and governmental organizations:

- Health-related applications. Hospitals can increase the efficiency of their services by monitoring patients' health and progress by performing automatic analysis of activities in their rooms [34]. They can also increase safety by, for example, only allowing authorized personnel and patients to have access to specific areas and devices. Health can be decentralized and made accessible at home through telecare and telehealth services in what it is commonly termed Ambient Assisted Living.
- Public transportation sector. Public transport can benefit from extra technology including satellite services, GPS-based spatial location, vehicle identification, image

processing and other technologies to make transport more fluent and hence more efficient and safe.

- Education services. Education-related institutions may use technology to create smart classrooms where the modes of learning are enhanced [35].
- Emergency services. Safety-related services like fire brigades can improve the reaction to a hazard by locating the place more efficiently and also by preparing the way to reach the place in connection with street services. The prison service can also quickly locate a place where a hazard is occurring or is likely to occur and prepare better access to it for security personnel.
- Production-oriented places. Companies can use RFID sensors to tag different products and track them along the production and commercialization processes. This allows identifying the product path from production to consumer and helps improving the process by providing valuable information for the company on how to react to favourable demand and unusual events like products that become unsuitable for sale [7]. Smart offices has been also the centre of attention and some interesting proposals aim at equipping offices with ways to assist their employees to perform their tasks more efficiently [36].

2.5 Social Implications

By the very definition of the fields of Ambient Intelligence and Smart Environments these systems are created to be immersed in a place where they will affect people's lives directly. Whether it is students in a classroom, people at home, pedestrians in a street or shoppers in a mall, their lives will be influenced by the technology deployed in those places. If the system works ideally, their life will be improved. The problem is that a computing system rarely works ideally and in this domain, given the complexity' of the environment, see some of the challenges listed in later sections, it is even more unlikely the system will work ideally. Hence there is scope for disappointment [37,38].

Privacy. Take video cameras as an example. They can be used to monitor streets so that street crime can be detected as soon as possible. They can also be used in a shopping centre to know more about shoppers' preferences. They can be used in smart homes to detect situations where somebody can be at risk [33,39], which is an extremely valuable safety net for anyone who is in a vulnerable position. For example, for elderly living alone, children with learning disabilities, public transport safety at night, people on their own after and surgery, etc.

Cameras are such a rich media. They can facilitate so much information which is relevant for the implementation of an intelligent environment. Still, leaving technical difficulties aside, like achieving understanding of what is captured by one or more cameras, they are fiercely resisted by users and researchers.

To illustrate the point think about extreme situations like having a camera in your own bathroom or bedroom at home. Sure there are many other situations where cameras can be used and indeed are being used. What is acceptable or not acceptable to share changes enormously with cultural values and the situation. Some users are happy to give up some degree of privacy in return for increased safety; some will never allow a camera recording their daily life activities.

Safety. Sensors record information about our daily activities and there is technology that can mine the recorded data to extract patterns of behaviour. The idea being that negative behaviours can be indentified and discouraged and positive ones encouraged and reinforced.

What happens when all that private information fall in the wrong hands? There have been many incidents where sensitive digital information from governments and military forces around the world has been forgotten in the pen drive, CD or laptop in an airport or train. How many unwanted calls do you get per week because a company (e.g., bank or electronics shop you bought something in instalments) stored your personal details in a PC and the company that do back-ups sells the information (most probably without the company's knowledge) to SPAM maker companies?

It is not unlikely then that the same can happen to sensitive private data about our habits and illnesses can be accessible to groups of people who are eager to take profit of that knowledge. Users will become more and more aware of this and extra measures have to be provided to bring peace of mind to the market. If the market is label as unsafe by the users then all those involved will lose a fantastic opportunity to benefit society.

3   The Future

The literature of the area is prolific and there is a growing body of research and developments reported in the recent technical literature (see for example: [40,41,42,43]).

Still we cannot claim these developments are being massively taken by society, there are some success stories in various areas and parts of the world but the systems produced are still too unreliable, expensive and difficult to use as to be embraced by society. So what are the current bottlenecks for the area preventing further progress?

3.1   Emotional and Social Intelligence

Cognitive intelligence is a hard goal in its own right and the area has a good deal of work ahead to provide robust and intelligent systems. Equally hard and still not so deeply investigated is the element of emotional intelligence. This is to some extent a less logical and predictable side of humans, it has to do with anger, fears, desires, pleasure, etc.

A system that is supposed to "... support people in their daily lives by assisting them in a sensible way" has to be aware of the user's preferences and has to know when is the right time to approach her/him and in which way, as well as to realize when it is better to stay silent. Think about a system that offers you help each 5 minutes over the whole day, or remind you of all the things you have listed as interesting but you do not have currently in the house or recommend you to have a box of chocolate when you are trying to lose weight.

Let us assume we accept that a system that can exhibit a level of subtle behaviour is what we need and let us do a little exercise to think how we can achieve that. How can we sense when the user is angry? Will it be because is shouting or cursing? Is it meaningful that the person is slamming doors?

Understanding all the subtle semantics of a dialogue to the extent to infer a particular state of mind is the terrain of spoken dialogue and natural language understanding, which still is a challenging area in computing. Detecting other states of mind like feeling tired, hungry, happy, or depressed can provide equally hard challenges.

3.2   Scaling Up from One to Many Users

Many current systems can provide some level of acceptable service in the case of one single user, for example the literature abounds on smart homes to support independent living which are based in the assumption only one person is the permanent resident in the house, or at least the only one that the house have to take care of.

When multiple occupants share the space and the house have some degree of responsibility for more than one of them then things are even harder. Funny examples are known where systems were not prepared for the complexity of a user having a pet wondering around the house triggering sensors here and there. Consider for example a family living under the same roof and a system that tries to provide services for all them. Choosing a T.V. program may be a situation of conflict, should the system stay away from such domestic rows or should it have a duty to advise and mediate? How the system should react when there are irreconcilable positions?

3.3   User Acceptance!

At the end of the day if these technologies want to be accepted and be as pervasive as a fridge or a washing machine are nowadays, then they have to achieve overall satisfaction. This will involve delivering adequate and reliable services which are judged to be good value for money.

Currently there are not standards or accepted measures of quality. The diversity of areas involved and the diversity of potential applications conspire against this. Still it is important for the area to achieve maturity that some sort of benchmark is agreed. See the Darmstadt Challenge [44] as an example of a step in this direction. Another interesting avenue is the possibility to provide users with the option to program the behaviour of the system [45,46,47,48].

4   Conclusions

The last section may have emphasized what the area is still missing and the hardship of working in a field which has ambitious practical aims. However, it is not all that gloomy. The same reasons used to say there is no guarantee of success can be used to argue there is no proof that the aims are unachievable. There are already good success stories and developments are gradually starting to appear in the form of smart homes, smart cars, smart classrooms, smart offices, etc. ([49]).

Patience and sustained work will be needed to extend the technical frontiers of this area bit by bit. To what extent these technologies will be taken by society it is to be discovered, meanwhile the potential benefits are such that it is worth trying. Researchers and developers should remember at all times that users are at the centre and that technology should be built for them.

References

1. Cook, D.J., Augusto, J.C., Jakkula, V.R.: Ambient intelligence: applications in society and opportunities for artificial intelligence. Pervasive and Mobile Computing (to appear, 2009)
2. Augusto, J.C., Shapiro, D. (eds.): Advances in Ambient Intelligence. Frontiers in Artificial Intelligence and Applications, vol. 164. IOS Press, Amsterdam (2007)
3. Ramos, C., Augusto, J.C., Shapiro, D.: Ambient intelligence: The next step for artificial intelligence. IEEE Intelligent Systems 23, 15–18 (2008)
4. Cook, D.J., Das, S.K.: Smart Environments: Technology, Protocols and Applications. Wiley-Interscience, Hoboken (2005)
5. Weiser, M.: The computer for the 21st century. Scientific American 265, 94–104 (1991)
6. Augusto, J.C.: Ambient intelligence: the confluence of ubiquitous/pervasive computing and artificial intelligence. In: Intelligent Computing Everywhere, pp. 213–234. Springer, London (2007)
7. Want, R.: RFID – a key to automating everything. Scientific American 290, 46–55 (2004)
8. i button, http://www.maxim-ic.com/products/ibutton/ibuttons/
9. Delapierre, G., Grange, H., Chambaz, B., Destannes, L.: Polymer-based capacitive humidity sensor. Sensors and Actuators 4, 97–104 (1983)
10. Wolffenbuttel, R.F., Mahmoud, K.M., Regtien, P.L.: Compliant capacitive wrist sensor for use in industrial robots. IEEE Transactions on Instrumentation and Measurements 39, 991–997 (1990)
11. Najafi, B., Aminian, K., Paraschiv-Ionescu, A., Loew, F., Bula, C., Robert, P.: Ambulatory system for human motion analysis using a kinematic sensor: Monitoring of daily physical activity in the elderly. IEEE Transactions on Biomedical Engineering 50, 711–723 (2003)
12. Stanford, V.: Biosignals offer potential for direct interfaces and health monitoring. IEEE Pervasive Computing 3, 99–103 (2004)
13. Pottie, G., Kaiser, W.: Wireless sensor networks. Communications of the ACM 43, 51–58 (2000)
14. Augusto, J.C., Liu, J., McCullagh, P., Wang, H., Yang, J.B.: Management of uncertainty and spatio-temporal aspects for monitoring and diagnosis in a smart home. International Journal of Computational Intelligence Systems 1, 361–378 (2008)
15. Benini, L., Poncino, M.: Ambient intelligence: a computational platform perspective, pp. 31–50. Kluwer Academic Publishers, Dordrecht (2003)
16. Manyika, J., Durrant-Whyte, H.: Data Fusion and Sensor Management: A Decentralized Information-Theoretic Approach. Ellis Horwood (1994)
17. Youngblood, G., Cook, D.J.: Data mining for hierarchical model creation. IEEE Transactions on Systems, Man, and Cybernetics, Part C: Applications and Reviews 37, 561–572 (2007)
18. Aztiria, A., Augusto, J.C., Izaguirre, A., Cook, D.J.: Learning accurate temporal relations from user actions in intelligent environments. In: Proceedings of the 3rd Symposium of Ubiquitous Computing and Ambient Intelligence, vol. 51, pp. 274–283 (2008)
19. Dey, A.: Understanding and using context. Personal and Ubiquitous Computing 5, 4–7 (2001)
20. Augusto, J.C., Nugent, C.D.: The use of temporal reasoning and management of complex events in smart homes. In: Proccedings of European Conference on AI (ECAI 2004), pp. 778–782. IOS Press, Amsterdam (2004)
21. Mozer, M.C.: Lessons from an adaptive home. In: Cook, D.J., Das, S.K. (eds.) Smart Environments: Technology, Protocols, and Applications, pp. 273–298. Wiley, Chichester (2004)
22. Cook, D.J., Youngblood, M., Das, S.K.: A multi-agent approach to controlling a smart environment. In: Designing Smart Homes. The Role of Artificial Intelligence, pp. 165–182. Springer, Heidelberg (2006)

23. Augusto, J., O'Donoghue, J.: Context-aware agents (the 6 ws architecture). In: Proceedings of the International Conference on Agents and Artificial Intelligence (ICAART), Porto, Portugal, INSTICC (2009)
24. de Ruyter, B., Aarts, E.: Social interactions in ambient intelligent environments. In: Augusto, J., Shapiro, D. (eds.) Proceedings of the 1st Workshop on Artificial Intelligence Techniques for Ambient Intelligence (AITAmI 2006), Riva del Garda (2006)
25. Matsumoto, O., Komoriya, K., Toda, K., Goto, S., Hatase, T., Nishimura, N.: Autonomous travelling control of the "tao aicle" intelligent wheelchair. In: 2006 IEEE/RSJ International Conference on Intelligent Robots and Systems (IROS 2006), pp. 4322–4327 (2006)
26. Pentland, A.: Perceptual environments. In: Cook, D.J., Das, S.K. (eds.) Smart Environments: Technology, Protocols and Applications. Wiley-Interscience, Hoboken (2005)
27. Majaranta, P., Räihä, K.J.: Text entry by gaze: Utilizing eye-tracking. In: MacKenzie, I., Tanaka-Ishii, K. (eds.) Text entry systems: Mobility, accessibility, universality, pp. 175–187. Morgan Kaufmann, San Francisco (2007)
28. Partala, T., Surakka, V., Vanhala, T.: Real-time estimation of emotional experiences from facial expressions. Interacting with Computers 18, 208–226 (2006)
29. Nakatsu, R.: Integration of multimedia and art for new human-computer communications. In: Ishizuka, M., Sattar, A. (eds.) PRICAI 2002. LNCS (LNAI), vol. 2417, pp. 19–28. Springer, Heidelberg (2002)
30. McTear, M., Raman, T.: Spoken Dialogue Technology: Towards the Conversational User Interface. Springer, Heidelberg (2004)
31. Ailisto, H., Kotila, A., Strommer, E.: Ubicom applications and technologies (2007), www.vtt.fi/~inf/pdf/tiedotteet/2003/T2201.pdf
32. Aghajan, H., Augusto, J., Delgado, R.L.C. (eds.): Human-Centric Interfaces for Ambient Intelligence. Elsevier, Amsterdam (2009)
33. Keshavarz, A., Tabar, A.M., Aghajan, H.: Distributed vision-based reasoning for smart home care. In: Proc. of ACM SenSys Workshop on DSC (2006)
34. O'Donoghue, J., Herbert, J.: An intelligent data management reasoning model within a pervasive medical environment. In: Augusto, J.C., Shapiro, D. (eds.) Proceedings of the 1st Workshop on Artificial Intelligence Techniques for Ambient Intelligence (AITAmI 2006), pp. 27–31 (2006)
35. Shi, Y., Xie, W., Xu, G., Shi, R.: The smart classroom: Merging technologies for seamless tele-education. IEEE Pervasive Computing, 47–55 (April-June 2003)
36. Marreiros, G., Santos, R., Ramos, C., Neves, J., Novais, P., Machado, J., Bulas-Cruz, J.: Ambient intelligence in emotion based ubiquitous decision making. In: Augusto, J.C., Shapiro, D. (eds.) Proceedings of the 2nd Workshop on Artificial Intelligence Techniques for Ambient Intelligence (AITAmI 2007), Hyderabad, India, pp. 86–91 (2007)
37. Dertouzos, M.: Human-centered systems. In: Denning, P. (ed.) The Invisible Future, pp. 181–192. McGraw-Hill, New York (2001)
38. Huuskonen, P.: Run to the hills! ubiquitous computing meltdown. In: Augusto, J.C., Shapiro, D. (eds.) Advances in Ambient Intelligence. Frontiers in Artificial Intelligence and Applications, vol. 164, pp. 157–172. IOS Press, Amsterdam (2007)
39. Aghajan, H., Augusto, J., Wu, C., McCullagh, P., Walkden, J.A.: Distributed vision-based accident management for assisted living. In: Okadome, T., Yamazaki, T., Makhtari, M. (eds.) ICOST. LNCS, vol. 4541, pp. 196–205. Springer, Heidelberg (2007)
40. Helal, S., Cook, D. (eds.): IE 2008: Proceedings of the 4th International Conference on Intelligent Environments, Seattle, USA, IET (2008)
41. Aarts, E., Crowley, J.L., de Ruyter, B., Gerhäuser, H., Pflaum, A., Schmidt, J., Wichert, R. (eds.): AmI 2008. LNCS, vol. 5355. Springer, Heidelberg (2008)
42. Augusto, J.C., Shapiro, D., Aghajan, H. (eds.): AITAmI2008: Proceedings of the 3rd Workhsop on Artificial Intelligence Techniques for Ambient Intelligence, ECCAI (2008)

43. PervHealth2008: 2nd International Conference on Pervasive Computing Technologies for Healthcare, Tampere, Finland. IEEE Explore, IEEE (2008)
44. Augusto, J., Bohlen, M., Cook, D., Flentge, F., Marreiros, G., Ramos, C., Qin, W., Suo, Y.: The darmstadt challenge (the turing test revisited). In: Proceedings of the International Conference on Agents and Artificial Intelligence (ICAART), Porto, Portugal, INSTICC (2009)
45. Muñoz, A., Vera, A., Botía, J.A., Gómez-Skarmeta, A.F.: Defining basic behaviours in ambient intelligence environments by means of rule-based programming with visual tools. In: Proceeding of the 1st Workshop on Artificial Intelligence Techniques for Ambient Intelligence, pp. 12–16 (2006)
46. Carolis, B.D., Cozzolongo, G., Pizzutilo, S.: An agent-based approach to personalized house control. In: Augusto, J., Shapiro, D. (eds.) Proceedings of the 1st Workshop on Artificial Intelligence Techniques for Ambient Intelligence (AITAmI 2006), Riva del Garda, Italy (2006)
47. Richard, N., Yamada, S.: Context-awareness and user feedback for an adaptive reminding system. In: Augusto, J.C., Shapiro, D. (eds.) Proceedings of the 2nd Workshop on Artificial Intelligence Techniques for Ambient Intelligence (AITAmI 2007), Hyderabad, India, pp. 57–61 (2007)
48. Chin, J., Callaghan, V., Clarke, G.: Soft-appliances: A vision for user created networked appliances in digital homes. Journal on Ambient Intelligence and Smart Environments (JAISE) 1, 69–75 (2009)
49. Nakashima, H., Aghajan, A., Augusto, J. (eds.): Handbook on Ambient Intelligence and Smart Environments. Springer, Heidelberg (2009)

PART I

Artificial Intelligence

Modelling Social Learning of Adolescence-Limited Criminal Behaviour

Tibor Bosse, Charlotte Gerritsen, and Michel C.A. Klein

Vrije Universiteit Amsterdam, Department of Artificial Intelligence
de Boelelaan 1081a, 1081 HV, The Netherlands
{tbosse,cg,mcaklein}@few.vu.nl

Abstract. Criminal behaviour exists in many variations, each with its own cause. A large group of offenders only shows criminal behaviour during adolescence. This kind of behaviour is largely influenced by the interaction with others, through social learning. This paper contributes a dynamical agent-based approach to simulate social learning of adolescence-limited criminal behaviour, illustrated for a small school class. The model is designed in such a way that it can be compared with data resulting from a large scale empirical study.

Keywords: Agent-based simulation, Social learning, Delinquent behaviour.

1 Introduction

Within Criminology, the analysis of the emergence of criminal behaviour is one of the main challenges [10]. An important mechanism behind the emergence of criminal behaviour is social learning [6]. To analyse this mechanism, this paper presents an agent-based approach to simulate social learning, which specifically addresses the mutual influence of peers, parents and school, with respect to delinquent behaviour.

To formalise and analyse the emergence of criminal behaviour through social learning, an artificial society has been modelled to represent a small school class. The models for the agents have been formally specified by executable temporal/causal logical relationships, using the modelling language TTL [4] and its executable sub-language LEADSTO [3]. This language allows the modeller to integrate both qualitative, logical aspects as quantitative, numerical aspects. Moreover, since the language has a formal logical semantics, simulation models created in TTL and LEADSTO can be formally analysed by means of logical analysis techniques.

In the field of Criminology, it is often quite difficult to perform experiments that involve changes in the real world. A model as the one presented in this paper can be used to study general patterns in the development of criminal behaviour. Simulation can help to answer what-if questions and to verify theories about the relation between different processes. Discussions with a team of criminologists taught us that the evidence provided by simulation models is already considered as useful knowledge about the relevance of criminological theories such as the *differential association theory,* which will be discussed below.

J. Filipe, A. Fred, and B. Sharp (Eds.): ICAART 2009, CCIS 67, pp. 19–32, 2010.
© Springer-Verlag Berlin Heidelberg 2010

In a next step of the research, we plan to validate the model using data of an existing empirical study e.g. [18]. In that study, the social networks of 1730 non-delinquent, minor delinquent and serious delinquent pupils at lower-level secondary schools in the Netherlands were analysed. This paper only reports about the first step, the model and simulations.

In Section 2 a summary from the literature on social learning is presented. Section 3 discusses the chosen modelling approach. The simulation model is presented in Section 4, and Section 5 discusses simulation results. In Section 6, these results are analysed using formal techniques. Section 7 presents related work. Finally, Section 8 concludes the paper.

2 Social Learning

According to [13], two types of delinquents can be distinguished: *life-course-persistent* offenders, who stay criminal throughout their entire life and *adolescence-limited* offenders, who only show antisocial behaviour during adolescence. Life-course-persistent anti-social behaviour is caused by neuropsychological problems during childhood that interact cumulatively with their criminogenic environments across development, which leads to a pathological personality. Adolescence-limited antisocial behaviour is caused by the gap between biological maturity and social maturity. It is learned from antisocial models that are easily mimicked, and it is sustained according to the reinforcement principles of learning theory. They peak sharply at about age 17 and drop fast in young adulthood. In the current paper, we explicitly focus on the adolescence-limited offenders.

An influential theory on the emergence of adolescence-limited criminal behaviour is the *differential association theory*, which was first proposed by [15] and later expanded by [6]. In short, this (informal) theory states that behaviour is learned through interaction with others. We learn most from the people we are in close contact with, like parents and peers. There are two basic elements to understanding the differential association theory. First, the content of what is learned is important (e.g., motives, attitudes and evaluations by others of the meaningful significance of each of these elements). Second, the process by which learning takes place is important, including the intimate informal groups and the collective and situational context where it occurs. Criminal behaviour itself is learned through assigning meaning to behaviour, experiences, and events during interaction with others.

According to [15], the extent to which delinquent behaviour is imitated is influenced by the *frequency*, *duration*, and *intensity* of the contact. Frequent, long and important or prestigious contacts have a larger influence. In addition, the *priority of learning* influences the social learning process: the earlier behaviour is learned, the more influential it is.

3 Modelling Approach

To formalise and analyse the emergence of criminal behaviour through social learning from an agent perspective, an expressive modelling language is needed. On the one hand, qualitative aspects have to be addressed, such as certain characteristics about the agents (e.g., their age), their social relationships (e.g., who are their parents and friends). On the other hand, quantitative aspects have to be addressed. For example, an agent's level of delinquency, which is the extent to which an agent exhibits delinquent behaviour, can best be described by a real number. The change of this delinquency can best be described by a mathematical formula. Another requirement of the chosen modelling language is its suitability to express on the one hand the basic mechanisms of social learning (for the purpose of simulation), and on the other hand more global properties of social learning (for the purpose of logical analysis and verification). For example, basic mechanisms of social learning involve decisions of individual agents to attach to their peers, whereas global properties are statements that consider the learning process over a longer period, like "eventually the delinquent pupils become less delinquent".

The predicate-logical Temporal Trace Language (TTL) [4] fulfils all of these desiderata. It integrates qualitative, logical aspects and quantitative, numerical aspects. This integration allows the modeller to exploit both logical and numerical methods for analysis and simulation. Moreover it can be used to express dynamic properties at different levels of aggregation, which makes it well suited both for simulation and logical analysis.

TTL is based on the assumption that dynamics can be described as an evolution of states over time. The notion of state as used here is characterised on the basis of an ontology defining a set of physical and/or mental (state) properties that do or do not hold at a certain point in time. These properties are often called *state properties* to distinguish them from dynamic properties that relate different states over time. A specific state is characterised by dividing the set of state properties into those that hold, and those that do not hold in the state. Examples of state properties are *'agent 1 has a delinquency level of 0.35'*, or *'agent 2 has an attachment to agent 3 of 0.5'*.

To formalise state properties, ontologies are specified in a (many-sorted) first order logical format: an *ontology* is specified as a finite set of sorts, constants within these sorts, and relations and functions over these sorts (sometimes also called signatures). The examples mentioned above then can be formalised by n-ary predicates (or proposition symbols), such as, for example, has_delinquency(agent1,0.35) or has_attachment_to(agent2, agent3, 0.5). Such predicates are called *state ground atoms* (or *atomic state properties*). For a given ontology Ont, the propositional language signature consisting of all ground atoms based on Ont is denoted by APROP(Ont). One step further, the *state properties* based on a certain ontology Ont are formalised by the propositions that can be made (using conjunction, negation, disjunction, implication) from the ground atoms. Thus, an example of a formalised state property is has_delinquency(agent1,0.35) & has_delinquency(agent2,0.45). Moreover, a *state* S is an indication of which atomic state properties are true and which are false, i.e., a mapping S: APROP(Ont) → {true, false}. The set of all possible states for ontology Ont is denoted by STATES(Ont).

To describe dynamic properties of complex processes such as the development of criminal behavior, explicit reference is made to *time* and to *traces*. A fixed time frame T is assumed which is linearly ordered. Depending on the application, it may be dense (e.g., the real numbers) or discrete (e.g., the set of integers or natural numbers or a finite initial segment of the natural numbers). Dynamic properties can be formulated that relate a state at one point in time to a state at another point in time. A simple example is the following (informally stated) dynamic property about the delinquency of agents:

> For all traces γ,
> there is a time point t such that
> all agents have a delinquency that is lower than d.

A *trace* γ over an ontology Ont and time frame T is a mapping γ : T → STATES(Ont), i.e., a sequence of states γ_t (t ∈ T) in STATES(Ont). The temporal trace language TTL is built on atoms referring to, e.g., traces, time and state properties. For example, 'in trace γ at time t property p holds' is formalised by state(γ, t) l= p. Here l= is a predicate symbol in the language, usually used in infix notation, which is comparable to the Holds-predicate in situation calculus. *Dynamic properties* are expressed by temporal statements built using the usual first-order logical connectives (such as ¬, ∧, ∨, ⇒) and quantification (∀ and ∃; for example, over traces, time and state properties). For example, the informally stated dynamic property introduced above is formally expressed as follows:

> ∀γ:TRACE ∃t:TIME ∀a:AGENT ∃x:REAL
> state(γ, t) l= has_delinquency(a, x) & x≤d

In addition, language abstractions by introducing new predicates as abbreviations for complex expressions are supported.

To be able to perform (pseudo-)experiments, only part of the expressivity of TTL is needed. To this end, the executable LEADSTO language [3] has been defined as a sublanguage of TTL, with the specific purpose to develop simulation models in a declarative manner. In LEADSTO, direct temporal dependencies between two state properties in successive states are modelled by *executable dynamic properties*. The LEADSTO format is defined as follows. Let α and β be state properties as defined above. Then, the notation α →⟩$_{e, f, g, h}$ β means:

If state property α holds for an interval with duration g, then after some delay between e and f state property β will hold for an interval with duration h.

As an example, the following executable dynamic property states that "if during 1 time unit the attachment between agent a1 and a2 is x1, and the difference in delinquency between both agents is x2, then for the next 5 time units (after a delay between 0 and 0.5 time units) the attachment between both agents will be β*x1+(1-β)*|x2|":

> ∀a1,a2:AGENT ∀x1,x2:REAL
> has_attachment_to(a1,a2,x1) ∧
> delinquency_difference(a1,a2,x2) →⟩$_{0, 0.5, 1, 5}$
> has_attachment_to(a1,a2,β*x1+(1-β)*|x2|)

Based on TTL and LEADSTO, two dedicated pieces of software have recently been developed. First, the LEADSTO Simulation Environment [3] takes a specification of executable dynamic properties as input, and uses this to generate simulation traces. Second, to automatically analyse the resulting simulation traces, the TTL Checker tool [4] has been developed. This tool takes as input a formula expressed in TTL and a set of traces, and verifies automatically whether the formula holds for the traces. In case the formula does not hold, the Checker provides a counter example, i.e., a combination of variable instances for which the check fails.

4 Simulation Model

To study the influence of social learning on delinquent behaviour, we modelled a school class with 10 pupils. There are three groups that influence the process of social learning, namely parents, school and peers. Therefore, each pupil is represented as an *agent*; the parents of the pupils and the school are modelled as *groups*. Each pupil is related to one parent group. The agents have a number of characteristics in our model (determined based on discussions with experts). We restricted our study to the characteristics that are collected in the empirical study [18]. The first property of an agent is its *age*. In our model the age is restricted to values between 12 and 17. The age is relevant for influence of peers on each other. The older an adolescent is (up to 17) the more his behaviour is influenced by peers. In addition, the *age difference* between peers is relevant, since older people are often more dominant in the relationship. The influence of school and parents tends to decrease as the adolescent gets older.

In addition, agents have a *basic level of influenceability*: this represents how easily they can be influenced. Oppositely, agents and groups have a level of *dominance*: this represents how easily they can influence others. For persons this is a character trait. Schools can also have a level of dominance. A dominant school can be seen as a strict school, while a school that is less strict could be considered to be less dominant.

The social relations between pupils in a school class are modelled via *attachment* relations. All agents are attached to each other with a specific *level of attachment*, representing the intensity of the contact as defined by [15]. The attachment relation is also used to model the attachment of pupils to their parents and to their school. We assume that a high attachment results in a higher influence of the attached agent or group on the behaviour of the pupil.

Finally, we model a level of *delinquency* for all agents and groups, also for parents and schools. The initial value for the delinquency of an agent could be based on a measurement of the number of delinquent acts of a pupil in the past. The interpretation of the delinquency of a school is indirect: the school has a low level of delinquency if it is a good school, i.e. teachers and other staff members have a low level of delinquency. When the atmosphere in the school is less positive, then it has a higher level of delinquency.

During the simulation, the levels of delinquency of the pupils change because of the influence of others. This process is depicted in Fig. 1, where the circles denote state properties and the arrows denote dynamic properties (relationships) between them. The age of each agent increases every year. Every agent starts with a basic influenceability; together with the age of the agent and the attachment to a specific group or agent, the *effective influenceability* of the agent by that agent or group is determined (denoted by has_influenceability in Fig. 1).

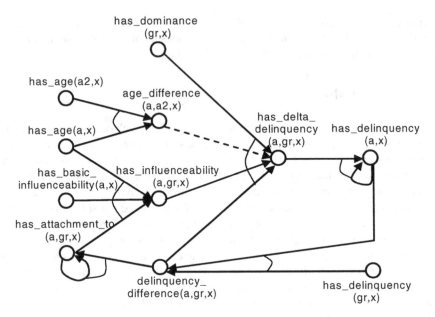

Fig. 1. Concepts and relations in the simulation model

This effective influenceability is combined with the level of dominance of the other party, the difference in delinquency between the agent and the other party, and - in case the other party is an agent - the age difference between two agents. This leads to the so-called *delta delinquency*. The delta delinquency represents all factors that influence the level of delinquency of an agent. In order to calculate the new delinquency of an agent, the delta delinquencies of all agents and groups in its environment are combined with the old delinquency (the delinquency the agent started out with).

In addition, the model is able to adapt the attachment between the agents. The idea behind this is that the strength of a relation is influenced by the overlap in values. If the difference in the level of delinquency is very high, then the attachment will decrease. However, because there are many other factors that influence the attachment as well, the difference in delinquency only causes a minor change in attachment.

The formal ontology used for the model is shown in Table 1. As can be seen, the concepts of influenceability, dominance, attachment, and delinquency are modelled as a real number between 0 and 1. Furthermore, the age is modelled as an integer between 12 and 17, and the delta delinquency as a real number between -1 and 1. For a complete overview of the simulation model, see the appendix[1].

The relationships between the concepts have been modelled in LEADSTO. Three example relationships (to determine the delta delinquency of groups, the new delinquency, and the new attachment to agents, respectively) are stated below. Here, the β's are decay factors, and the w's are weight factors. Note that these relationships correspond to (conjunctions of) arrows in Figure 1.

---

[1] http://human-ambience.few.vu.nl/docs/ICAART09.pdf

Table 1. Formal ontology

Formal predicate	Informal description	$X \in$
has_basic_influenceability (A:AGENT, X:REAL)	agent A has a basic influenceability of X (i.e., a static characteristic of an agent)	[0..1]
has_attachment_to(A:AGENT, G:GROUP, X:REAL)	agent A has an attachment to group G of strength X	[0..1]
has_attachment_to(A1:AGENT, A2:AGENT, X:REAL)	agent A1 has an attachment to agent A2 of strength X	[0..1]
has_age(A:AGENT, X:REAL)	agent A has age X	[12..17]
has_influenceability(A:AGENT, G:GROUP, X:REAL)	group G has an influence on agent A with strength X	[0..1]
has_influenceability(A1:AGENT, A2:AGENT, X:REAL)	agent A2 has an influence on agent A1 with strength X	[0..1]
age_difference(A1:AGENT, A2:AGENT, X:REAL)	the age difference between agent A1 and agent A2 is X	[-5..5]
has_delinquency(A:AGENT, X:REAL)	agent A has a delinquency of X	[0..1]
has_delinquency(G:GROUP, X:REAL)	group G has a delinquency of X	[0..1]
delinquency_difference(A:AGENT, G:GROUP, X:REAL)	the delinquency difference between agent A and group G is X	[-1..1]
delinquency_difference(A1:AGENT,A2:AGENT,X:REAL)	the delinquency difference between agent A1 and agent A2 is X	[-1..1]
has_dominance(G:GROUP, X:REAL)	group G has dominance X (static value)	[0..1]
has_dominance(A:AGENT, X:REAL)	agent A has dominance X (static value)	[0..1]
has_delta_delinquency(A:AGENT, G:GROUP, X:REAL)	the amount of change of the delinquency of agent A caused by group G is X	[-1..1]
has_delta_delinquency(A1:AGENT,A2:AGENT,X:REAL)	the amount of change of the delinquency of Agent A1 caused by agent A2 is X	[-1..1]
has_gender(A:AGENT, G:GENDER)	agent A has gender G	

Delta Delinquency Determination (for Groups)
$\forall$a:AGENT $\forall$g:GROUP $\forall$x1,x2,x3:REAL
delinquency_difference(a,g,x1) $\land$ has_influenceability(a,g,x2) $\land$
has_dominance(g,x3) $\rightarrow$
has_delta_delinquency(a,g,β2*(β1*x1+(1-β1)*x1* (w4*x2+w5*x3)))

New Delinquency Determination
$\forall$a1:AGENT $\forall$g:GROUP $\forall$d,s,p,x1,...,x10:REAL
has_old_delinquency(a1,d) $\land$ has_delta_delinquency(a1,school,s) $\land$
has_delta_delinquency(a1,g,p) $\land$ are_parents_of(g,a1) $\land$
has_delta_delinquency(a1,agent1,x1) $\land$ ...
has_delta_delinquency(a1,agent10,x10) $\rightarrow$
has_delinquency(a, d+ (s+p+x1+...+x10)/12))

New Attachment Determination (for Agents)
$\forall$a1,a2:AGENT $\forall$x1,x2:REAL
has_attachment_to(a1,a2,x1) $\land$ delinquency_difference(a1,a2,x2) $\rightarrow$
has_attachment_to(a1,a2, β3*x1+(1-β3)*abs(x2))

5   Simulation Results

A number of simulation experiments have been performed to see whether the behaviour of the model was as expected for some common scenarios. A thorough evaluation will be performed later when the results will be compared with data of an empirical study.

In the **first scenario** there is one bad guy with criminal parents in an otherwise reasonable school class. We are interested in the question whether the criminal boy makes the other boys bad or whether the group is able to straighten out the delinquent. In this scenario agent 1 has a delinquency of 0.8 while the other agents have a delinquency of 0.3. All agents are male[2] and are 12 years old at the start of the simulation. They have a basic influenceability with a value of 0.4, a level of dominance of 0.6 and a mutual attachment of 0.3. The attachments are stable in this simulation. Every agent has parents with a dominance of 0.7 and a delinquency of 0.2, except for agent 1, whose parents have a delinquency of 0.8.

The resulting trace is shown in Figure 2. Here, time is on the horizontal axis and the level of delinquency is on the vertical axis. The three graphs show the combined delinquencies of all pupils, the delinquency of agent 1 and the delinquency of the other agents (that all show the same behaviour; agent 10 is just taken as an example), respectively. The two lines in the first graph correspond to the lines in the second and third graph, respectively, where a more detailed scale is used. The results show that the interaction between the agents leads to a decreased delinquency of agent 1. The delinquency of the other agents increases slightly to 0.31 and from this point on it decreases to 0.255 at time point 100. From time point 70 on, there is a more or less stable difference in delinquency between the agent with criminal parents and the others.

In a **second scenario** (Figure 3), the influence of the school is examined by increasing its delinquency to 0.8. The level of delinquency of the agents and their parents were identical to the settings in the previous scenario. The results show that the increased delinquency of the school causes an increased level of delinquency of all the agents. This influence appeared to be larger than the influence of individual agents, because it propagates through to pupils, who again influence each other.

In the **third scenario**, half of the pupils (and their parents) have a high delinquency. The other pupils (and their parents) have the same level of delinquency as in scenario 1. In this case all agents influence each other and their delinquencies grow towards each other, while a difference remains because of the influence of the parents (see Fig. 4).

Finally, the **fourth scenario** represents a school class with two groups (3 delinquent pupils with a high mutual attachment, 3 extremely non-delinquent pupils with a high mutual attachment) and 4 individuals with a high basic influenceability. One of these 'group-less pupils' has a high attachment to a person in the criminal group, one to a person in the non-criminal group, and the others had no specific relations. The attachments can change over time. The goal of this scenario is to see whether a pupil will be incorporated in a group if he has a strong relationship with one of them. Figure 5 shows the resulting delinquencies.

Interestingly, we see that all group-less pupils reach a level of delinquencies that is close to that of the pupils in the 'good group', even for the pupils that have a strong relation to a pupil in one of the groups. This observation can be explained by the fact

---

[2] Note that the model does not incorporate a direct influence of gender. Difference between male and female pupils can be modeled indirectly by giving the males higher initial delinquencies.

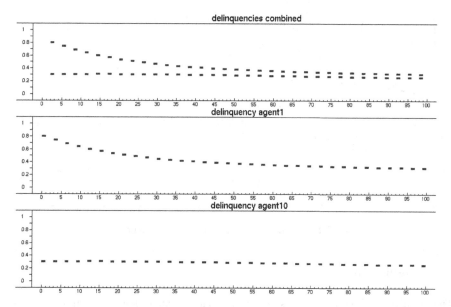

Fig. 2. Delinquency in a school class with one bad guy

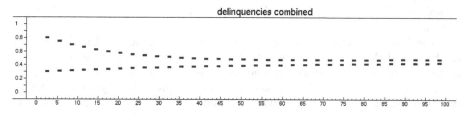

Fig. 3. Influence of a bad school

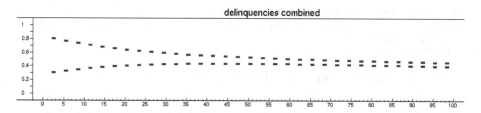

Fig. 4. Delinquency in a school class with half of the pupils being criminal

that the delinquency of the parents of the group-less pupils is close to the delinquency of the parents in the good group. However, if we look closely at the delinquencies of the group-less people (lower graph in Figure 5), we see that they develop slightly differently (notice the different scale). Apparently, the delinquency of the pupil with a friend in the bad (good) group initially grows faster (slower), but eventually it reaches the same level as the other group-less pupils.

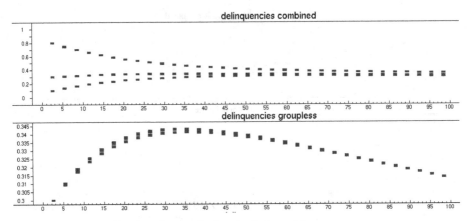

Fig. 5. Delinquencies in school class with two groups

6   Formal Analysis

The detailed settings and results of ten simulation experiments (including the ones described in Section 5) are shown in the appendix. Among the different experiments, various parameter settings were varied, in particular the initial delinquencies of agents, parents, and school, the initial attachment between agents, and several weight factors.

To analyse the resulting simulation traces in more detail, the TTL Checker tool [4] has been used. As mentioned earlier, this tool takes as input a TTL formula and a set of traces, and verifies automatically whether the formula holds for the traces. For the current domain, a number of hypotheses have been expressed as dynamic properties in TTL, which were inspired by relevant questions in Criminology (see Sections 1 and 2). To give a simple example, consider the following dynamic property (P1), which expresses that the delinquency of an agent keeps on decreasing over time:

P1   Strict Monotonic Decrease of Delinquency
For all time points t1 and t2, if t2 is later than t1, then the agent's delinquency at t2 is lower than at t1.
P1(γ:TRACE, a:AGENT) ≡
$\forall$t1,t2:TIME $\forall$d1,d2:REAL
[state(γ, t1) |= has_delinquency(a, d1) &
state(γ, t2) |= has_delinquency(a, d2) & t1<t2] $\Rightarrow$ d1>d2

Note that this formula comprises two free variables (the trace γ and the agent a), for which different values can be instantiated. For example, in order to check whether agent 1 satisfies the criterion of strict monotonic decrease of delinquency in simulation trace 5, the formula P1(trace1, agent1) should be checked. Similarly, it is possible to check whether the property holds for all agents and all traces, or for a certain percentage of them.

Besides checking whether the delinquency of agents keeps on decreasing, also other properties can be verified. A relevant question in Criminology is what the relative influences of (respectively) parents, peers, and school on the development of a person's delinquency are. For example, might it be the case that the biggest contribution is provided by parents and school only, and that the influence of class-mates can almost be neglected? To analyse these kinds of hypotheses, properties like the following have been established:

P2 Agent converges to Parents and School

At the end of the trace, the delinquency of agent a lies within a margin δ of the average of the delinquencies of its parents and the school at the start of the trace.

P2(γ:TRACE, a:AGENT) $\equiv$
$\forall$d1,d2,d3:REAL $\forall$p:AGENT
[state(γ, start_time) |= has_delinquency(p, d1) &
  state(γ, start_time) |= has_delinquency(school, d2) &
  state(γ, end_time) |= has_delinquency(a, d3) &
  are_parents_of(p,a) &]
$\Rightarrow$ d3-δ < (d1+d2)/2 < d3+δ

If this property were true (for a small δ), this would indicate that the development of a pupil could be predicted by taking into account the delinquency of the parents and the school only. Some initial checks have pointed out that the lowest δ for which the property satisfies all generated traces is 0.22. In other words, for all of the traces the influence of parents and school was relatively high. In addition to P2, a property was created to compare the change in delinquency between two agents a1 and a2.

P3 Bigger change in Delinquency

During the whole trace, agent a1 made a bigger change in delinquency than agent a2.

P3(γ:TRACE, a1,a2:AGENT) $\equiv$
$\forall$d1,d2,d3,d4:REAL
[state(γ, start_time) |= has_delinquency(a1, d1) &
  state(γ, start_time) |= has_delinquency(a2, d2) &
  state(γ, end_time) |= has_delinquency(a1, d3) &
  state(γ, end_time) |= has_delinquency(a2, d4)]
$\Rightarrow$ |d1-d3| > |d2-d4|

This property can be used, for example, to find out whether in a school class with many "good" pupils and one "bad" guy (see scenario 1), the bad pupil tends to move towards the good ones, or vice versa. In our simulation traces, such a bad pupil indeed turned out to converge towards his classmates.

To summarise, a number of TTL properties have been checked against the gener-ated simulation traces, as a first pilot study of the applicability of the approach. Al-though no real conclusions can be drawn as yet, these checks pointed out that the traces satisfy basic properties that were inspired by criminological theories, such as property P2 and P3.

Finally, it is important to note that, in addition to simulated traces, the TTL Checker can also take empirical traces as input. In future work, several properties as those introduced here will be verified against empirical traces that are constructed on the basis of experiments in real classrooms

7 Related Work

With respect to related work, the research presented in this paper on the one hand has commonalities with literature from the social and behavioural sciences (in particular, the area of Criminology), and on the other hand with literature in AI and Computer Science (among others, agent-based simulation).

Concerning the criminological and psychological area, first of all the current paper is related to early articles from the 60's and 70's such as [1], [6] and [15], which were the first to formulate (different variants of) the social learning theory. Here, the theory put forward by [1] is more generic, whereas the other two focus specifically on social learning in Criminology. For an overview of these theories, see [11]. In fact, these theories formed the basis of the research questions addressed in this paper. Based on these theories, [14] identified a number of (informal) properties that are expected to hold for social learning in Criminology, such as "the more frequently persons show deviant behaviour, the more frequently they will have contact with patterns of deviant behaviour". Although a detailed verification (using larger-scale experiments and statistical techniques) is left for future work, an initial analysis provides evidence that our model indeed satisfies these properties. Next, a number of papers in Criminology propose more refined models for social learning, often focusing on specific aspects of the learning. For example, [16] compared three theoretical models of the interrelations among associations between delinquent peers, delinquent beliefs, and delinquent behaviour. A main difference with our work is that these models are not computational. Nevertheless, their conclusions are in agreement with the initial results found in this paper. Finally, several authors have performed empirical studies on social learning of delinquent behaviour in schools [5] and [18]. Our model was designed explicitly with the purpose of reproducing such data.

Concerning the literature in AI and Computer Science, we are not aware of approaches using multi-agent technology to simulate delinquent behaviour of individuals in a group. However, various papers have similarities to the work proposed here. First, [9] present a model that is rather similar to ours, but which uses differential equations to describe the development of juvenile criminal behaviour. Another difference with our model is that they aim for an integration of multiple criminological theories (namely social learning theory, career theory, and rational choice theory), whereas we focus (in more detail) on the former only. Moreover, several authors have created models that address social learning and criminal behaviour at a more global level. For example, [7] presents an economic model for social learning, although not explicitly focussed on learning of delinquent behaviour. Similarly, [19] presents an agent-based economic model for the market for offenses. This model addresses the global development of delinquency in a population. These models differ from our model in the sense that they are situated at a macroscopic level, thereby abstracting from differences between individuals. An approach that does consider individual differences, but that addresses a different domain, is presented by [17]. They present a simulation model of the dynamics of terrorist networks, based on networks of non-deterministic finite automata. Furthermore, a large number of approaches address simulation of the environmental aspects of criminal behaviour, such as the displacement of crime and the emergence of "hot spots", e.g., [12] and [2]. Finally, relevant work is put forward by [8]. They identify a number of (cognitive) factors that are relevant in social learning in general. However, in contrast to our work, they do not provide a computational model.

8 Conclusions

This paper presented an agent-based approach to simulate and formally analyse the process of social learning of delinquency during adolescence. The general mechanism of change by influences of peers is possibly also useful in other domains in which social learning is relevant. In this paper, however, we focused on learning of delinquent behaviour. Inspired by criminological literature, the approach incorporates the influences of three types of groups, namely peers, parents, and school. Various relevant factors were identified, such as influenceability, dominance, and attachment, and their mutual relationships were formalised by means of the hybrid modelling language LEADSTO. Moreover, it was shown how the approach can be used to generate simulation traces, and how such traces can be automatically verified against relevant properties, expressed in the language TTL. Although preliminary, the first results are promising. Firstly, they provide evidence that the proposed model is a useful experimental tool to give insight in social learning processes as described in the criminological literature. Secondly, some interesting patterns have already been found. For example, the simulation results suggest that the influence of the school on delinquency is relatively high (scenario 3), that the impact of attachment is relatively low (scenario 4), and that every individual learning process approaches a final delinquency near the average of the delinquencies of parents, school, and peers.

In the current paper, no detailed empirical validation of the model has been presented. However, as mentioned in the introduction, various empirical studies have been performed, of which large data sets are available [5] and [18]. The model has been explicitly designed with the objective of using such data sets for validation in the future. Currently, some initial steps in this direction are taken. During such a validation, several questions are addressed, such as "is it realistic that the average delinquency almost always decreases?", or "is it realistic to have a relatively stable delinquency for school and parents?". When these questions are solved, the model can be further fine-tuned, in particular by choosing realistic values for all parameter settings and weight factors involved.

References

1. Bandura, A.: Social Learning Theory. Prentice-Hall, Englewood Cliffs (1977)
2. Bosse, T., Gerritsen, C.: Agent-Based Simulation of the Spatial Dynamics of Crime: On the Interplay between Criminal Hot Spots and Reputation. In: Proceedings of the 7th International Joint Conference on Autonomous Agents and Multi-Agent Systems, AAMAS 2008, pp. 1129–1136. ACM Press, New York (2008)
3. Bosse, T., Jonker, C.M., van der Meij, L., Treur, J.: A Language and Environment for Analysis of Dynamics by Simulation. International Journal of AI Tools 16(3), 435–464 (2007)
4. Bosse, T., Jonker, C.M., van der Meij, L., Sharpanskykh, A., Treur, J.: Specification and Verification of Dynamics in Cognitive Agent Models. In: Proceedings of the 6th International Conference on Intelligent Agent Technology, IAT2006, pp. 247–254. IEEE Computer Society Press, Los Alamitos (2006); Extended version: International Journal of Cooperative Information Systems 18, 167–193 (2009)

5. Bruinsma, G.J.N.: Crime as Social Learning Process. A Test of the Differential Association Theory in the version of K.-D. Opp (in Dutch). Gouda Quint, Arnhem. (1985)
6. Burgess, R., Akers, R.L.: A Differential Association-Reinforcement Theory of Criminal Behavior. Social Problems 14, 363–383 (1966)
7. Chamley, C.P.: Rational Herds: Economic Models of Social Learning. Cambridge University Press, New York (2003)
8. Conte, R., Paolucci, M.: Intelligent Social Learning. Journal of Artificial Societies and Social Simulation 4(1) (2001)
9. van Dijkum, C., Landsheer, H.: Experimenting with a Nonlinear Dynamic Model of Juvenile Criminal Behavior. Simulation & Gaming 31, 479–490 (2000)
10. Gottfredson, M., Hirschi, T.: A General Theory of Crime. Stanford University Press (1990)
11. Lanier, M.M., Henry, S.: Essential Criminology. Westview Press, Boulder (1998)
12. Liu, L., Wang, X., Eck, J., Liang, J.: Simulating Crime Events and Crime Patterns in RA/CA Model. In: Wang, F. (ed.) Geographic Information Systems and Crime Analysis, pp. 197–213. Idea Group, Singapore (2005)
13. Moffitt, T.E.: Adolescence-Limited and Life-Course-Persistent Antisocial Behavior: A Developmental Taxonomy. Psychological Review 100(4), 674–701 (1993)
14. Opp, K.D.: The Economics of Crime and the Sociology of Deviant Behaviour - A Theoretic Confrontation of Basic Propositions. Kyklos 42(3), 405–430 (1989)
15. Sutherland, E.H., Cressey, D.R.: Principles of Criminology, 7th edn. J.B. Lippincott, Philadelphia (1966)
16. Thornberry, T.P., Lizotte, A.J., Krohn, M.D., Farnworth, M., Jang, S.J.: Delinquent Peers, Beliefs, and Delinquent Behavior: A Longitudinal Test of Interactional Theory. Criminology 32, 47–83 (1994)
17. Tsvetovat, M., Carley, K.M.: Structural Knowledge and Success of Anti-Terrorist Activity: The Downside of Structural Equivalence. Journal of Social Structure 6 (2005)
18. Weerman, F.M., Bijleveld, C.C.J.H.: Birds of Different Feathers. European Journal of Criminology 4(4), 357–383 (2007)
19. Winoto, P.: An Agent-Based Simulation of the Market for Offenses. In: AAAI Workshop on Multi-Agent Modeling and Simulation of Economic Systems, Edmonton, Canada (2002)

How Do Emotions Induce Dominant Learners' Mental States Predicted from Their Brainwaves?

Alicia Heraz and Claude Frasson

HERON Lab, University of Montreal, CP 6128 succ., Centre Ville, Montreal QC, Canada
{herazali,frasson}@iro.umontreal.ca

Abstract. In this paper we discuss how learner's electrical brain activity can be influenced by emotional stimuli. We conducted an experiment in which we exposed 17 learners to a set of pictures from the International Affective Picture System (IAPS) while their electrical brain activity was recorded. We got 33.106 recordings. In an exploratory study we examined the influence of 24 picture categories from the IAPS on the amplitude variations of the 4 brainwaves frequency bands: δ, φ, α and β. We used machine learning techniques to track the amplitudes in order to predict the dominant frequency band which inform about the learner mental and emotional states. Correlation and regression analyses show a significant impact of the emotional stimuli on the amplitudes of the brainwave frequency bands. Standard classification techniques were used to assess the reliability of the automatic prediction of the dominant frequency band. The reached accuracy was 90%.

Keywords: Electrical Brain Activity, Machine Learning Techniques, Learner Brainwaves Model.

1 Introduction

Innovative Research is rapidly expanding the level of control that is achievable in Human-Machine Interactions. Scientists have been experimenting with non-invasive brain-computer interfaces that read brain signals with an electroencephalogram (EEG). EEG-based brain-computer interfaces use sensors placed on the head to detect brainwaves and feed them into a computer as input [17]. To close the performance gap between the user and the computer, many research focused on the user modeling [5], [13].

Most of the work in this field has focused on identifying the user's emotions as they interact with computer systems such as tutoring systems [10] or educational games [6], [7]. The importance of the systematic study of emotions has become more present in several disciplines [9], [14], [18], [19] since it was largely ignored until the late 20th century.

Kort, Reilly and Picard (2001) proposed a comprehensive four-quadrant model that explicitly links learning and affective states; this model has not yet been supported by empirical data from human learners. Conati (2002) has developed a probabilistic system that can reliably track multiple emotions of the learner during interactions

J. Filipe, A. Fred, and B. Sharp (Eds.): ICAART 2009, CCIS 67, pp. 33–43, 2010.

with an educational game. Their system relies on dynamic decision networks to assess the affective states of joy, distress, admiration, and reproach. The performance of their system has been measured on the basis of learner self reports [6] and inaccuracies that were identified have been corrected by updating their model [7]. D'Mello (2005) study reports data to integrate affect-sensing capabilities into an intelligent tutoring system with tutorial dialogue, namely AutoTutor. They identified affective states that occur frequently during learning. They applied various classification algorithms towards the automatic detection of the learners affect from the dialogue patterns manifested in AutoTutor's log files.

Unfortunately, many of these of systems lack precision because they are based on learner self reports, or use tools to analyze the learner external behavior like facial expression [10], vocal tones [8] or gesture recognition [13]. In addition, one affective state is not sufficient to encompass the whole gamut of learning [5].

Our previous work [11], [12] indicated that an EEG is an efficient info source to detect emotions. Results show that the student's affect (Anger, Boredom, Confusion, Contempt, Curious, Disgust, Eureka, and Frustration) can be accurately detected (82%) from brainwaves [11]. We have also conducted an experimentation in which we explored the link between brainwaves and emotional assessment on the SAM scale (pleasure, arousal and domination). Results were promising, with 73.55%, 74.86% and 75.16% for pleasure, arousal and dominance respectively [12]. Those results support the claim that all rating classes for the three emotional dimensions (pleasure, arousal and domination) can be automatically predicted with good accuracy through the nearest neighbour algorithm.

As a contrast to the learner self reports and use tools to analyze the learner external behaviour; our previous work is directed towards measuring emotions from the learner brainwave activity to track the learner's emotional states transitions. But what is the influence of feeling emotions on Brainwaves? What impact has emotional stimuli on the amplitudes of the brainwaves frequency bands?

In this paper, we focus on 4 different frequency bands: delta, theta, alpha and beta. We measure their amplitudes to identify the predominant learner mental state corresponding to the highest amplitude. We use the International Affective Picture System to induce emotions and we aim to how these effects the brainwaves amplitudes. Innovative Research is rapidly expanding the level of control that is achievable in Human-Machine Interactions. Scientists have been experimenting with non-invasive brain-computer interfaces that read brain signals with an electroencephalogram (EEG). EEG-based brain-computer interfaces use sensors placed on the head to detect brainwaves and feed them into a computer as input [17]. To close the performance gap between the user and the computer, many research focused on the user modelling [5], [13].

2   Brainwaves and Electroencephalogram

In the human brain, each individual neuron communicates with the other by sending tiny electrochemical signals. When millions of neurons are activated, each contributing its small electrical current, they generate a signal that is strong enough to be detected by an electroencephalogram (EEG) device [1], [4].

The EEG used in this experimentation is Pendant EEG. Commonly, Brainwaves are categorized into 4 different frequency bands, or types, known as delta, theta, alpha, and beta waves. Each of these wave types often correlates with different mental states. Table 1 lists the different frequency bands and their associated mental states.

Table 1. Brainwaves Categories

Brainwave	Frequency	Mental State
Delta (δ)	0-4 Hz	Deep sleep
Theta (θ)	4-8 Hz	Creativity, dream sleep, drifting thoughts
Alpha (α)	8-12 Hz	Relaxation, calmness, abstract thinking
Beta (β)	+12 Hz	Relaxed focus, high alertness, agitation

Delta frequency band is associated with deep sleep. Theta is dominant during dream sleep, meditation, and creative inspiration. Alpha brainwave is associated with tranquillity and relaxation. By closing one's eyes can generate increased alpha brainwaves. Beta frequency band is associated with an alert state of mind, concentration, and mental activity [17].

The electrical signal recorded by the EEG is sampled, digitized and filter to divide it into 4 different frequency bands: Beta, Alpha, Theta and Delta (Figure 1).

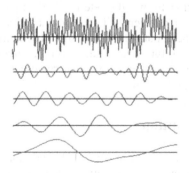

Fig. 1. A raw EEG sample and its filtered component frequencies. Respectively (from the top): Beta, Alpha, Theta and Delta Brainwaves [17].

3 Picture Categories in the International Affective Picture System

The International Affective Picture System (IAPS) is a large colored bank of pictures. It provides the ratings of emotions. It includes contents across a wide range of semantic categories. IAPS is developed and distributed by the NIMH Center for Emotion and Attention (CSEA) at the University of Florida in order to provide standardized database that are available to researchers in the study of emotion and attention. IAPS has been characterized primarily along the dimensions of valence, arousal, and dominance. Even though research has shown that the IAPS is useful in the study of discrete emotions, the categorical structure of the IAPS has not been characterized thoroughly.

Mickels (2005) experiment consisted of collecting descriptive emotional category data on subsets of the IAPS in an effort to identify pictures that elicit one discrete emotion more than others. Results revealed multiple emotional categories for the pictures and indicated that this picture set has great potential in the investigation of discrete emotions [16]. Table 2 shows the categories identified by Mikel's study.

Table 2. Mikel's categories for the IAPS

Category	Description
A	Anger
D	Disgust
F	Fear
U	Undifferentiated
S	Sadness
Am	Amusement
Aw	Awe
C	Contentment
U	Undifferentiated
†	Pictures that are outside two standard deviations from the overall mean and may thus be blends of positive and negative emotions.

This study provided categorical data that allows the IAPS to be used more generally in the study of emotion from a discrete categorical perspective. In accord with previous reports [3], gender differences in the emotional categorization of the IAPS images were minimal. These data show that there are numerous images that elicit single discrete emotions and, furthermore, that overall, a majority of the images elicit either single discrete emotions or emotions that represent a blend of discrete emotions, also in accord with previous reports.

4   Experiment Description

In our experimentation we use Pendant EEG [15], a portable wireless electroencephalograph. Electrode placement was determined according to the \10-20 International System of Electrode Placement." This system is based on the location of the cerebral cortical regions Electrodes were placed on PCz, A1 and A2 [17]. Pendant EEG sends the electrical signals to the machine via an infrared connection. Light and easy to carry, it is not cumbersome and can easily be forgotten within a few minutes. The learner wearing Pendant EEG is completely free of his movements: no cable connects them to the machine. The experiment included 17 learners selected from the Computer Science Department of University of Montreal. In order to induce the emotions which occur during learning, we use IAPS.

Participant is connected to Pendant EEG. The duration of the experimentation for each participant varies between 15 and 20 minutes. This one is free to stop when he wishes. He's invited to indicate his emotions any time, whenever it changes (figure 2).

The purpose of the experiment is to record the emotions at each change of brainwave amplitude. The recording set size is 33106.

Fig. 2. A learner wearing Pendant EEG

5 Data Treatment

Before using the database as an input to several learning algorithms, preliminary treatments of formatting, cleaning and selection had to be applied to it. The initial database was composed of 33106 tuples that contained the user id and the picture category from IAPS. The first treatment that was applied to the database was to extract a dataset of tuples that contain the picture category and the transition from two vectors $Amp_t(\delta_1, \theta_1, \alpha_1, \beta_1)$ and $Amp_{t+\Delta t}(\delta_2, \theta_2, \alpha_2, \beta_2)$, where $Amp_t(\ )$ is the amplitudes recorded at instant t and $Amp_{t+\Delta t}(\ )$ is the amplitude at $t + \Delta t$. Δt (in sec) is the time between each modification in one of the 4 brainwaves amplitudes.

We also applied some few data cleaning with respect to picture categories frequencies by removing every picture categories that had a frequency inferior to 6. The most represented image category in the dataset is U (Undifferentiated) which is more than 4 times more frequent than S (Sadness), which is the next most frequent one, but since undifferentiated images appear in the case of transitions from emotion 5 (disgust) to another, we decided to keep that category in the dataset. Figure 3 shows the repartition of pictures categories in the dataset.

The empty categories were: AwAwC, ADF and AS. They were removed. Most of pictures that the learners saw were in the categories: U(13282), S(4058), D(2432), DF(2026) and AwE(1910).

In addition, we created the class $do\min ance$. It gives the order of the brainwaves amplitudes. Since we have 4 types of brainwaves frequency bands, $do\min ance$ takes 4! =24 different values from the set $\{dtab, dtba,..., btad\}$. The value $btda$ means that the first highest amplitude recorded is for beta brainwave, the second is for theta, the third is for delta and the fourth one is for alpha. This means that the

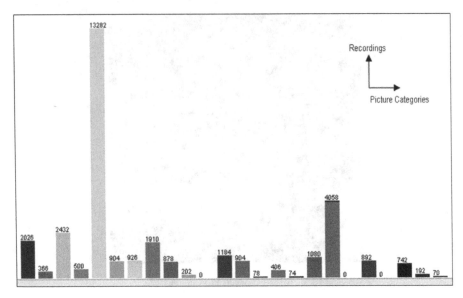

Fig. 3. Repartition of picture categories in the dataset; three empty categories were removed

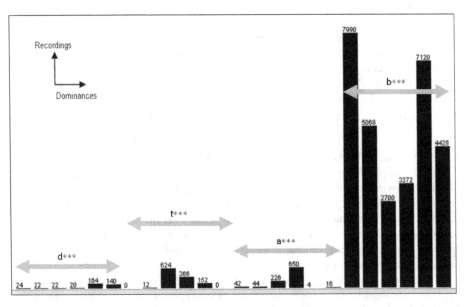

Fig. 4. Dominance Values Repartition

predominant mental state is the one associated to beta. Most of time, $b***$ values are predominant, figure 4 shows that fact.

The percentage of predominance of Delta, Theta, Alpha and Beta on the 33106 recordings were respectively: 1,2%, 3,5%, 3% and 92,2%.

6 Prediction Results

Determining the impact of emotional stimuli on the brainwaves is a multi-class classification problem. The mapping function is:

$$f : (\delta, \theta, \alpha, \beta, pictureCat) \to do\min ance$$

For classification we used Weka a collection of machine learning algorithms for solving data mining problems implemented in Java and open sourced under the GPL [22].

Many classification algorithms were tested. Best results were given by Naïve Bayes, K-Nearest Neighbor and Decision trees [20]. Table 4 shows the overall classification results using k-fold cross-validation5 (k = 10). In k-fold cross-validation the data set (N) is divided into k subsets of approximately equal size (N/k). The classifier is trained on (k-1) of the subsets and evaluated on the remaining subset. Accuracy statistics are measured. The process is repeated k times. The overall accuracy is the average of the k training iterations. The various classification algorithms were successful in detecting the new dominant value from the four brainwaves amplitudes and the picture category. Classification accuracy varies from 78.02% to 93.82%. Kappa statistic measures the proportion of agreement between two rates with correction for chance. Kappa scores ranging from 0.4 – 0.6 are considered to be fair, 0.6 – 0.75 are good, and scores greater than 0.75 are excellent (Robson, 1993). In the case of the algorithms we tested Kappa scores vary from 0.73 to 0.92 (good to excellent). Results are shown on table 3.

Table 3. The Best Results

Algorithm	Accuracy	Kappa
Naïve Bayes	78.02%	0.73
k-NN (k=1)	92.52%	0.91
Decision Tree	93.82%	0.93

For the decision tree Algorithm, table 4 shows the details of classification accuracy among the 24 values of the class Dominance.

For the decision tree algorithm and according to table 4, we calculated the Youden's J-index to increase the weight to the rating classes with minority instances [23] as the following formula:

$$JIndex = Card(RC)^{-1} \sum_{e \in RC} \Pr ecision_e$$

With $Card(RC)$ is the cardinality of rating classes list and is 22 (24-2; we removed the 2 classes *tdab* and *tadb* since they are empty). The JIndex value is 73.32% which is less (but still good) than the classification prediction shown in table 4 (93.82%). This result supports the claim that all rating classes for the 22 classes can be automatically detected with good accuracy (73.32%) through the decision tree algorithm. Figure 5 shows the Confusion Matrix.

The highest classification rates appear on the Matrix Diagonal. Two classes were removed: G=*tdab* and L=*tadb*. They are empty.

Table 4. Detailed Accuracy by Class

Precision	Recall	F-Measure	Class
0.783	0.75	0.766	dtab
0.708	0.773	0.739	dtba
0.6	0.682	0.638	datb
0.737	0.7	0.718	dabt
0.873	0.841	0.857	dbat
0.831	0.771	0.8	dbta
0	0	0	~~tdab~~
0.818	0.75	0.783	tdba
0.904	0.893	0.898	tbad
0.898	0.885	0.891	tbda
0.938	0.895	0.916	tabd
0	0	0	~~tadb~~
0.976	0.952	0.952	atdb
0.907	0.886	0.897	atbd
0.925	0.872	0.897	abdt
0.89	0.871	0.88	abtd
0.5	0.5	0.5	adtb
0.938	0.938	0.938	adbt
0.954	0.957	0.956	btad
0.943	0.94	0.942	btda
0.927	0.922	0.924	bdat
0.93	0.935	0.933	bdta
0.944	0.946	0.945	batd
0.934	0.942	0.938	badt

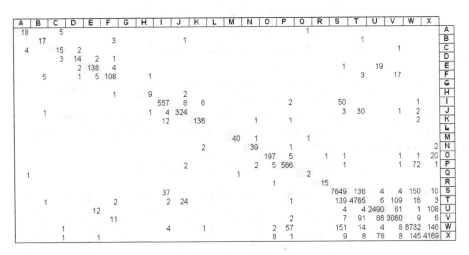

Fig. 5. The confusion Matrix

7 Implementation

In our previous works, we conceived the Architecture of a multi agent System (MAS) for 3 agents that assess emotional parameters from brainwaves. Via the JADE (Java Agent Development Framework) platform [2] and according to the communication language FIPA-ACL, these agents communicate with the planner located in the tutoring module of an ITS. They send to the latter the predicted emotional state. To complete this work, we aim by doing this experimentation to extend our MAS in the future and add the Brainwave Dominance Predictor (BDP) Agent. BDP Agent will induce emotional stimuli to regulate the Brainwave Activity. New pedagogical strategies will be implemented and suggested to an ITS to improve the learning conditions. Figure 6 shows the overall architecture.

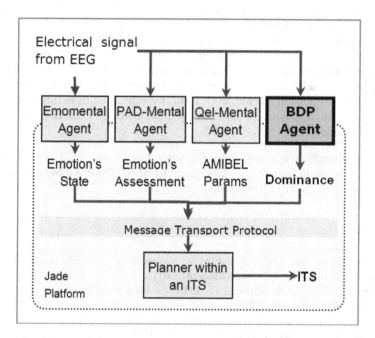

Fig. 6. Extended Architecture of the Multi-Agent Brain-Sensitive System

The BDP Agent will be implemented within the MAS in the future.

8 Conclusions

This study has presented machine learning techniques to follow and track the learner's brainwaves frequency bands amplitudes. It completes many previous works that assess emotional parameters from brainwaves by using an EEG. This can be useful for some particular learners as taciturn, impassive and disabled learners. We do not consider the whole cases of disabled learners. We will consider only disabled learner who cannot express facial emotions or body gestures due to an accident or a surgery and also those who lost their voice or cannot talk. Here we are talking about physical disability and not mental disability. This procedure allowed us to record the brainwaves amplitudes of the learners exposed to emotional stimuli from the International Picture System. These data were used to predict the future dominant amplitude knowing the picture category and the actual brainwaves frequency band amplitudes.

We acknowledge that the use of EEG has some potential limitations. In fact, any movement can cause noise that is detected by the electrodes and interpreted as brain activity by Pendant EEG. Nevertheless, we gave a very strict instructions to our participants. They were asked to remain silent, immobile and calm. We believe that the instructions given to our participants, their number (17) and the database size (33106 records) can considerably reduce this eventual noise. Results are encouraging, a potential significant impact of emotional stimuli and the brainwave amplitudes. The decision tree analyses resulted in accurate predictions 93.82% and the Yuden's J-Index is 73.22%. If the method described above proves to be effective in tracking the learner's brainwaves amplitudes, we can direct our focus to a second stage. An ITS would select an adequate pedagogical strategy that adapt to certain learner's mental states correlated to the brainwaves frequency bands in addition to cognitive and emotional states. This adaptation would increase the bandwidth of communication and allow an ITS to respond at a better level. If this hypothesis holds in future replication, then it would give indications on how to help those learners to induce positive mental states during learning.

Acknowledgements. We acknowledge the support of the FQRSC (Fonds Québécois de la Recherche sur la Société et la Culture) and NSERC (National Science and Engineering Research Council) for this work.

References

1. Bear, M.F., Connors, B.W., Paradiso, M.A.: Neuroscience: Exploring the Brain, 2nd edn. Lippincott Williams & Williams, Baltimore (2001)
2. Bellifemine, F., Poggi, A., Rimassa, G.: JADE - A FIPA-compliant Agent Framework. In: PAAM 1999, London, UK (1999)
3. Bradley, M.M., Codispoti, M., Cuthbert, B.N., Lang, P.J.: Emotion and motivation: Defensive and appetitive reactions in picture processing. Emotion 1 (2001)
4. Cantor, D.S.: An overview of quantitative EEG and its applications to neurofeedback. In: Evans, J.R., Abarbanel, A. (eds.) Introduction to Quantitative EEG and Neurofeedback. Academic Press, London (1999)
5. Conati, C.: Probabilistic assessment of user's emotions in educational games. Journal of Applied Artificial Intelligence (2002)
6. Conati, C.: How to evaluate models of user affect? In: André, E., Dybkjær, L., Minker, W., Heisterkamp, P. (eds.) ADS 2004. LNCS (LNAI), vol. 3068, pp. 288–300. Springer, Heidelberg (2004)
7. Conati, C., Mclaren, H.: Data-driven Refinement of a Probabilistic Model of User Affect. In: Ardissono, L., Brna, P., Mitrović, A. (eds.) UM 2005. LNCS (LNAI), vol. 3538, pp. 40–49. Springer, Heidelberg (2005)

8. D'Mello, S.K., Craig, S.D., Gholson, B., Franklin, S., Picard, R.W., Graesser, A.C.: Integrating Affect Sensors in an Intelligent Tutoring System. In: Affective Interactions: The Computer in the Affective Loop Workshop at 2005 International conference on Intelligent User Interfaces. ACM Press, New York (2005)
9. Ekman, P.: Are there basic emotions? Psychological Review (1992)
10. Fan, C., Sarrafzadeh, A., Overmyer, S., Hosseini, H.G., Biglari-Abhari, M., Bigdeli, A.: A fuzzy approach to facial expression analysis in intelligent tutoring systems. In: Méndez-Vilas, A., Mesa González, J.A (eds.) (2003)
11. Heraz, A., Frasson, C.: Predicting the three major dimensions of the learner's emotions from brainwaves. International Journal of Computer Science (2008)
12. Heraz, A., Razaki, R., Frasson, C.: Using machine learning to predict learner emotional state from brainwaves. In: 7th IEEE conference on Advanced Learning Technologies: ICALT 2007, Niigata, Japan (2007)
13. Kort, B., Reilly, R., Picard, R.: An affective model of interplay between emotions and learning: Reengineering educational pedagogy—building a learning companion. In: Okamoto, T., Hartley, R., Kinshuk, Klus, J.P. (eds.) Proceedings IEEE International Conference on Advanced Learning Technology (2001)
14. Mandler, G.: Emotion. In: Bly, B.M., Rumelhart, D.E. (eds.) Cognitive science. Handbook of perception and cognition, 2nd edn. Academic Press, San Diego (1999)
15. McMilan, B.: (2006), http://www.pocket-neurobics.com
16. Mikels, J.A., Fredrickson, B.L., Larkin, G.R., Lindberg, C.M., Maglio, S.J., Reuter-Lorenz, P.A.: Emotional category data on images from the International Affective Picture System. Behav. Res. Methods 37 (2005)
17. Palke, A.: Brainathlon: Enhancing Brainwave Control Through Brain-Controlled Game Play. Master thesis, Mills College, Oakland, California, USA (2004)
18. Panksepp, J.: Affective neuroscience: The foundations of human and animal emotion. Oxford University Press, New York (1998)
19. Picard, R.W.: Affective computing. MIT Press, Cambridge (1997)
20. Quinlan, R.: C4.5: Programs for Machine Learning. Morgan Kaufmann Publishers, San Mateo (1993)
21. Robson, C.: Real word research: A resource for social scientist and practitioner researchers. Blackwell, Oxford (1993)
22. Witten, I.H., Frank, E.: Data Mining: Practical Machine Learning Tools and Techniques with Java Implementations. Morgan Kaufmann, San Francisco (2005)
23. Youden, W.J.: How to evaluate accuracy. Materials Research and Standards, ASTM (1961)

A Multiagent Semantics for the Game Description Language

Stephan Schiffel and Michael Thielscher

Technical University of Dresden
Dresden, Germany
{stephan.schiffel,mit}@inf.tu-dresden.de

Abstract. The Game Description Language (GDL) has been developed for the purpose of formalizing game rules. It serves as the input language for general game players, which are systems that learn to play previously unknown games without human intervention. In this paper, we show how GDL descriptions can be interpreted as multiagent domains and, conversely, how a large class of multiagent environments can be specified in GDL. The resulting specifications are declarative, compact, and easy to understand and maintain. At the same time they can be fully automatically understood and used by autonomous agents who intend to participate in these environments. Our main result is a formal characterization of the class of multiagent domains that serve as formal semantics for—and can be described in—the Game Description Language.

1 Introduction

A novel and challenging research problem for Artificial Intelligence, General Game Playing is concerned with the development of systems that learn to play a previously unknown game solely on the basis of the rules. The Game Description Language (GDL) [1] has been developed to formalize the rules of any finite, information-symmetric n-player game in such a way that the description can be automatically processed by a general game player [2]. As a declarative language, GDL supports specifications that are modular and easy to develop, understand, and maintain. While the basic semantics for GDL is grounded in standard logic, the language uses several pre-defined predicates as keywords, whose intended meaning is only informally described in [1].

In this paper, we show that GDL can be understood as a specification language for a large class of multiagent environments. This allows for formalizing the physics and laws that govern an arbitrary domain in such a way that agents can automatically understand the rules and thus know how to participate in this environment. There is a variety of potential applications for machine processable descriptions of multiagent environments: the rules of an e-marketplace can be made accessible to agents, the interface of interactive Internet platforms for software agents can be formally described, and agent competitions can be run without revealing detailed problem specifications in advance. In each of these cases, an autonomous agent—or a team of agents—can learn how to participate in a new or modified environment without the need to be (re-)programmed for each specific case. Because GDL uses a decidable subset of logic programming,

J. Filipe, A. Fred, and B. Sharp (Eds.): ICAART 2009, CCIS 67, pp. 44–55, 2010.

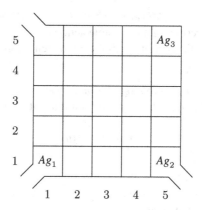

Fig. 1. A simple multiagent domain: two "guard" agents Ag_1 and Ag_2 shall cooperatively try to catch Ag_3, whose goal in turn is to escape via one of the three exits at locations $(1, 1)$, $(5, 1)$, and $(1, 5)$. All agents act synchronously and can move horizontally or vertically to an adjacent position. Ag_3 is caught when it ends up in the same location as Ag_1 or Ag_2, or when it crosses path with one of them in a simultaneous move.

autonomous agents require just a simple, standard reasoning module to be able to understand and effectively process a given set of rules. Moreover, if an agent environment is specified in GDL, successful general game playing systems such as [3,4,5,6] can be readily employed as intelligent agents for these environments.

The main result in this paper is the definition of a formal class of multiagent environments which can be expressed in GDL and, conversely, which can be used to provide a semantics for any GDL game description. As a by-product we thus obtain a formal semantics for the special, pre-defined keywords in GDL.

The rest of the paper is organized as follows. In Section 2, we formally define the class of deterministic, synchronous multiagent environments. In Section 3, we show how these can be axiomatically described in GDL, and in the section that follows, we present the converse result by showing how all GDL games can be interpreted as a deterministic, asynchronous multiagent environment. We conclude in Section 5.

2   Multiagent Environments

As the running example in this paper, we will consider the multiagent domain depicted in Figure 1. As a discrete environment it can be formally described as a finite state transition system. However, even though it is obviously just a toy-size example, it has a considerably large state space, rendering an explicit encoding difficult—and practically impossible for even slightly larger environments. Fortunately it is possible to exploit the fact that any natural and realistic multiagent environment has an internal structure, which allows one to describe its dynamics with the help of symbols that represent individual components. Our example domain, for instance, can be formally described using the following symbolic expressions: Ag_1, Ag_2, Ag_3, representing the three agents; $At(r, x, y)$, where $r \in \{Ag_1, Ag_2, Ag_3\}$ and $x, y \in \{1, \ldots, 5\}$, representing the position of each agent; and *Move(d)*, where $d \in \{North, South, East, West\}$, along with *Stay* and *Exit*, representing the possible actions.

Based on a suitable collection of symbols, multiagent domains can be formally described as follows.

Definition 1. *Let* Σ *be a countable set of ground (i.e., variable-free) symbolic expressions. A (discrete, synchronous, deterministic)* multiagent environment *is a structure*

$$(R, s_1, t, l, u, g)$$

where

- $R \subseteq \Sigma$ *finite (the agents, or* roles*);*
- $s_1 \subseteq \Sigma$ *finite (the* initial state*);*
- $t \subseteq 2^\Sigma$ *(the terminal states);*
- $l \subseteq R \times \Sigma \times 2^\Sigma$ *(the action preconditions, or* legality relation*);*
- $u : (R \mapsto \Sigma) \times 2^\Sigma \mapsto 2^\Sigma$ *(the transition function, or* update function*);*
- $g \subseteq R \times \mathbb{N} \times 2^\Sigma$ *(the* utility, *or* goal relation*).*

Here, 2^Σ *denotes the set of all* finite *subsets of* Σ, *and for any* $r \in R$ *and* $S \in 2^\Sigma$, $l(r, a, S)$ *holds for finitely many* $a \in \Sigma$.

This definition deserves some explanation. For the sake of simplicity, the symbolic expressions are not categorized—there is no formal distinction between symbols for objects, state components, actions, etc. For practical purposes, it is important that states are finitely representable; hence, while possibly infinitely many symbols give rise to infinitely many states, a state itself is an element of the set of all *finite* subsets of the given symbols. The legality relation $l(r, a, S)$ defines a to be a legal action for agent r in state S. Again for the sake of practical usability, it is assumed that every agent in every state has only finitely many possible actions. The update function takes an action for each agent and (synchronously) applies the joint actions to a current state, resulting in the updated state. For the sake of simplicity, we take natural numbers $n \in \mathbb{N}$ as the utility of a state S for agent r in the goal relation $g(r, n, S)$.

For illustration, consider a formalization of the multiagent environment of Figure 1 using the symbols introduced above.

- $R = \{Ag_1, Ag_2, Ag_3\}$;
- $s_1 = \{At(Ag_1, 1, 1), At(Ag_2, 5, 1), At(Ag_3, 5, 5)\}$;
- t contains all states

$$\{At(Ag_1, x_1, y_1), At(Ag_2, x_2, y_2)\}$$

(that is, where Ag_3 has escaped) along with all states

$$\{At(Ag_1, x_1, y_1), At(Ag_2, x_2, y_2), At(Ag_3, x_3, y_3)\}$$

in which $x_1 = x_3 \land y_1 = y_3$ or $x_2 = x_3 \land y_2 = y_3$ (that is, where Ag_3 has been caught);

- l is defined as follows: each agent can always *Stay*; in every non-terminal state each agent can *Move(d)* in any direction d unless this would lead outside the physical environment; Ag_3 can *Exit* from any of the locations $(1,1)$, $(5,1)$, or $(1,5)$, provided it has not been caught.
- u is defined as follows: actions *Stay*, *Exit*, and *Move(d)* have the expected effects on the individual locations of the agents, with the exception that when the paths of Ag_1 and either of Ag_1 or Ag_2 (or both) cross in a simultaneous move, then Ag_3 ends up (caught) in the same location as Ag_1 or Ag_2, respectively. For illegal actions or states that are not reachable, u may be arbitrarily defined.
- g shall be defined as true for $n = 100$ and $r = Ag_3$ in terminal states in which this agent has escaped; conversely, g holds for $n = 100$ and both Ag_1 and Ag_2 in terminal states in which Ag_3 got caught. In all other states the goal relation gives value 0 for all three agents.

We have thus obtained a formal, symbolic description of the example multiagent environment. However, this specification is not yet amenable to automatic processing by an autonomous agent, because it uses natural language to describe some of the components. If this were to be translated into an explicit enumeration of the transition function, this would again yield too large a description to be of any practical use. In the following section, we show how the Game Description Language can be readily used to provide a fully axiomatic, compact description of arbitrary multiagent environments; a description that on the one hand is declarative and easy to understand and maintain by humans, and on the other hand can be fully automatically processed by artificial, autonomous agent systems.

3   Axiomatizing Multiagent Environments as Game Descriptions

The Game Description Language (GDL) has been developed to formalize the rules of any finite game with complete information in such a way that the description can be automatically processed by a general game player. In this section, we first recapitulate the GDL syntax from [1] and then show how the multiagent environments defined in the preceding section can be formally described in this language.

3.1   General GDL Syntax

GDL is based on the standard syntax of logic programs, including negation. A logic program is a set of clauses according to the following definition (see, for example, [7]).

Definition 2.

- *A* term *is either a variable, or a function symbol applied to terms as arguments (a* constant *is a function symbol with no argument);*
- *An* atom *is a predicate symbol applied to terms as arguments;*
- *A* literal *is an atom or its negation;*
- *A* clause *is an implication* $h \Leftarrow b_1 \wedge \ldots \wedge b_n$ *where* head h *is an atom and* body $b_1 \wedge \ldots \wedge b_n$ *a conjunction of literals* $(n \geq 0)$.

Table 1. GDL keywords

role(R)	R is a player
init(P)	P holds in the initial position
true(P)	P holds in the current position
legal(R, M)	player R has legal move M
does(R, M)	player R does move M
next(P)	P holds in the next position
terminal	the current position is terminal
goal(R, N)	player R gets goal value N in the current position

We adopt the Prolog convention according to which variables are denoted by uppercase letters and predicate and function symbols start with a lowercase letter. (The interested reader may take a peek at Figure 2 at this point to see some example clauses, which in fact constitute a complete GDL axiomatization of our running example domain.) GDL imposes some general restrictions on a set of clauses, with the intention to ensure finite derivability.

Definition 3. *The* dependency graph *for a set G of clauses is a directed, labeled graph whose nodes are the predicate symbols that occur in G and where there is a* positive *edge $p \xrightarrow{+} q$ if G contains a clause $p(\bar{s}) \Leftarrow \ldots \wedge q(\bar{t}) \wedge \ldots$, and a* negative *edge $p \xrightarrow{-} q$ if G contains a clause $p(\bar{s}) \Leftarrow \ldots \wedge \neg q(\bar{t}) \wedge \ldots$.*

To constitute a valid *GDL specification, a set of clauses G and its dependency graph Γ must satisfy the following.*

1. *There are no cycles involving a negative edge in Γ (this is also known as being* stratified *[8,9]);*
2. *Each variable in a clause occurs in at least one positive atom in the body (this is also known as being* allowed *[10]);*
3. *If p and q occur in a cycle in Γ and G contains a clause*

$$p(s_1, \ldots, s_m) \Leftarrow b_1(\bar{t}_1) \wedge \ldots \wedge q(v_1, \ldots, v_k) \wedge \ldots \wedge b_n(\bar{t}_n)$$

then for every $i \in \{1, \ldots, k\}$,
   - *v_i is variable-free, or*
   - *v_i is one of $s_1, \ldots, s_m$, or*
   - *v_i occurs in some $\bar{t}_j$ ($1 \leq j \leq n$) such that b_j does not occur in a cycle with p in Γ.*

Stratified logic programs are known to admit a specific *standard model*; we refer to [8] for details and just mention the following properties.

1. To obtain the standard model, clauses with variables are replaced by their (possibly infinitely many) ground instances.
2. Clauses are interpreted as reverse implications.
3. The standard model is minimal while interpreting negation as non-derivability (the "negation-as-failure" principle [11]);

The second and third restriction in Definition 3 essentially guarantee that a logic program entails a *finite* number of ground atoms via its standard model. This is necessary to enable agents to make effective use of a set of game rules.

3.2   GDL Keywords

As a tailor-made specification language, GDL uses a few pre-defined predicate symbols.
These are shown in Table 1 together with their informal meaning. A further, standard
predicate is $\mathtt{distinct}(X, Y)$ to express (syntactic) inequality of two terms.[1]

GDL imposes additional restrictions on the use of these keywords.

Definition 4. *A valid GDL specification is a set of clauses G that, in addition to the
restrictions in Definition 3, satisfies the following conditions.*

- $\mathtt{role}$ *only appears in the head of clauses that have an empty body;*
- $\mathtt{init}$ *only appears as head of clauses and is not connected, in the dependency
  graph for G, to any of* $\mathtt{true}, \mathtt{legal}, \mathtt{does}, \mathtt{next}, \mathtt{terminal}, \mathtt{goal};$
- $\mathtt{true}$ *only appears in the body of clauses;*
- $\mathtt{does}$ *only appears in the body of clauses and is not connected, in the dependency
  graph for G, to any of* $\mathtt{legal}, \mathtt{terminal}, \mathtt{goal};$
- $\mathtt{next}$ *only appears as head of clauses.*

According to the informal semantics given in [1], a GDL specification G is to be
understood as follows. The derivable instances of $\mathtt{role}(R)$ define the players. The
initial state is composed of the derivable instances of $\mathtt{init}(P)$. In order to determine
the legal moves of a player in any given state, this state has to be encoded first, using
the keyword $\mathtt{true}$. More precisely, let $S = \{p_1, \ldots, p_n\}$ be a state (e.g., the derivable
instances of $\mathtt{init}(P)$ at the beginning), then G is extended by the clauses

$$\mathtt{true}(p_1) \Leftarrow$$
$$\ldots \qquad\qquad (1)$$
$$\mathtt{true}(p_n) \Leftarrow$$

Those instances of $\mathtt{legal}(R, A)$ which are derivable from this extended program define
all legal actions A for player R in state S. In the same way, the clauses for $\mathtt{terminal}$
and $\mathtt{goal}(R, N)$ define termination and goalhood (of value N for player R) *relative*
to the encoding of a given state. Determining a state transition, finally, requires the
encoding of the current state along with clauses representing a joint move. Specifically,
if players $r_1, \ldots, r_n$ make moves $a_1, \ldots, a_n$, then

$$\mathtt{does}(r_1, a_1) \Leftarrow$$
$$\ldots \qquad\qquad (2)$$
$$\mathtt{does}(r_n, a_n) \Leftarrow$$

must be added to G, and then the derivable instances of $\mathtt{next}(P)$ compose the updated
state.

---

[1] The semantics of this predicate is given by tacitly assuming the addition of the clause

$$\mathtt{distinct}(s, t) \Leftarrow$$

for every pair s, t of syntactically different ground terms.

3.3 Multiagent Environments in GDL

GDL provides all necessary features for declarative, formal descriptions of arbitrary multiagent environments as defined in Section 2. Of course there are many possible ways in which any specific environment can be axiomatized. We therefore define two sets of GDL clauses as *logically equivalent* if for any finite set of ground clauses (1) and (2) added, the two standard models of the two resulting logic programs agree on the interpretation of all GDL keywords. Before we can show how to formalize multiagent environments in GDL, we need the following syntactic definitions.

For any finite subset $S = \{p_1, \ldots, p_n\} \subseteq \Sigma$ of a set of ground terms, the following conjunction axiomatizes S as the current state:

$$S^{\texttt{true}} \overset{\text{def}}{=} \texttt{true}(p_1) \wedge \ldots \wedge \texttt{true}(p_n) \wedge \neg p_S \tag{3}$$

Here, p_S is an auxiliary predicate, one for every finite $S \subseteq \Sigma$, whose purpose is to ensure that the conjunction does not hold for states that are strict supersets of S:

$$\begin{aligned} p_S \Leftarrow \;& \texttt{true}(X) \wedge \\ & \texttt{distinct}(X, p_1) \wedge \ldots \wedge \texttt{distinct}(X, p_n) \end{aligned} \tag{4}$$

Hence, p_S is true for any state in which at least one state component X is true that differs syntactically from any of $p_1, \ldots, p_n$, that is, the elements of S. This ensures that the conjunction defined in (3) is an exact axiomatization of state S.

Furthermore, for any function $A : \{r_1, \ldots, r_n\} \mapsto \Sigma$, where $r_1, \ldots, r_n \in \Sigma$, the following conjunction axiomatizes A as a joint action:

$$A^{\texttt{does}} \overset{\text{def}}{=} \texttt{does}(r_1, A(r_1)) \wedge \ldots \wedge \texttt{does}(r_n, A(r_n)) \tag{5}$$

We are now ready to show how GDL can be used to axiomatize multiagent domains.

Definition 5. *Let $E = (R, s_1, t, l, u, g)$ be a multiagent environment based on ground symbolic expressions Σ, then any valid set of GDL clauses is an axiomatic description of E if it is logically equivalent to the following.*

- $\texttt{role}(r) \Leftarrow$ *for each $r \in R$;*
- $\texttt{init}(p) \Leftarrow$ *for each $p \in s_1$;*
- $\texttt{terminal} \Leftarrow S^{\texttt{true}}$ *for each $S \in t$;*
- $\texttt{legal}(r, a) \Leftarrow S^{\texttt{true}}$ *for each $(r, a, S) \in l$;*
- $\texttt{next}(p) \Leftarrow A^{\texttt{does}} \wedge S^{\texttt{true}}$ *for each $p \in u(A, S)$ and $A : R \mapsto \Sigma$, $S \subseteq \Sigma$;*
- $\texttt{goal}(r, n) \Leftarrow S^{\texttt{true}}$ *for each $(r, n, S) \in g$.*

It is important to realize that this direct axiomatization, where all relations and functions are encoded explicitly, is used solely to define the intended semantics. In practice, of course, a domain can be described in a much more compact manner, using variables, logical equivalence, and possibly auxiliary predicates. As an example, Figure 2 depicts a complete GDL specification of the multiagent environment introduced in Section 2. It is not too difficult to verify that this is a valid set of clauses according to Definition 3 and 4 and that it is indeed a correct axiomatic description of this domain according to Definition 5.

$\text{role}(\text{ag}_1) \Leftarrow$
$\text{role}(\text{ag}_2) \Leftarrow$
$\text{role}(\text{ag}_3) \Leftarrow$

$\text{init}(\text{at}(\text{ag}_1, 1, 1)) \Leftarrow$
$\text{init}(\text{at}(\text{ag}_2, 5, 1)) \Leftarrow$
$\text{init}(\text{at}(\text{ag}_3, 1, 5)) \Leftarrow$

$\text{terminal} \Leftarrow \text{true}(\text{at}(\text{ag}_1, \text{X}, \text{Y})) \wedge \text{true}(\text{at}(\text{ag}_3, \text{X}, \text{Y}))$
$\text{terminal} \Leftarrow \text{true}(\text{at}(\text{ag}_2, \text{X}, \text{Y})) \wedge \text{true}(\text{at}(\text{ag}_3, \text{X}, \text{Y}))$
$\text{terminal} \Leftarrow \neg\text{remain}$

$\text{remain} \Leftarrow \text{true}(\text{at}(\text{ag}_3, \text{X}, \text{Y}))$

$\text{legal}(\text{R}, \text{stay}) \Leftarrow \text{true}(\text{at}(\text{R}, \text{X}, \text{Y}))$
$\text{legal}(\text{ag}_3, \text{exit}) \Leftarrow \neg\text{terminal} \wedge \text{true}(\text{at}(\text{ag}_3, 1, 1))$
$\text{legal}(\text{ag}_3, \text{exit}) \Leftarrow \neg\text{terminal} \wedge \text{true}(\text{at}(\text{ag}_3, 5, 1))$
$\text{legal}(\text{ag}_3, \text{exit}) \Leftarrow \neg\text{terminal} \wedge \text{true}(\text{at}(\text{ag}_3, 1, 5))$
$\text{legal}(\text{R}, \text{move}(\text{D})) \Leftarrow \neg\text{terminal} \wedge \text{true}(\text{at}(\text{R}, \text{U}, \text{V})) \wedge \text{adjacent}(\text{U}, \text{V}, \text{D}, \text{X}, \text{Y})$

$\text{adjacent}(\text{X}, \text{Y}_1, \text{north}, \text{X}, \text{Y}_2) \Leftarrow \text{co}(\text{X}) \wedge \text{succ}(\text{Y}_1, \text{Y}_2)$
$\text{adjacent}(\text{X}, \text{Y}_1, \text{south}, \text{X}, \text{Y}_2) \Leftarrow \text{co}(\text{X}) \wedge \text{succ}(\text{Y}_2, \text{Y}_1)$
$\text{adjacent}(\text{X}_1, \text{Y}, \text{east}, \text{X}_2, \text{Y}) \Leftarrow \text{co}(\text{Y}) \wedge \text{succ}(\text{X}_1, \text{X}_2)$
$\text{adjacent}(\text{X}_1, \text{Y}, \text{west}, \text{X}_2, \text{Y}) \Leftarrow \text{co}(\text{Y}) \wedge \text{succ}(\text{X}_2, \text{X}_1)$

$\text{co}(1) \Leftarrow \quad \text{co}(2) \Leftarrow \quad \text{co}(3) \Leftarrow \quad \text{co}(4) \Leftarrow \quad \text{co}(5) \Leftarrow$
$\text{succ}(1, 2) \Leftarrow \quad \text{succ}(2, 3) \Leftarrow \quad \text{succ}(3, 4) \Leftarrow \quad \text{succ}(4, 5) \Leftarrow$

$\text{next}(\text{at}(\text{R}, \text{X}, \text{Y})) \Leftarrow \text{does}(\text{R}, \text{stay}) \wedge \text{true}(\text{at}(\text{R}, \text{X}, \text{Y}))$
$\text{next}(\text{at}(\text{R}, \text{X}, \text{Y})) \Leftarrow \text{does}(\text{R}, \text{move}(\text{D})) \wedge \text{true}(\text{at}(\text{R}, \text{U}, \text{V})) \wedge \text{adjacent}(\text{U}, \text{V}, \text{D}, \text{X}, \text{Y}) \wedge$
$\qquad\qquad\qquad\qquad \neg\text{capture}(\text{R})$
$\text{next}(\text{at}(\text{ag}_3, \text{X}, \text{Y})) \Leftarrow \text{true}(\text{at}(\text{ag}_3, \text{X}, \text{Y})) \wedge \text{capture}(\text{ag}_3)$

$\text{capture}(\text{ag}_3) \Leftarrow \text{true}(\text{at}(\text{ag}_3, \text{X}, \text{Y})) \wedge \text{true}(\text{at}(\text{R}, \text{U}, \text{V})) \wedge \text{does}(\text{ag}_3, \text{move}(\text{D}_1)) \wedge$
$\qquad\qquad\qquad \text{does}(\text{R}, \text{move}(\text{D}_2)) \wedge \text{adjacent}(\text{X}, \text{Y}, \text{D}_1, \text{U}, \text{V}) \wedge \text{adjacent}(\text{U}, \text{V}, \text{D}_2, \text{X}, \text{Y})$

$\text{goal}(\text{R}, 0) \Leftarrow \text{role}(\text{R}) \wedge \neg\text{terminal}$
$\text{goal}(\text{R}, 0) \Leftarrow \text{role}(\text{R}) \wedge \text{distinct}(\text{R}, \text{ag}_3) \wedge \text{terminal} \wedge \neg\text{remain}$
$\text{goal}(\text{R}, 100) \Leftarrow \text{role}(\text{R}) \wedge \text{distinct}(\text{R}, \text{ag}_3) \wedge \text{terminal} \wedge \text{true}(\text{at}(\text{ag}_3, \text{X}, \text{Y}))$
$\text{goal}(\text{ag}_3, 0) \Leftarrow \text{terminal} \wedge \text{true}(\text{at}(\text{ag}_3, \text{X}, \text{Y}))$
$\text{goal}(\text{ag}_3, 100) \Leftarrow \text{terminal} \wedge \neg\text{remain}$

Fig. 2. A complete, formal description of the multiagent environment of Figure 1

The specification of the Game Description Language in [1] lacks a fully formal definition of the intended meaning of a specification. This is why there are no formal grounds on which it could actually be proved that Definition 5 yields a correct description of a multiagent environment. In fact, we can and will use our formal concept of a multiagent domain to provide just this precise semantics for GDL in terms of a transition system.

4   A Multiagent Semantics for GDL

In the preceding section, we have shown how GDL provides a declarative, compact language to formally describe a large class of multiagent environments in a machine processable fashion. In this section, we show how the abstract model of a multiagent environment can in turn be used to provide a formal semantics for GDL in terms of a transition system. In this way we make precise what is only informally described in [1].

Any valid game description G in GDL contains a finite set of function symbols, including constants, which implicitly determines a (usually infinite) set of ground terms. This set constitutes the symbol base Σ in the transition-based semantics for G. The syntactic restrictions in GDL ensure finite derivability, so that each state, the set of roles, etc. are all finite subsets of Σ. The following definition of the semantics of a GDL description is straightforwardly obtained by reversing the mapping from a multiagent environment into GDL (cf. Definition 5). To this end, we redefine the abbreviations S^{true} and A^{does} as logic program facts (rather than conjunctions as in (3) and (5)). This allows to add them to G in order to determine terminal states, legal moves, updates, and goalhood:

$$S^{\text{true}} \overset{\text{def}}{=} \{ \text{true}(p_1) \Leftarrow$$
$$\ldots$$
$$\text{true}(p_n) \Leftarrow \}$$
$$A^{\text{does}} \overset{\text{def}}{=} \{ \text{does}(r_1, A(r_1)) \Leftarrow$$
$$\ldots$$
$$\text{does}(r_n, A(r_n)) \Leftarrow \}$$

It is worth mentioning that auxiliary predicate p_S (cf. (4)) is not needed in the axiomatization of a state as a set of facts, because the principle of negation-as-failure and S^{true} imply $\neg\text{true}(p)$ for any $p \notin S$.

Definition 6. *Let G be a valid GDL specification, whose signature determines the set of ground terms Σ. The semantics of G is the multiagent environment (R, s_1, t, l, u, g) where[2]*

- $R = \{r \in \Sigma : G \models \text{role}(r)\}$;
- $s_1 = \{p \in \Sigma : G \models \text{init}(p)\}$;
- $t = \{S \in 2^{\Sigma} : G \cup S^{\text{true}} \models \text{terminal}\}$;
- $l = \{(r, a, S) : G \cup S^{\text{true}} \models \text{legal}(r, a)\}$, *where $r \in R$, $a \in \Sigma$, and $S \in 2^{\Sigma}$*;
- $u(A, S) = \{p \in \Sigma : G \cup A^{\text{does}} \cup S^{\text{true}} \models \text{next}(p)\}$, *for all $A : (R \mapsto \Sigma)$ and $S \in 2^{\Sigma}$*;
- $g = \{(r, n, S) : G \cup S^{\text{true}} \models \text{goal}(r, n)\}$, *where $r \in R$, $n \in \mathbb{N}$, and $S \in 2^{\Sigma}$*.

This definition provides a formal semantics for GDL in terms of abstract multiagent environments. Finite derivability in valid GDL specifications implies that the

---

[2] Below, entailment ($\models$) is via the standard model of a set of clauses.

entailment relation is decidable, which in turn ensures that the definition of the semantics is effective.

In the preceding section we have seen that one and the same multiagent environment can be axiomatically described in many different ways. With the help of Definition 6 it is now easy to verify that two logically equivalent GDL descriptions (as defined in Section 3.3) describe exactly the same environment.

Proposition 1. *The semantics of two logically equivalent, valid GDL descriptions coincide.*

Proof. By definition, two logically equivalent GDL descriptions agree on the interpretation of all GDL keywords for all finite additions of clauses (1) and (2). It is easy to see, then, that the various components of their semantics according to Definition 6 must be identical.

Based on this result it is also straightforward to prove that Definition 6 indeed provides the complement to the encoding of a multiagent environment in GDL.

Proposition 2. *Let E be a multiagent environment and G any axiomatic description thereof, then the semantics of G is E.*

Proof. Consider the generic encoding of E given in Definition 5. It is easy to verify that the standard model for this set of clauses, augmented by any finite set of facts about relations `true` and `does` (cf. clauses (1) and (2), respectively, in Section 3.2), determines a semantics (R, s_1, t, l, u, g) via Definition 6 which equals E. The claim follows from Proposition 1 and the fact that any GDL encoding for E is logically equivalent to the generic clauses given in Definition 5.

5   Discussion

We have shown how the Game Description Language, developed in the context of General Game Playing, can be understood as a declarative language to provide compact and machine processable specifications of a large class of multiagent environments. This can be applied to formalize the rules, for example, of an e-marketplace, of publicly accessible agent platforms on the Internet, of problem domains used in agent competitions, etc. By automatically processing these specifications, autonomous agents can fully automatically learn how to participate in a new or modified environment without the need to be (re-)programmed. Moreover, successful off-the-shelf general game playing systems can be readily employed as intelligent agents for these environments.

It is interesting to note that GDL has been originally developed as problem specification language for a competition [2], much like the Planning Domain Description Language (PDDL) [12], which today is a quasi standard for the specification of planning domains. GDL can be viewed as a generalization of PDDL to domains with multiple agents, because solving a planning problem can be understood as playing a single-player game. Indeed, most features of current versions of PDDL can be expressed in GDL, though with one notable exception: sensing actions are not included in the current version of GDL. Although a GDL specification leaves agents with uncertainty about how the world evolves (an agent can decide on its own actions but not on those of all other agents), the language has been written for games without information asymmetry. An important research issue for the near future is to extend the Game Description Language so as to support descriptions of games with asymmetric information and sensing actions, which is a typical feature of card games, for instance. This would then provide a suitable formalization language for an even larger class of multiagent environments than considered in this paper.

In the second part of the paper, we have used the concept of a multiagent environment to provide a formal, transition-based semantics for GDL. With this we have made precise what is only informally described in [1]. Our semantics for GDL in terms of multiagent environments is related to an existing formal characterization of GDL by a game structure [13]. The main difference of the latter in comparison to our work are:

- It is restricted to propositional GDL;
- It puts further restrictions on GDL, such as not allowing predicate `init` to occur in clause with non-empty bodies;
- It uses an inductive definition of the set of all states in order to obtain only those which are reachable from the initial state. Since it is possible to give valid GDL specifications of games that do not terminate, this definition would be undecidable in the general setting.[3]

These restrictions have been imposed because the focus in [13] lies on the use of Temporal Logic for the purpose of verifying properties of games, such as termination or winnability. In contrast to this, the semantics given in the present paper covers full GDL.

Acknowledgements. This research was partially supported by *Deutsche Forschungsgemeinschaft* under Contract TH 541/16-1.

References

1. Love, N., Hinrichs, T., Haley, D., Schkufza, E., Genesereth, M.: General Game Playing: Game Description Language Specification. Technical Report LG–2006–01, Stanford Logic Group, Computer Science Department, Stanford University, 353 Serra Mall, Stanford, CA 94305 (2006), games.stanford.edu
2. Genesereth, M., Love, N., Pell, B.: General game playing: Overview of the AAAI competition. AI Magazine 26, 62–72 (2005)

---

[3] It is worth noting that this does not contradict the finite derivability property of valid GDL specifications, which just implies that all *local* reasoning problems are decidable. More specifically, given a particular state it is decidable whether an action is possible, and given a joint action it is also decidable what properties hold in the updated state, etc. On the other hand, GDL is expressive enough to describe any Turing machine as a "game" using clauses like

$$\text{init}(\text{head}(0)) \Leftarrow$$
$$\text{next}(\text{head}(\text{succ}(X))) \Leftarrow \text{true}(\text{head}(X)) \wedge$$
$$\text{does}(\text{tm}, \text{move\_forward})$$

Hence, reachability of states is generally undecidable in GDL.

3. Kuhlmann, G., Dresner, K., Stone, P.: Automatic heuristic construction in a complete general game player. In: Proceedings of the AAAI National Conference on Artificial Intelligence, pp. 1457–1462. AAAI Press, Boston (2006)
4. Clune, J.: Heuristic evaluation functions for general game playing. In: Proceedings of the AAAI National Conference on Artificial Intelligence, Vancouver, pp. 1134–1139. AAAI Press, Menlo Park (2007)
5. Schiffel, S., Thielscher, M.: Fluxplayer: A successful general game player. In: Proceedings of the AAAI National Conference on Artificial Intelligence, Vancouver, pp. 1191–1196. AAAI Press, Menlo Park (2007)
6. Finnsson, H., Björnsson, Y.: Simulation-based approach to general game playing. In: Proceedings of the AAAI National Conference on Artificial Intelligence, Chicago, pp. 259–264. AAAI Press, Menlo Park (2008)
7. Lloyd, J.: Foundations of Logic Programming, 2nd extended edn. Series Symbolic Computation. Springer, Heidelberg (1987)
8. Apt, K., Blair, H.A., Walker, A.: Towards a theory of declarative knowledge. In: Minker, J. (ed.) Foundations of Deductive Databases and Logic Programming, pp. 89–148. Morgan Kaufmann, San Francisco (1987)
9. van Gelder, A.: The alternating fixpoint of logic programs with negation. In: Proceedings of the 8th Symposium on Principles of Database Systems, ACM SIGACT-SIGMOD, pp. 1–10 (1989)
10. Lloyd, J., Topor, R.: A basis for deductive database systems II. Journal of Logic Programming 3, 55–67 (1986)
11. Clark, K.: Negation as failure. In: Gallaire, H., Minker, J. (eds.) Logic and Data Bases, pp. 293–322. Plenum Press (1978)
12. McDermott, D.: The 1998 AI planning systems competition. AI Magazine 21, 35–55 (2000)
13. van der Hoek, W., Ruan, J., Wooldridge, M.: Strategy logics and the game description language. In: Proceedings of the Workshop on Logic, Rationality and Interaction, Beijing, China (2007)

Verifying Context-Dependent Reduction Relations for Knowledge Specifications

Alexei Sharpanskykh and Jan Treur

Vrije Universiteit Amsterdam, Department of Artificial Intelligence
De Boelelaan 1081a, 1081 HV Amsterdam, The Netherlands
{sharp,treur}@cs.vu.nl

Abstract. Knowledge can be specified at different levels of conceptualisation or abstraction. In this paper, lessons learned on the philosophical foundations of cognitive science are discussed, with a focus on how the relationships of cognitive theories with specific underlying (physical/biological) makeups can be dealt with. It is discussed how these results can be applied to relate different types of knowledge specifications. More specifically, it is shown how different knowledge specifications can be related by means of reduction relations, similar to how specifications of cognitive theories can be related to specifications within physical or biological contexts. By the example of a specific reduction approach, it is shown how the process of reduction can be automated, including mapping of specifications of different types and checking the fulfilment of reduction conditions.

Keywords: Reduction relations, Automated mapping of specifications, Cognitive science.

1 Introduction

Specification languages play a major role in the development of knowledge models, as a means to describe specific functionalities aimed at. Functionalities can be described at different levels of conceptualisation and abstraction, and often different languages are available to specify them, varying from symbolic, logical languages to algorithmic, numerical languages. The question in how far such different types of specifications can be related to each other has not a straightforward general answer yet. Specifications of different types can just be used without explicitly relating them, as part of a heterogeneous specification. In a particular case relationships can be defined of the type that output of one functionality specification is related to input for another specification. However, it may be useful when general methods are available to relate the contents of different specifications as well. The aim of this paper is to explore possibilities for such general methods, inspired by recent work in the philosophical foundations of Cognitive Science.

Within the philosophical literature the position of Cognitive Science has often been debated; e.g., [1]. Recent developments have provided more insight in the specific characteristics of Cognitive Science, and how it relates to other sciences. A main issue

J. Filipe, A. Fred, and B. Sharp (Eds.): ICAART 2009, CCIS 67, pp. 56–69, 2010.

that had to be clarified is the role of the specific (physical or biological) makeup of individuals (or species) in Cognitive Science. Cognitive theories have a nontrivial dependence on the context(s) of these specific makeups. Due to this context-dependency, for example, regularities or relationships between cognitive states are not considered genuine universal laws and cannot be directly related to general physical or biological laws, as they simply can be refuted by considering a different makeup. The classical approaches to reduction that provide means to relate properties (or laws) of one level of conceptualization to properties (or laws) of another level (e.g., bridge law reduction [11], functional reduction [9] and interpretation mappings [15] do not address this context-dependency properly. In this paper context-dependent refinements of these approaches are used (as introduced in [16]) that provide a way to clarify in which sense regularities in a cognitive theory relate on the one hand to general physical/biological laws and on the other hand to specific makeups or mechanisms. Using theorem proving techniques and tools it is shown how such contect-dependent reduction relations can be worked out in more formal detail and used as a basis for automated verification of such relations between different knowledge specifications.

In this paper, first the lessons learned about the philosophical foundations of Cognitive Science are briefly summarised in Section 2. Section 3 shows how these findings can be applied to relate different knowledge specifications. This is illustrated for an example of adaptive functionality, for which two different types of knowledge specifications are given: one logical specification, and one algorithmic, numerical specification. Section 4 describes how different types of reduction relations can be defined to relate the two types of knowledge specification. Furthermore, in Section 5 it is shown in this example how the interpretation mapping approach to reduction can be automated, including checking the fulfilment of reduction conditions. The paper concludes with a discussion in Section 6.

2 Some of Main Issues

In this section some of the motivations behing context-dependent reduction approaches are briefly discussed, following [16]. The status of Cognitive Science has since long been the subject of debate within the philosophical literature; e.g. [2, 8, 9]. Among the issues questioned are the existence and status of higher-level cognitive laws, and the connection of a higher-level specification to reality. Within the philosophical literature on reduction since a long time much effort has been invested to address these issues, with partial success; e.g., [11]. In response to the severe criticisms, alternative views have been explored.

In recent years much attention has been paid to explore the possibilities of the notion of *mechanism* within Philosophy of Science; e.g., [3, 6]. One of the issues addressed by mechanisms is how a certain (higher-level) capability is realised by organised (lower-level) operations. This paper shows how certain aspects addressed by mechanisms can also be addressed by refinements of approaches to reduction, such as the bridge law approach, the functional approach, and the interpretation mapping approach.

Before going into the details, first some of the central claims from the literature in Philosophy of Mind are illustrated for an example case study:

(a) Cognitive laws are not genuine laws but depend on circumstances, for example, in the form of an organism's makeup.
(b) Cognitive laws can not be related (in a truth-preserving manner) to physical or biological laws.
(c) Cognitive concepts and laws cannot be related to reality in a principled manner, but, if at all, in different manners depending on circumstances.

A central issue in these claims is the observation that the relationship between a higher-level conceptualisation and reality has a dependency on the context of the physical or biological makeup of individuals and species, and this dependency remains unaddressed and hidden in the classical reduction approaches. Perhaps one of the success factors of the approaches based on mechanisms is that referring to a mechanism can be viewed as a way to make this context-dependency explicit.

To get more insight in the issue, an example case study is used concerning functionality for adaptive behaviour, as occurs, in conditioning processes in the sea hare *Aplysia*. For *Aplysia* underlying neural mechanisms of learning are well understood, based on long term changes in the synapses between neurons; see, for example, [5]. *Aplysia* is able to learn based on the (co)occurrence of certain stimuli; for example; see [5].

The example functionality for adaptive behaviour is described from a global external viewpoint as follows. Before a learning phase a tail shock leads to a response (contraction), but a light touch on its siphon is insufficient to trigger such a response. Suppose a training period with the following protocol is undertaken: in each trial the subject is touched lightly on its siphon and then immediately shocked on its tail. After a number of trials the behaviour has changed: the subject also shows a response (contraction) to a siphon touch. From an external viewpoint, the overall behaviour can be summarised by the specification of a relationship between stimuli and (re)actions involving a number of time points:

> If a number of times a siphon touch occurs, immediately followed by a tail shock, and after that a siphon touch occurs, then contraction will take place.

\To obtain a higher-level description of the functionality of this adaptive behaviour, a sensitivity state for stimulus-action pairs s-a is assumed that can have levels low, medium and high, where high sensitivity entails that stimulus s results in action a, and lower sensitivities do not entail this response:

> If s-a sensitivity is high and stimulus s occurs, then action a occurs.
> If stimulus stim1 and stimulus stim2 occur and stim1-a sensitivity is high, and stim2-a sensitivity is not high, then stim2-a sensitivity becomes one level higher.

As a next step, it is considered how the mechanism behind the higher-level description works at the biological level for *Aplysia*. The internal neural mechanism for *Aplysia*'s conditioning can be depicted as in Fig. 1, following [5].

A tail shock activates a sensory neuron SN1. Activation of SN1 activates the motoneuron MN via the synapse S1; activation of MN makes the sea hare move. A siphon touch activates the sensory neuron SN2. Activation of SN2 normally is not sufficient to activate MN, as the synapse S2 is not strong enough. After learning, the synapse S2 has become stronger and activation of SN2 is sufficient to activate MN. During the learning SN2 and MN are activated simultaneously, and the strength of the synapse S2 increases. This description is on the one hand based on the specific makeup of Aplysia's neural system, but on the other hand makes use of general neurological laws. A (simple) neurological theory consisting of the following laws explains the mechanism:

Activations of neurons propagate through connections via synapses with high strength. Simultaneous activation of two connected neurons increases the strength of the synapse connecting them. When an external stimulus occurs that is connected to a neuron, then this neuron will be activated. When a neuron is activated that is connected to an external action, then this action will occur.

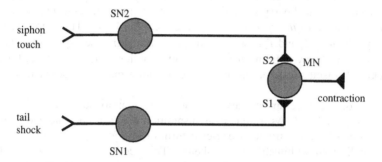

Fig. 1. A neural mechanism for adaptive functionality

Claims (a) and (b) discussed above are illustrated by the Aplysia case as follows. The neurological laws considered are general laws, independent of any specific makeup; they are (assumed to be) valid for any neural system. In contrast, the validity of the higher-level specification not only depends on these laws but also on the makeup of the specific type of neural system; for example, if some of the connections of Aplysia's neural system are absent (or wired differently), then the higher-level specification will not be valid for this organism. As the neurological laws do not depend on this makeup, the higher-level specification can not be related (in a truth-preserving manner) to the neurological laws. Claim (c) can be illustrated by considering other species than Aplysia as well, with different neural makeup, but showing similar conditioning processes. A central issue shown in this illustration of the claims is the notion of makeup, which provides a specific context of realisation of the higher-level specification. Indeed, the classical approaches to reduction ignore this aspect, whereas the approaches based on the notion of mechanism explicitly address it. However, variants of these classical approaches can be defined that also explicitly take into account this aspect of context-dependency, and thus provide support for the claims (a) to (c) instead of ignoring them. This will be addressed in Section 3.

3 Context-Dependent Reduction Relations

Reduction addresses relationships between descriptions of two different levels, usually indicated by a higher-level theory T_2 (e.g., a cognitive theory) and a lower-level or base theory T_1 (e.g., a neurological theory). A specific reduction approach provides a particular *reduction relation*: a way in which each higher-level property or law a (an expression in T_2) can be related to a lower-level property or law b (an expression in T_1), this b is often called a *realiser* for a. Reduction approaches differ in how these relations are defined. Within the traditional philosophical literature on reduction, three approaches play a central role. In the classical approach, following Nagel [11] reduction relations are based on (biconditional) bridge principles $a \leftrightarrow b$ that relate the expressions a in the language of a higher-level theory T_2 to expressions b in the language of the lower-level or base theory T_1. In contrast to Nagel's *bridge law reduction*, *functional reduction* (e.g., [9]) is based on functionalisation of a state property a in terms of its causal task C, and relating it to a state property b in T_1 performing this causal task C. From the logical perspective two closely related notions to formalise reduction relations are *(relative) interpretation mappings* (e.g., [14, 10]. These approaches relate the two theories T_2 and T_1 based on a mapping φ relating the expressions a of T_2 to expressions b of T_1, by defining $b = \varphi(a)$. Within philosophical literature, for example, Bickle [2] discusses a variant of the interpretation mapping approach with roots in [7].

For each of the three approaches to reduction as mentioned a context-dependent variant will be defined. As a source of inspiration [8] is used, where it is briefly sketched how a local or structure-restricted form of bridge law reduction can handle multiple realisation within different makeups. This section shows how this idea of context-dependent reduction can be worked out for each of three approaches, thus obtaining variants making the dependency on a specific makeup.

In context-dependent reduction the aim is to identify multiple context-specific sets of realisers. When contexts are defined in a sufficiently fine-grained manner, within one context the set of realisers can be taken to be unique. The contexts may be chosen in such a manner that all situations in which a specific type of realisation occurs are grouped together and described by this context. In Cognitive Science such a grouping could be based on species. When within each context one unique set of realisers exists, from an abstract viewpoint contexts can be seen as a form of parameterisation of the different possible sets of realisers.

In context-dependent reduction approaches, a context can be taken a description S (of an organism or system with a certain structure) by a set of statements within the language of the lower-level theory T_1. For a given context S as a parameter, for each expression of T_2 there exists a realiser within the language of T_1. Context-dependent reduction as sketched by Kim ([8], pp. 233-236), assumes that the contexts all are specified within the same base theory T_1. However, if mental state properties (for example, having certain sensory representations) are assumed that can be shared between, for example, biological organisms and robot-like architectures, it may be useful to allow contexts that are described within different base theories. In the multi-theory-based multi-context reduction approach developed below, a collection of lower-level theories $\mathcal{T}_1$ is assumed, and for each theory T in $\mathcal{T}_1$ a set of contexts C_T, such that each organism or system is described by a specific theory T in $\mathcal{T}_1$ together with a specific context or makeup S in C_T; these contexts S are assumed to be descriptions in the language of T and consistent with T. For the case that within one context only one realisation is possible, the theories T in $\mathcal{T}_1$ and contexts S in C_T can be used to parameterise the different sets of realisers that are possible. Below it is shown how contexts can be incorporated in the three reduction approaches discussed above, adopted from [16].

Context-dependent Bridge Law Reduction

For this approach, a unique set of realisers is assumed within each context S for a theory T in $\mathbf{\mathcal{T}_I}$; this is expressed by context-dependent biconditional bridge laws. Such context-dependent bridge laws are parameterised by the theory T in $\mathbf{\mathcal{T}_I}$ and context S in $\mathbf{C}_T$, and can be specified by

$a_1 \leftrightarrow b_{1,T,S}, ..., a_k \leftrightarrow b_{k,T,S}$

Here a_i is an expression specified in the language of theory T_2, and b_i is an expression in the language of theory T_1 corresponding to a_i. Given such a parameterised specification, the criterion of context-dependent bridge law reduction for a law $L(a_1, ..., a_k)$ of T_2 can be formulated (in two equivalent manners) by:

(i) $T_2 \vdash L(a_1, ..., a_k) \Rightarrow$

$\forall T \in \mathbf{\mathcal{T}_I} \forall S \in \mathbf{C}_T T \cup S \cup \{a_1 \leftrightarrow b_{1,T,S}, ..., a_k \leftrightarrow b_{k,T,S}\} \vdash L(a_1, ..., a_k)$

(ii) $T_2 \vdash L(a_1, ..., a_k) \Rightarrow \forall T \in \mathbf{\mathcal{T}_I} \forall S \in \mathbf{C}_T T \cup S \vdash L(b_{1,T,S}, ..., b_{k,T,S})$

Here $T \vdash A$ denotes that A is derivable in T. Note that this notion of context-dependent bridge law reduction implies unique realisers (up to equivalence) per context: from a $\leftrightarrow b_{T,S}$ and a $\leftrightarrow b'_{T,S}$ it follows that $b_{T,S} \leftrightarrow b'_{T,S}$. So the idea is that to obtain context-dependent bridge law reduction in cases of multiple realisation, the contexts are defined with such a fine grain-size that within one context unique realisers exist.

Context-dependent Functional Reduction

For a given collection of context theories $\mathbf{\mathcal{T}_I}$ and sets of contexts $\mathbf{C}_T$, for context-dependent functional reduction a first criterion is that a joint causal role specification $C(P_1, ..., P_k)$ can be identified such that it covers all relevant state properties of theory T_2. As an example, consider the case discussed in ([8], pp. 105-107). Here the joint causal role specification C(alert, pain, distress) for three related mental state properties is described by:

For any x,
if x suffers tissue damage and is normally alert, x is in pain
if x is awake, x tends to be normally alert
if x is in pain, x winces and groans and goes into a state of distress
if x is not normally alert or is in distress, x tends to make typing errors

By a Ramseification process [13] the following joint causal role specification is obtained. There exist properties P_1, P_2, P_3 such that $C(P_1, P_2, P_3)$ holds, where $C(P_1, P_2, P_3)$ is

For any x,
if x suffers tissue damage and has P_1, x has P_2
if x is awake, x has P_1
if x has P_2, x winces and groans and has P_3
if x has not P_1 or has P_3, x tends to make typing errors

The state property 'being in pain' of an organism is formulated in a functional manner as follows:

There exist properties P_1, P_2, P_3 such that $C(P_1, P_2, P_3)$ holds and the organism has property P_2.

Similarly, '*being alert*' is formulated as:

There exist properties P_1, P_2, P_3 such that $C(P_1, P_2, P_3)$ holds and the organism has property P_1.

A first criterion for context-dependent functional reduction is that for each theory T in $\mathbfit{T}_1$ and context S in $\mathbf{C}_T$ at least one instantiation of it within T exists:

$\forall T \in \mathbfit{T}_1 \ \forall S \in \mathbf{C}_T \ \exists P_1, ..., P_k \ \ T \cup S \vdash C(P_1, ..., P_k)$.

The second criterion for context-dependent functional reduction, concerning laws or regularities L is

$T_2 \vdash L(a_1, ..., a_k)$
$\Rightarrow \ \forall T \in \mathbfit{T}_1 \ \forall S \in \mathbf{C}_T \ \forall P_1, ..., P_k \ [T \cup S \vdash C(P_1, ..., P_k) \ \Rightarrow \ T \cup S \vdash L(P_1, ..., P_k)]$

In general this notion of context-dependent functional reduction may still allow multiple realisation within one theory and context. However, by choosing contexts with an appropriate grain-size it can be achieved that within one given theory and context unique realisation occurs. This can be done by imposing the following additional criterion expressing that for each T in $\mathbfit{T}_1$ and context S in $\mathbf{C}_T$ there exists a unique set of instantiations (parameterised by T and S) realising the joint causal role specification $C(P_1, ..., P_k)$:

$\forall T \in \mathbfit{T}_1 \ \forall S \in \mathbf{C}_T \ \exists P_1, ..., P_k \ [T \cup S \vdash C(P_1, ..., P_k) \ \&$
$\quad \forall Q_1, ..., Q_k \ [\ T \cup S \vdash C(Q_1, ..., Q_k)$
$\quad\quad\quad\quad \Rightarrow \ T \cup S \vdash P_1 \leftrightarrow Q_1 \ \& \ ... \ \& \ P_k \leftrightarrow Q_k \ ]]$

This *unique realisation criterion* guarantees that for all systems with theory T and context S any basic state property in T_2 has a unique realiser, parameterised by theory T in $\mathbfit{T}_1$ and context S in $\mathbf{C}_T$. When also this third criterion is satisfied, a form of reduction is obtained that we call *strict context-dependent functional reduction*. Based on the unique realisation criterion, the universally quantified form for relations between laws is equivalent to the following existentially quantified variant:

$T_2 \vdash L(a_1, ..., a_k) \ \Rightarrow$
$\forall T \in \mathbfit{T}_1 \ \forall S \in \mathbf{C}_T \ \exists P_1, ..., P_k \ [T \cup S \vdash C(P_1, ..., P_k) \ \& \ T \cup S \vdash L(P_1, ..., P_k)]$

Context-dependent Interpretation Mappings

To obtain a form of context-dependent interpretation, the notion of interpretation mapping can be generalised to a multi-mapping, parameterised by contexts. A *context-dependent interpretation* of a theory T_2 in a collection of theories $\mathbfit{T}_1$ with sets of contexts $\mathbf{C}_T$ specifies for each theory T in $\mathbfit{T}_1$ and context S in $\mathbf{C}_T$ an appropriate mapping $\varphi_{T,S}$ from the expressions of T_2 to expressions of T. When both the higher and lower level theories are specified using a sorted predicate language, then such a multi-mapping can be defined on the basis of mappings of each predicate symbol from the language of T_2 and of its arguments – terms of the language of T_2 – to formulae in the language of T_1. Mappings of sorts, constants, variables and functions may be specified to define mappings of terms. Mappings of composite formulae in the language of T_2 are defined as follows:

$$\varphi_{T,S}(A_1 \,\&\, A_2) \quad = \quad \varphi_{T,S}(A_1) \,\&\, \varphi_{T,S}(A_2)$$

$$\varphi_{T,S}(\neg\, A) \quad = \quad \neg\, \varphi_{T,S}(A)$$

$$\varphi_{T,S}(\exists x.\, A) \quad = \quad \exists \varphi_{T,S}(x)\, \varphi_{T,S}(A)$$

Here A, A_1 and A_2 are formulae in the language of T_2. A multi-mapping $\varphi_{T,S}$ is a context-dependent interpretation mapping when it satisfies the property that if a law (or regularity) L can be derived from T_2, then for each T in $\mathcal{T}_1$ and context S in $\mathbf{C}_T$ the corresponding $\varphi_{T,S}(L)$ can be derived from $T \cup S$:

$$T_2 \vdash L \;\Rightarrow\; \forall T \in \mathcal{T}_1\; \forall S \in \mathbf{C}_T \quad T \cup S \vdash \varphi_{T,S}(L)$$

Note that also here within one theory T in $\mathcal{T}_1$ and context S in $\mathbf{C}_T$ multiple realisation is still possible, expressed as the existence of two essentially different interpretation mappings $\varphi_{T,S}$ and $\varphi'_{T,S}$, i.e., such that it does not always hold that $\varphi_{T,S}(a) \leftrightarrow \varphi'_{T,S}(a)$. An additional criterion to obtain unique realisation per context is: when for a given theory T in $\mathcal{T}_1$ and context S in $\mathbf{C}_T$ two interpretation mappings $\varphi_{T,S}$ and $\varphi'_{T,S}$ are given, then for all formulae a in the language of T_2 it holds that

$$T \cup S \vdash \varphi_{T,S}(a) \leftrightarrow \varphi'_{T,S}(a)$$

When for each theory and context this additional criterion is satisfied as well, the interpretation is called a *strict context-dependent interpretation*.

4 Case Study

In this section the applicability of the context-dependent reduction approaches described in Section 3 is illustrated for a case study involving adaptive functionality inspired by the conditioning processes in *Aplysia* (see Section 2) which is worked out in much formal detail.

To formalise both the lower and higher level theories the reified temporal predicate language *RTPL* [14] has been used, a many-sorted temporal predicate logic language that allows specification and reasoning about the dynamics of a system. To express state properties of a system ontologies are used. An ontology is a signature specified by a tuple $<S_1,..., S_n,..., C, f, P, arity>$, where S_i is a sort for $i=1,.., n$, C is a finite set of constant symbols, f is a finite set of function symbols, P is a finite set of predicate symbols, arity is a mapping of function or predicate symbols to a natural number. In *RTPL* state properties (that can be represented by formulae within the state language) are used as terms (denoting objects). The sort STATPROP contains the names of all state properties. The set of function symbols of *RTPL* includes $\wedge, \vee, \rightarrow, \leftrightarrow$: STATPROP x STATPROP $\rightarrow$ STATPROP; not: STATPROP $\rightarrow$ STATPROP, and $\forall, \exists$: SVARS x STATPROP $\rightarrow$ STATPROP, of which the counterparts in the state language are Boolean propositional connectives and quantifiers. To represent dynamics of a system sort TIME (a set of time points) and the ordering relation $>$: TIME x TIME are introduced in *RTPL*. To indicate that some state property holds at some time point the relation at: STATPROP x TIME is introduced. The terms of *RTPL* are constructed by induction in a standard way from variables, constants and function symbols typed with all before-mentioned sorts. The set of well-formed *RTPL* formulae is defined inductively in a standard way using

Boolean connectives and quantifiers over variables of *RTPL* sorts. The language *RTPL* has the semantics of many-sorted predicate logic.

In the following a specification of the higher-level model *HM* for conditioning (as in *Aplysia*) is provided formalised in *RTPL* using the state ontology from Table 1. Time is assumed to be discrete in this example, and sort TIME contains natural numbers.

Table 1. State ontology for the higher-level model HM

Sort	Elements
STIMULUS	stim1, stim2
ACTION	contraction
DEGREE	low, medium, high
Predicate	**Description**
sensitivity: STIMULUS x ACTION x DEGREE	Describes the sensitivity degree of a stimulus-action relation
observesstimulus: STIMULUS	Describes the observation of a stimulus
performsaction: ACTION	Describes an action being performed

In the following formalization a and s are variable names.

HMP1 Action performance

For any time point, if the sensitivity of a relation s-a is high and the stimulus s is observed, then at some later time point action a will be performed.
Formally:

$\forall$t1:TIME [at(sensitivity(s, a, high) $\wedge$ observesstimulus(s), t1) $\Rightarrow$ $\exists$t2:TIME t2 > t1 & at(performsaction(a), t2)]

HMP2 Sensitivity increase

For any time points t1 and t2, such that t1+1 < t2 ≤ t1+c5+1
if stimulus stim1 is observed at t1
   and the sensitivity of relation stim1-a is high
   and stimulus stim2 is observed at t2
   and the sensitivity of relation stim2-a is v,
   and v' is the value-successor of v,
then at t2+2 the sensitivity of relation stim2-a will become v'. Formally:

$\forall$t1, t2:TIME $\forall$v, v':DEGREE [t1+1 < t2 ≤ t1+c5+1 & at(observesstimulus(stim1) $\wedge$ sensitivity(stim1, a, high), t1) & at(observesstimulus(stim2) $\wedge$ sensitivity(stim2, a, v) $\wedge$ has_successor(v, v'), t2) $\Rightarrow$ at(sensitivity(stim2, a, v'), t2+2)]

Both has_successor(low, medium) and has_successor(medium, high) are always TRUE.

HMP3 Unconditional persistency of the high sensitivity value
Formally:

$\forall$t4:TIME [at(sensitivity(s, a, high), t4) $\Rightarrow$ at(sensitivity(s, a, high), t4+1)]

HMP4 Conditional persistency of the sensitivity value other than high

For any time point t5,
if the sensitivity value of the relation stim2-a is v≠high and
  and not
     stimulus stim2 was observed at time point t5-1,

and there exists time point t6 t5-1 > t6 ≥ t5 - c5 -1 such that stimulus stim1 was observed at t6
then at the next time point the sensitivity value of the relation stim2-a stays the same.
Formally:

∀t5:TIME ∀v:DEGREE [at(sensitivity(stim2, a, v) ∧ v≠high, t5) &
¬(at(observesstimulus(stim2), t5-1) & ∃t6 t5-1 > t6 ≥ t5 − c5 − 1 & at(observesstimulus(stim1),
t6)) ⇒ at(sensitivity(stim2, a, v), t5+1)]

A lower-level model *LM* for the same adaptive functionality is formalised below as a
neurological makeup *NM* together with the general neurological activation rules *NA*.
For the formalisation the ontology from Table 2 were used.

Table 2. State ontology for formalising the lower-level model LM

Sort	Elements
NEURON	sn1, sn2, mn
SYNAPSE	S1, S2
VALUE	natural numbers
Predicate	**Description**
stimulusconnection: STIMULUS x NEURON	Describes a connection between a stimulus and a (sensory) neuron
occurs: STIMULUS, occurs: ACTION	Describes an occurrence of a stimulus/action
activated: NEURON	Describes the activation of a neuron
connectedvia: NEURON x NEURON x SYNAPSE	Describes a connection between two neurons by a synapse
has_strength: SYNAPSE x VALUE	Describes the strength of a synapse
actionconnection: NEURON x ACTION	Describes a connection between a (preparatory) neuron and an action

LMP1 Neuron activation based on a stimulus
For any time point,
if a stimulus occurs,
then the neuron connected to this stimulus will be activated for c5 following time points.
Formally:

∀t5:TIME ∀st:STIMULUS ∀y:NEURON [at(stimulusconnection(st, y) ∧ occurs(st), t5)
⇒ ∀t2:TIME t5 < t2 ≤ t5+c5 & at(activated(y), t2)]

LMP2 Propagation of neuron activations
For any time point, if a neuron is activated, and this neuron is connected to some other neuron
by a synapse with strength higher than B2,
then the other neuron will be also activated at the next time point. Formally:

∀t1:TIME ∀x, y:NEURON ∀s:SYNAPSE ∀v:VALUE [at(connectedvia(x, y, s) ∧ activated(x)
∧ has_strength(s, v) ∧ v > B2, t1) ⇒ at(activated(y), t1+1)]

LMP3 Increase of the synapse's strength
For any time point,
if two neurons connected by a synapse with strength v are activated
and at the previous time point both neurons were not activated,
then at the next time point the strength of the synapse will be v+d(v). Formally:

∀t3:TIME ∀x,y:NEURON ∀s:SYNAPSE ∀v:VALUE [at(activated(x) ∧ activated(y) ∧ connectedvia(x, y, s) ∧ has_strength(s, v), t3) & at(not(activated(x) ∧ activated(y)), t3-1) ⇒
at(has_strength(s, v+d(v)), t3+1)]

LMP4 Conditional persistency of the strength value of a synapse
For any time point,

if the value of a synapse is v,
and not
both neurons are activated and
at the previous time point both neurons were not activated,
then the synapse's strength remains the same. Formally:

∀t4:TIME ∀x,y:NEURON ∀s:SYNAPSE ∀v:VALUE [at(connectedvia(x, y, s) ∧ has_strength(s, v), t4) & ¬(at(activated(x) ∧ activated(y)), t4) & at(not(activated(x) ∧ activated(y)), t4-1)) ⇒ at(has_strength(s, v), t4+1)]

LMP5 Occurrence of an action
For any time point,
if a neuron is not activated
and at the previous time point the neuron was activated,
then after c4 time points the action related to the neuron will be performed. Formally:

∀t7: TIME ∀x:NEURON [at(not(activated(x)), t7) & at(actionconnection(x, a) ∧ activated(x), t7-1)
⇒ at(occurs(a), t7+c4)]

The neurological makeup *NM* is assumed to be stable in this example and is specified more formally as follows (inspired by *Aplysia*'s makeup shown in Fig. 1):

∀t:TIME at(stimulusconnection(stim1, SN1) ∧ stimulusconnection(stim2,SN2) ∧ connectedvia(SN,MN, S1) ∧ connectedvia(SN2, MN, S2) ∧ actionconnection(MN, contraction) ∧ has_strength(S1,v1) ∧ has_strength(S2,v2),t)

Applying the context-dependent Reduction Approaches

An interpretation mapping from the higher-level model *HM* to the lower-level model *LM* can be defined as follows. The variables and constants of sorts ACTION, STIMULUS, TIME, VALUE are mapped without changes.

$\varphi_{NA,NM}$(v:DEGREE)   = v:VALUE, where v is a variable.

Suppose within the context of makeup *NM*, stimulus s is connected to the motoneuron MN via a path passing synapse S, then:

$\varphi_{NA,NM}$(sensitivity(s, a, low)) = has_strength (S, v) ∧ v < B1
$\varphi_{NA,NM}$(sensitivity(s, a, medium)) = has_strength (S, v) ∧ B1 ≤ v ∧ v ≤ B2
$\varphi_{NA,NM}$(sensitivity(s, a, high)) = has_strength (S, v) ∧ v > B2
$\varphi_{NA,NM}$(sensitivity(s, a, v)) = has_strength (S, v),
where v is a variable

To avoid clashes between names of variables, every time when a new variable is introduced by a mapping, it should be given a name different from the names already used in the formula.

Note that the reduction relation depends on the context *NM*. Within context *NM* sensitivity for stimulus stim1 relates to synapse S1 and sensitivity for stimulus stim2 to synapse S2. Therefore, for example,

$\varphi_{NA,NM}$(sensitivity(stim1, a, high)) = has_strength (S1, v) ∧ v > B2
$\varphi_{NA,NM}$(sensitivity(stim2, a, high)) = has_strength (S2, v) ∧ v > B2.

Here v is a variable of sort VALUE.

Observation and action predicates are mapped as follows:

$\varphi_{NA,NM}$(observesstimulus(s))   = occurs(s)
$\varphi_{NA,NM}$(performsaction(a))   = occurs(a)
$\varphi_{NA,NM}$(has_successor(v, v'))   = v'=v + d(v)

All other function and predicate symbols of the language of *HM* are mapped without changes.

Based on the mapping $\varphi_{NA,NM}$ as defined for basic state properties, by compositionality the mapping of more complex relationships is made as described in Section 3, for example:

$\varphi_{NA,NM}(\forall t1:TIME$ [at(sensitivity(s, a, high) $\wedge$ observesstimulus(s), t1) $\Rightarrow$
$\exists t2:TIME$ t2 > t1 & at(performsaction(a), t2)])

$= \forall t1:TIME$ [at($\varphi_{NA,NM}$(sensitivity(s, a, high) $\wedge$ observesstimulus(s), t1) $\Rightarrow$
$\exists t2:TIME$ t2 > t1 & at($\varphi_{NA,NM}$(performsaction(a)), t2)]

$= \forall t1:TIME$ [at($\varphi_{NA,NM}$(sensitivity(s, a, high)) $\wedge$ $\varphi_{NA,NM}$(observesstimulus(s)), t1) $\Rightarrow$
$\exists t2:TIME$ t2 > t1 & at($\varphi_{NA,NM}$(performsaction(a)), t2)]

$= \forall t1:TIME$ [at(has_strength (syn, v) $\wedge$ v > B2 $\wedge$ occurs(s), t1) $\Rightarrow$
$\exists t2:TIME$ t2 > t1 & at(occurs(a), t2)]

Here s and a are the variable names and variable syn corresponds to s.

This and other regularities derivable from the higher-level specification *HM* can be mapped automatically as described below in Section 5 onto regularities that are derivable from $NA \cup NM$, which illustrates the criterion for interpretation mapping.

In similar manners the other two context-based approaches can be applied to the case study. For example, context-dependent bridge principles for *NA* and context *NM* can be defined by (where the path from stimulus s to neuron *MN* is via synapse S):

sensitivity(s, a, low) $\leftrightarrow$ has_strength (S, v) $\wedge$ v < B1
sensitivity(s, a, medium) $\leftrightarrow$ has_strength (S, v) $\wedge$ B1 $\leq$ v $\wedge$ v $\leq$ B2
sensitivity(s, a, high) $\leftrightarrow$ has_strength (S, v) $\wedge$ v > B2
observesstimulus(s) $\leftrightarrow$ occurs(s)
performsaction(a) $\leftrightarrow$ occurs(a)
has_successor(v, v') $\leftrightarrow$ v'=v + d(v)
v:DEGREE $\leftrightarrow$ v:VALUE,

where v is a variable.

Context-dependent functional reduction can be applied by taking the joint causal role specification for sensitivity(stim2, a, low), sensitivity(stim2, a, medium), sensitivity(stim2, a, high) assuming that the sensitivity of relation stim1-a is high as follows:

C(P1, P2, P3) = def
[$\forall t1, t2:TIME$ [t1+1 < t2 $\leq$ t1+c5+1 & at(observesstimulus(stim1), t1) &
at(observesstimulus(stim2) $\wedge$ P1, t2)
$\Rightarrow$ at(P2, t2+2)]] & [$\forall t1, t2:TIME$ [t1+1 < t2 $\leq$ t1+c5+1 & at(observesstimulus(stim1), t1) &
at(observesstimulus(stim2) $\wedge$ P2, t2)
$\Rightarrow$ at(P3, t2+2)]] &
[$\forall t1:TIME$ [at(P3 $\wedge$ observesstimulus(s), t1)
$\Rightarrow$ $\exists t2:TIME$ t2 > t1 & at(performsaction(a), t2)]] &
[$\forall t4:TIME$ at(P3, t4) $\Rightarrow$ at(P3, t4+1)] &
[$\forall t5:TIME$ $\forall v:DEGREE$ [at(P1, t5) & $\neg$(at(observesstimulus(stim2),t5-1) & $\exists t6$ t5-1> t6 $\geq$ t5 $-$
c5-1 at(observesstimulus(stim1), t6)) $\Rightarrow$ at(P1, t5+1)]] &
[$\forall t5:TIME$ $\forall v:DEGREE$ [at(P2, t5) & $\neg$(at(observesstimulus(stim2), t5-1) & $\exists t6$ t5-1>t6 $\geq$ t5$-$
c5- 1 at(observesstimulus(stim1), t6)) $\Rightarrow$ at(P2, t5+1)]]

5 Implementation

To perform an automated context-dependent mapping of a higher level model specification to a lower level model specification, a software tool has been implemented in Java™ based on the mapping principles described in Sections 3 and 4. As input for this tool a higher level model specification in sorted predicate logic is provided together with a set of mappings of basic elements of the ontology used for formalisation of the higher level specification. While a mapping is being performed on any higher-level formula, the tool traces possible clashes of variable names and renames new variables when needed. As a result, a specification in the lower level specification language is generated.

The context-dependent interpretation mapping should satisfy the reduction conditions described in Section 3. For the case considered, these conditions have the form: if a law (or property) L is derived from HM, then the corresponding $\varphi_{NA,NM}(L)$ should be derived from NA $\cup$ NM: HM $\vdash$ L $\Rightarrow$ NA $\cup$ NM $\vdash \varphi_{NA,NM}(L)$. This will be applied to the properties in the specification HM.

Since both HM and NA $\cup$ NM are specified using the reified temporal predicate language, to establish if a formula can be derived from a set of formulae, the theorem prover Isabelle for many-sorted higher-order logic has been used [12]. As input for Isabelle a theory specification is provided. A simple theory specification consists of a declaration of ontologies, lemmas and theorems to prove. Sorts are introduced using the construct datatype (e.g., datatype neuron = sn1| sn2| mn). Furthermore, sorts for higher-order logics can be defined: e.g., sort STATPROP is defined for the case study:

datatype statprop= stimulusconnection event neuron| activated neuron | occurs event | connectedvia neuron neuron synapse | actionconnection neuron event |has_strength synapse nat

Here each element of statprop refers to a state property, expressed using the state ontology. The elements of the state ontology should be also defined in the theory : e.g., activated:: "neuron $\Rightarrow$ statprop"; stimulusconnection:: "event $\Rightarrow$ neuron $\Rightarrow$ statprop". The formulae of the state language are imported into the reified language using the predicate at:: "statprop $\Rightarrow$ nat $\Rightarrow$ bool".

The first theory specification defines the following lemma expressing the criterion for the mapping of the property HMP1 (Action Performance), which expresses

NA $\cup$ NM $\vdash \forall$t1:TIME [at(has_strength (syn, v) $\wedge$ v > B2 $\wedge$ occurs(s), t1)
  $\Rightarrow \exists$t2:TIME t2 > t1 & at(occurs(a), t2)]

To enable the automated proof of this lemma the implication introduction rule [12] is applied, which moves the part $\forall$t1:TIME $\forall$s:STIMULUS [at(has_strength (syn, v) $\wedge$ v > B2 $\wedge$ occurs(s), t1) to the assumptions. Then, the lemma is proved automatically by the *blast* method, which is an efficient classical reasoner. Note that for the actual proof only the relevant part of NA $\cup$ NM has been used.

The second specification defines the lemma for the mapping of the property HMP2 (Sensitivity increase), which expresses

NA $\cup$ NM $\vdash \forall$t1, t2:TIME $\forall$v, v':VALUE [t1+1 < t2 $\leq$ t1+c5+1 &
at(occurs(stim1) $\wedge$ has_strength (S1, var) $\wedge$ var > B2, t1) &
at(occurs(stim2) $\wedge$ has_strength (S2, v) $\wedge$ v'=v + d(v), t2) $\Rightarrow$ at(has_strength (S2, v'), t2+2)]

For the proof of this lemma the same strategy has been used as for the previous example. The proofs of both examples have been performed in a fraction of a second.

6 Discussion

Within Cognitive Science, cognitive theories provide higher-level descriptions of the functioning of specific neural makeups. The concepts and relationships used in the descriptions do not have a direct one-to-one relationship to reality such as concepts and relationships used within Physics or Chemistry have. Due to the nontrivial dependence of cognitive theories on the context of specific (neural) makeups of individuals or species, relationships between cognitive states are not considered genuine universal laws; by changing the specific makeup they simply can be refuted. Therefore they cannot have a direct truth-preserving relationship to general physical/biological laws. The classical approaches to reduction do not take into account this context-dependency in an explicit manner. Therefore, refinements of these classical reduction approaches are used in this paper that incorporate the context-dependency in an explicit manner. These context-dependent reduction approaches make explicit how laws or regularities in a cognitive theory depend on lower-level laws on the one hand and specific makeups on the other hand. The detailed formalised definitions of the approaches described in this paper enable practical application to higher-level and lower-level knowledge specification. As in the case of cognitive theories, here the context-dependent reduction approaches make explicit how concepts and relationships in higher-level specifications relate to lower-level specifications. Using these formalized relations reduction approaches can be automated. In particular, this paper illustrates how the interpretation mapping approach can be automated, including mapping of specifications and checking the fulfilment of reduction criteria. In the example considered the mapping of basic ontological elements was assumed to be given. In the future research approaches to identify basic ontological mappings will be developed.

References

1. Bennett, M.R., Hacker, P.M.S.: Philosophical Foundations of Neuroscience. Blackwell, Malden (2003)
2. Bickle, J.: Psychoneural Reduction: The New Wave. MIT Press, Cambridge (1998)
3. Craver, C.F.: Role Functions, Mechanisms and Hierarchy. Philosophy of Science 68, 31–55 (2001)
4. Galton, A.: Operators vs Arguments: The Ins and Outs of Reification. Synthese 150, 415–441 (2006)
5. Gleitman, H.: Psychology. W.W. Norton & Company, New York (2004)
6. Glennan, S.S.: Mechanisms and the Nature of Causation. Erkenntnis 44, 49–71 (1996)
7. Hooker, C.: Towards a General Theory of Reduction. Dialogue 20, 38–59, 201–236, 496–529 (1981)
8. Kim, J.: Philosophy of Mind. Westview Press (1996)
9. Kim, J.: Physicalism, or Something Near Enough. Princeton University Press, Princeton (2005)
10. Kreisel, G.: Models, translations and interpretations. In: Skolem, T.A.l. (ed.) Mathematical Interpretation of Formal Systems, pp. 26–50. North-Holland, Amsterdam (1955)
11. Nagel, E.: The Structure of Science. Harcourt, Brace & World, New York (1961)
12. Nipkow, T., Paulson, L.C., Wenzel, M.: Isabelle/HOL. LNCS, vol. 2283. Springer, Heidelberg (2002)
13. Ramsey, F.P.: Theories. Ramsey, F.P (1931); The Foundations of Mathematics and Other Essays. In: Braithwaite, R. B. (eds.). Routledge and Kegan Paul, London (1929)
14. Schoenfield, J.R.: Mathematical Logic. Addison-Wesley, Reading (1967)
15. Tarski, A., Mostowski, A., Robinson, R.M.: Undecidable Theories. North-Holland, Amsterdam (1953)
16. Treur, J.: Laws and Makeups in Context-Dependent Reduction Relations. In: Love, B.C., McRae, K., Sloutsky, V.M. (eds.) Proc. of the 30th Annual Conference of the Cognitive Science Society, CogSci 2008, pp. 1752–1757. Cognitive Science Society, Austin (2008)

Combining Artificial Intelligence Techniques for the Training of Power System Control Centre Operators*

Luiz Faria, António Silva, Zita Vale, and Carlos Ramos

Engineering Institute - Porto Polytechnic Institute, Porto, Portugal

Abstract. Control Centre operators are essential to assure a good performance of Power Systems. Operators actions are critical in dealing with incidents, especially severe faults, like blackouts. In this paper we present an Intelligent Tutoring approach for training Portuguese Control Centre operators in incident analysis and diagnosis, and service restoration of Power Systems, offering context awareness and an easy integration in the working environment.

1 Introduction

Power Systems have a very complex structure and require sophisticated and precise operation and control. The most important real-time decisions concerning Power System operation are taken in Control Centres (CC) by specially trained operators. Although Power System reliability has been increasing during the last decades, incidents still occur, resulting sometimes in blackout situations, with very high economic and social impact.

Blackouts have been a major concern in Power Systems specially since the occurrence of the 9th November 1965 Northeast Blackout in USA. In recent years, several blackouts assumed catasthrophic proportions. On 14th August 2003 a major blackout paralyzed most of the Northeast of USA and large areas of the Southeast of Canada. This incident affected directly more than 50 million people and took more than 4 days to recover from. More recently, on the 4th October 2006, the UCTE (Union for the Coordination of Transmission of Electricity) European Network experienced a quasi blackout situation affecting 9 European countries and North Africa and about 10 million consumers.

This last incident was due to a set of unforeseen circumstances and the difficulty felt by the CC operators in interacting quickly and in an effective manner. Moreover, the restoration process was hampered by the lack of coordination between the different entities involved. All this points to the creation of conditions allowing the restoration plans' training of the personnel belonging to the different utilities in a coordinated way [1].

Control Center operators performance is essential to minimize the incident consequences. The need of a good response of Control Centres to severe faults, is even more

---

* The authors would like to thank FCT foundation and FEDER, PEDIP, POSI, POSC, and PTDC programmes for their support in several research projects leading to the development of the work described here.

J. Filipe, A. Fred, and B. Sharp (Eds.): ICAART 2009, CCIS 67, pp. 70–83, 2010.

important nowadays, due to the widespread adoption of the Electricity Markets [2]. As Power Systems reliability increased, the number of incidents offering occasion for operator on-the-job training has decreased. The consequences of incorrect operator behaviour are all the more severe during a serious incident [3]. Operator training is vital for overcoming these problems, as well as the availability of adequate decision support tools.

Intelligent Tutoring Systems (ITS) have been chosen as the tool to address the operator's training requirements in diagnosis and restoration tasks. The reasons for that choice can be described as such:

1. They provide a way to represent domain knowledge in a structured and efficient way, allowing the inference of new knowledge.
2. They use a model of the trainee, showing a non-monotonous behavior and adapting better to the trainees characteristics and evolution.
3. With the right didactic knowledge they can select different pedagogical approaches in the various phases of the learning process.
4. They are able to monitor the trainees performance and evolution, gathering information to guide the system's adaptation.
5. They typically require very little intervention from the training staff, and can be used in the working environment without disturbing the normal routines.

In this paper we present an ITS used for the training of Control Centre operators. Several Artificial Intelligence techniques are used to make this system able to minimize experts'efforts in the training sessions preparation and to enable on the job and cooperative effective training. The ITS that has been developed for the Control Centre operators involves two main areas: one devoted to the training of fault diagnosis skills (DiagTutor) and another dedicated to the training of power system restoration techniques (CoopTutor). The Figure 1 shows this tutoring environment architecture.

2 Why Operators' Task Is So Demanding?

During the analysis of alarm messages lists, CC operators must look for the group of relevant messages that describe each type of fault. The same group of messages can show up in the reports of different types of faults. So CC operators have to analyze the arrival of additional information, whose presence or absence determines the final diagnosis. Additionally, operators have to deal with uncertain, incomplete and inconsistent information, due to data loss or errors occurred in the data gathering system.

In fact, if we consider all the messages that are generated during the period of an incident, including not only the messages originated in the plants involved in the incident but also in other plants of the Power System, operators can be forced to consider several hundreds of messages in just a few minutes. It is important to note that an incident usually causes the generation of not only the messages that are relevant to the analysis of this particular incident but also a lot of other messages that are not important in that context. However, on different contexts, these messages could be important, which stresses the need for a contextual interpretation of the information.

On the other hand, several incidents can take place almost at the same time, and one incident can have consequences in much more than two plants, resulting on a much

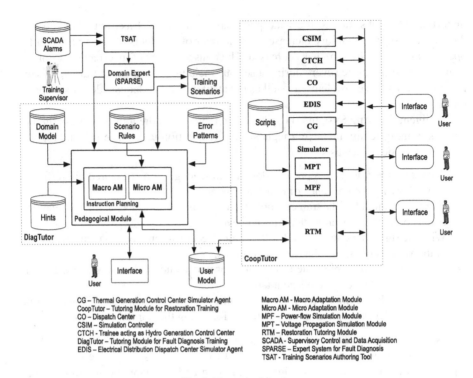

CG – Thermal Generation Control Center Simulator Agent
CoopTutor – Tutoring Module for Restoration Training
CO – Dispatch Center
CSIM – Simulation Controller
CTCH – Trainee acting as Hydro Generation Control Center
DiagTutor – Tutoring Module for Fault Diagnosis Training
EDIS – Electrical Distribution Dispatch Center Simulator Agent

Macro AM - Macro Adaptation Module
Micro AM - Micro Adaptation Module
MPF – Power-flow Simulation Module
MPT – Voltage Propagation Simulation Module
RTM – Restoration Tutoring Module
SCADA - Supervisory Control and Data Acquisition
SPARSE – Expert System for Fault Diagnosis
TSAT - Training Scenarios Authoring Tool

Fig. 1. Tutor Architecture

more complex interpretation of the situation. If we also take into account the need to consider missing information, we can have an idea of the difficulties CC operators face.

The occurrence of an incident demands a quick response from CC operators in order to minimize the consequences of that incident, avoiding the propagation of the incident to other power plants. It requires the start and completion of the service restoration as fast as possible, minimizing the period of service unavailability. Such goals require immediate CC operator reaction in two phases: first, he must make the correct incident diagnosis, and second, he must select and execute the correct maneuvers in order to restore power. This second step requires a close coordination between the teams responsible for power generation, transmission and distribution. Their actions should be based on a careful planning and guided by adequate strategies but those plans are often derailed by unforeseen circumstances. The correct execution of the restoration plan is very important because an incorrect maneuver can often leave the power system in a unstable state or even expand the area affected by the original blackout.

All these facts stress the need for preparing the CC operators to face the difficulties presented by the Power System operation. This paper presents a tutoring environment that addresses the two main areas of training: fault diagnosis skills (section 3) and power system restoration techniques in a cooperative environment (section 4).

3 Tutoring Module for Fault Diagnosis Training

This tutoring module, named DiagTutor, is focused on Fault Diagnosis Training. In order to illustrate how a training session is conducted and the interaction between the operator and the tutor, this section presents a very simplified diagnosis problem containing a DmR (monophase tripping with reclosure) incident, occurred in panel 204 of SED substation. The relevant SCADA[1] messages related to this incident are included in Table 1. These SCADA messages correspond to the following events: breaker tripping, breaker moving and breaker closing [3]. In a real training scenario the operator is faced with a huge amount of messages, typically several hundreds.

Table 1. Incident in panel 204 of SED substation

14-12-2003 04:24:45.200	SED	204	CCL,2	TRIPPING	0 1
14-12-2003 04:24:45.240	SED	204	CCL,2	-BK BREAKER	0 1
14-12-2003 04:24:45.860	SED	204	CCL,2	-BK BREAKER	0 1

3.1 Reasoning about Operator Answers

The interaction between the trainee and the tutor is performed by means of prediction tables (Figure 2) where the operator selects a set of premises and the corresponding conclusion. The premises represent events (SCADA messages), temporal constraints between events or previous conclusions [4].

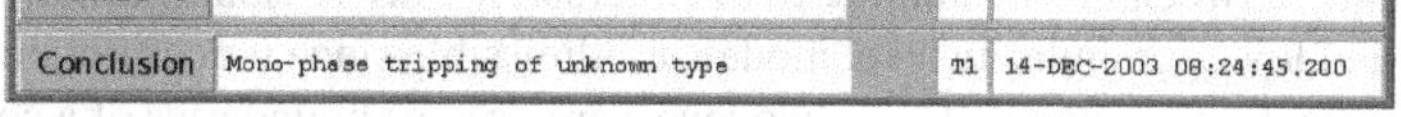

Fig. 2. Prediction Table

DiagTutor does not require the operators reasoning to follow a predefined set of steps, as in other implementations of the model tracing technique [5]. In order to evaluate this reasoning, the tutor will compare the prediction tables content with the specific situation model. This model has been obtained by matching the domain model with the inference produced by the SPARSE expert system [3]. With this process, the system is able to identify the errors that reveal the operators misconceptions, to provide assistance on each problem solving action, to monitor the trainee's knowledge evolution and to offer different learning opportunities.

---

[1] Supervisory Control and Data Acquisition system.

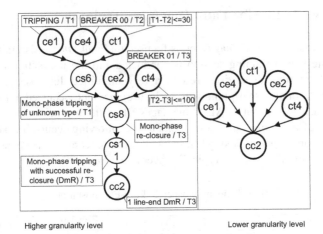

Fig. 3. Higher and lower granularity levels of the situation specific model

The identified errors are used as opportunities to correct the faults in the operators reasoning. The operators entries in prediction tables cause immediate responses from the tutor. In case of error, the operator can ask for help which is supplied as hints. Hinting is a tactic that encourages active thinking structured within guidelines dictated by the tutor [6]. The first hints are generic, becoming more detailed if the help requests are repeated.

The situation specific model generated by the tutoring system for the problem presented is shown in the left frame of Figure 3. It displays high granularity since it includes all the elementary steps used to get the problem solution. The tutor uses this model to detect errors in the operator reasoning by comparing the situation specific model with the set of steps used by the operator. This models granularity level is adequate to a novice trainee but not to an expert operator. The right frame of Figure 3 represents a model used by an expert operator, including only concepts representing events, temporal constraints between events and the final conclusion. Any reasoning model between the higher and lower granularity level models is admissible since it does not include any violation to the domain model. These two levels are used as boundaries of a continuous cognitive space.

The DiagTutor behavior during a training session includes two activities: the inference of the trainee reasoning during problem solving, in order to detect misconceptions; and the reaction to trainee errors and assistance through presentation of hints. These functions of DiagTutor are implemented by the Micro Adaptation Module (in Figure 1).

3.2   Adapting the Curriculum to the Operator

The Macro Adaptation Module, included in the Pedagogical Module (Figure 1), is responsible to adapt the Curriculum Plan to the current trainee needs. In particular, this module is able to select, from the Training Scenarios library, a problem fitting the trainee needs.

The preparation of the tutoring sessions learning material is typically a time-consuming task. In industrial environments, there is not usually a staff exclusively dedicated to training tasks. Specifically, in the electrical sector, the preparation of training sessions is done by the most experienced operators which are often already overloaded with power system operation tasks [7]. In order to overcome this difficulty, we developed two separate tools. The first one generates and classifies training scenarios from previously stored real cases. As these may not cover all the situations that CC operators must be prepared to face, another tool is used to create new training scenarios or to adapt already existing ones [7].

The process used by the Macro Adaptation Module to define the problems features involves two phases. First, the tutor must define the difficulty level of the problem, using heuristic rules. These rules relate parameters like the trainees performance in previous problems and his overall level of knowledge. In the second phase, the tutor uses the user models contents to select the most suitable type of incidents to be included in the problem, taking into account the domain concepts involved in each type of incident and the corresponding trainees expertise.

3.3 Difficulty Level Selection

To evaluate the problems difficulty level, we need to identify the cases characteristics that increase their complexity. The parameters involved here are the number of incidents contained in the case, the variety of incident types, the number of affected plants and the existence of chronological inversion in SCADA messages.

The choice of the difficulty level depends on two factors contained in the trainees model: the trainees global knowledge and a global acquisition factor. The first parameter is a measure of the trainee knowledge level in the whole range of domain concepts and is calculated using the mean of his knowledge level in each domain concept. The Macro Adaptation Module needs appropriate thresholds for deciding on the next problem difficulty level. The opinion of the trainees, regarding their personal evolution as the problems difficulty level is changed, can be used to adjust these thresholds.

The acquisition factors record how well trainees learn new concepts. When a new concept is introduced, the tutor monitors the trainee performance on the first few problems, namely how well and how quickly he solves them. This analysis determines the trainees acquisition factor. The procedure used to determine the trainees acquisition in each domain concept is based on the number of times the trainees knowledge level about the concept has increased, considering the three first applications of the concept.

The mechanism used to define the difficulty level of the problems is based on the following rule:

If the global knowledge level and the global acquisition factor change in opposite directions
  Then the problem difficulty level does not change
  Else the problem difficulty level changes in the same direction of the global knowledge level.

3.4 Problem Type Adequacy to the Trainee Cognitive Status

The mechanism used to classify each kind of incident in terms of adequacy to the trainee is based on a neural network (Figure 4).

The nodes belonging to the input layer correspond to the concepts (included in the domains knowledge base) to be assimilated by the trainees. Each node represents the application of a concept in a specific context. For instance, the nodes ce1/T1 and ce1/T5 represent two instances of the same concept and characterize the application of the concept of breaker tripping in the situations of first tripping and tripping after an automatic reclosure. The input vector contains an estimate of the trainees expertise level for each concept or its application and is obtained from the user model. Therefore, this vector represents an estimate of the trainees domain knowledge.

The output layer units represent the adequacy of an incident type to the current learners knowledge status. The number of units corresponds to the number of incident types (DS, DtR, DmR, DtD, DmD). Each output layers node, representing a type of

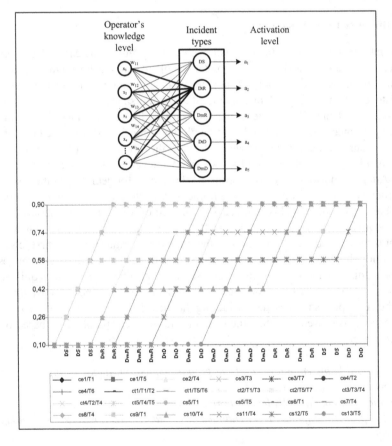

Fig. 4. Classification Mechanism

incident, is connected only to the input nodes corresponding to concepts involved with that incident type. These connections are done with links of weight w_{ij}.

The values used as weights are $w_{ij} = \{1, 0, N\}$ where N is used to indicate that there is no connection between node i of the output layer and the input node j. This means that concept j is not involved in an incident type i.

Each output neuron activation level is computed using the input vector and its weight vector. The activation is defined by the Euclidean distance, given by (1).

$$a_i = \sqrt{\sum_{j=1}^{n}(w_{ij} - x_j)^2} \qquad (1)$$

We can see that a neuron with a weight vector (w) similar to the activation level vector of the input nodes (x) will have a low activation level and vice versa. The output layers node with the lowest activation will be the winner.

In Figure 4 we illustrate a situation where all the model variables are set to their minimum value (0.1) and achieve a maximum value of 0.9. It is also assumed that the ideal operator applies correctly all the domain concepts involved in the problem and that the updating rate is constant.

It can be observed that, after the third iteration, the concepts used in the DS incident type overcome the medium level (0.5), leading to a new type of incident (DtR) in the next iteration. After the fourth iteration, some concepts that are not used in DS but are involved in DtR incident overtake the minimum level for the first time.

We observed that an early introduction of new concepts can contribute to increase the instructional process efficiency. The problem selection mechanism ensures that the sequence of problems is not monotonous, tending to stimulate the operators performance with new kinds of incidents.

4   Tutoring Module for Restoration Training

4.1   Restoration Training Issues

The management of a power system involves several distinct entities, responsible for different parts of the network. The power system restoration process requires a close coordination between generation, transmission and distribution personnel, whose actions must be based on a careful planning and guided by adequate strategies [8].

In the specific case of the Portuguese transmission network, four main entities can be identified: the National Dispatch Centre (CG); the Operational Centre (CO); the Hydroelectric Control Centres (CTCH); and the Distribution Dispatch (EDIS).

The power restoration process is conducted by these entities in such a way that the parts of the grid they are responsible for will be led to their normal state, by performing the actions specified in detailed operating procedures and fulfilling the requirements defined in previously established protocols. This process requires frequent negotiation between entities, agreement on common goals, and synchronization.

The purpose of the training tutor is to allow the training of the established restoration procedures and the drilling of some basic techniques. Power system utilities have built detailed plans containing the actions to be executed and the procedures to be followed in case of incident. In the case of the Portuguese network, there are specific plans for the system restoration following several cases of sectorial blackouts as well as national blackouts, with or without loss of interconnection with the Spanish network.

One typical restoration plan, to be applied after a regional blackout induced by a fault on the 150kV bus of the SRA substation, can be described as follows:

1. Notify Distribution Dispatch Center about the incident and expected restoration time. Wait for the fault on the SRA 150 kV bus to be repaired.
2. Feed the 150kV to SRA bus using the 400/150 kV autotransformers.
3. Switch SVI substation to manual.
4. Energize the lines fed by the 150 kV bus of SRA with priority to lines connected to substations SOR and SRU and to power plants CCD and CVN.
5. Contact the Hydroelectric Power Plants CC, asking for the restoration of their lines with priority for the ones between CCD and CAR and between CCD and SVI/CVF.
6. Wait for the automatic operators of SCV and SGR substations to restore the 150/60 kV transformers, if no voltage is available in 60 kV buses.
7. Wait for SOR substation automatic operator to restore the service, including the line to SVI.
8. Finish the restoration of 150 kV line between substations SRA and SED.
9. Check if the automatic operators work is concluded and finish the restoration if it has not been done automatically.
10. Notify Distribution Dispatch Center about the end of the restoration process.

Our Restoration training subsystem, the CoopTutor, is based on a Multi-Agent system [9]. These agents can be seen as virtual entities that possess knowledge about the domain. As the real operators, they have assigned tasks, goals to achieve and beliefs about the simulated reality and others agents activity. They work asynchronously, performing their different duties simultaneously and synchronizing their activities only when the need arises. Therefore, the system needs a facilitator entity (Simulator in Figure 1) that supervises the process, ensuring that the simulation stays coherent and convincing. In our system, the trainee can select which of the available roles he wants to play, leaving to the tutor the responsibility of supplying the virtual agents that will simulate the other would be participants.

4.2   Trainees Model

The representation method used to model the trainees knowledge about the domain knowledge is a variation of the Constraint-Based Modelling (CBM) technique [10].

This student model representation technique is based on the assumption that diagnostic information is not extracted from the sequence of students actions but rather from the situation, also described as problem state, that the student arrived at. Hence, the student model should not represent the students actions but the effects of these actions. Because the space of false knowledge is much greater than the one for the correct one, it was suggested the use of an abstraction mechanism based on constraints. In this representation, a state constraint is an ordered pair (Cr,Cs) where Cr stands for relevance condition, and Cs for satisfaction condition. Cr identifies the class of problem states in which this condition is relevant and Cs identifies the class of relevant states that satisfy Cs.

Under these assumptions, domain knowledge can be represented as a set of state constraints. Any correct solution for a problem cannot violate any of the constraints. A violation indicates incomplete or incorrect knowledge and constitutes the basic piece of information that allows the Student Model to be built on. This CBM technique does not require an expert module and is computationally undemanding because it reduces student modelling processing to basic pattern matching. One example of a state constraint, as used in our system, can be found below:

If there is a request to CTCH to restore the lines under its responsibility
Then the lines that supply the hydroelectric power plants must have already been restored
Otherwise an error has occurred

Each violation to a state constraint like the one above allows the tutor to intervene both immediately or at a later stage, depending on the seriousness of the error or the pedagogical approach that was chosen. This technique gives the tutor the flexibility needed to address trainees with a wide range of experience and knowledge, tailoring, in a much finer way, the degree and type of support given, and, at the same time, spared us the exhaustive monitoring and interpretation of students errors during an extended period, which would be required by alternative methods.

Nevertheless, it was found the need for a metaknowledge layer in order to adapt the CBM method to an essentially procedural, time-dependent domain like the power system restoration field. This layer is composed of rules that control the constraints application, depending on several parameters: the phase of the restoration process in which the trainee is; the previously satisfied constraints; and the set of constraints that can be simultaneously triggered.

These rules establish a dependency network between constraints that can be represented by a graph (Figure 5). The relationships between constraints expressed by this graph can be of precedence, mutual exclusion or priority [11].

The constraint evaluation is the responsibility of the Tutoring Module (RTM in Figure 1). Its mechanism is triggered by the arrival of messages relating certain actions performed by the trainee on the simulated reality or other external events that should force the verification of past trainee's actions.

The range of events that are so monitored include the change of status of circuit breakers, the change of status of a substation (from automatic to manual mode or vice-versa), the sending of some special messages and, finally, events automatically triggered after certain time intervals. The output of the Constraint Evaluation module will be one or more pedagogical actions.

The Constraint Evaluation mechanism on Figure 6 is responsible for for checking which constraints (if any) are triggered by the event. This behavior is modulated by the

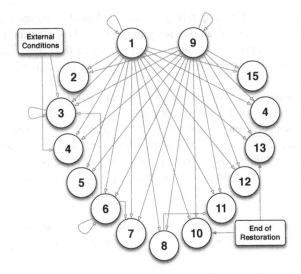

Fig. 5. Constraint Dependency Graph

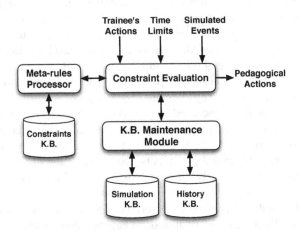

Fig. 6. Constraint Evaluation Mechanism

Meta-rules Processor that states which constraints are available for evaluation in each situation, as explained before in this section.

4.3 The Cooperative Learning Environment

The CoopTutor is prepared to offer two different modes of training: it can train individual operators as if they belonged to a team, interacting with virtual operators, but is also capable of dealing with the interaction between several trainees engaged in a cooperative process, providing specialized agents capable of fulfilling the roles of the missing operators.

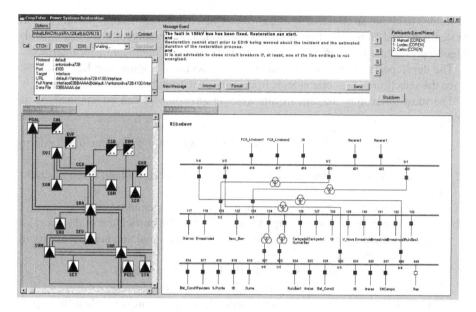

Fig. 7. CoopTutor Interface

At the same time, when several trainees are are working in cooperative mode, it monitors their interaction, stepping in when a serious imbalance is detected. The tutor can therefore be used as a distance learning tool, with several operators being trained at different locations. This simplifies considerably the logistics of the training sessions preparation.

In order to support the monitoring activities of the cooperative discussion and decision processes, the core data contained in the student model has been complemented with information concerning the quantity and characteristics of the interactions detected between trainees. The tutor gathers this data by loosely monitoring the interaction patterns and performing a surface level analysis of the messages' contents.

The tutor will intervene in the cooperative discussion process only if it detects a clear imbalance between participants. It may be called to step in though by the trainees themselves, if they are not able to agree on a course of action or if they find themselves in an impasse. In the first case, the tutor will use the knowledge contained in the CBM module to evaluate the different proposals. In the second case, it will combine the constraint satisfaction data previously gathered with procedural knowledge representing the sequence of the specific restoration plan, in order to issue recommendations about the next step to perform. The general aspect of the CoopTutor interface is shown in Figure 7.

5   Conclusions

This paper describes how an Intelligent Tutoring System can be used for the training of Power Systems Control Centre operators in two main tasks: Incident Analysis and Diagnosis; and Service Restoration. Several Artificial Intelligence techniques were combined to obtain an effective Intelligent Tutoring environment, namely: Multi-Agent Systems, Neural Networks, Constraint-based Modelling, Intelligent Planning, Knowledge Representation, Expert Systems, User Modelling, and Intelligent User Interfaces.

The developed system is used in the training Electrical Engineering BSc students, since the selection of new operators is frequently done from this kind of students.

The most interesting features of this training environment can be summarized as follows:

1. The connection with SPARSE, a legacy Expert System used for Intelligent Alarm Processing [3].
2. The use of prediction tables and different granularity levels for fault diagnosis training.
3. The use of the model tracing technique to capture the operators reasoning.
4. The development of tools to help the adaptation of the curriculum to the operator - one that generates training scenarios from real cases and another that assists in creating new scenarios.
5. The automatic assignment of the difficulty level to the problems.
6. The identification of the operators knowledge acquisition factors.
7. The use of Neural Networks to automatically select the next problem to be presented.
8. The use of the Multi-Agent Systems paradigm to model the interaction of several operators during system restoration.
9. The use of the Constraint-based Modelling technique in restoration training.

References

1. UCTE: Union for the co-ordination of transmission of electricity - system disturbance on 4 november 2006 - final report (2007), http://www.ucte.org
2. Praça, I., Ramos, C., Vale, Z., Cordeiro, M.: Mascem: A multiagent system that simulates competitive electricity markets. IEEE Intelligent Systems- Special Issue on Agents and Markets 18, 54–60 (2003)
3. Vale, Z., Moura, A., Fernandes, M., Marques, A., Rosado, A., Ramos, C.: Sparse: An intelligent alarm processor and operator assistant. IEEE Expert - Special Track on AI Applications in the Electric Power Industry 12, 86–93 (1997)
4. Faria, L., Vale, Z., Ramos, C.: Diagnostic tasks training based on a model tracing approach. International Journal of Engineering Intelligent Systems for Electrical Engineering & Communications (CRL) 13, 223–230 (2005)
5. Anderson, J., Corbett, A., Koedinger, K., Pelletier, R.: Cognitive tutors: Lessons learned. The Journal of the Learning Sciences 4, 167–207 (1995)
6. Razzaq, L., Heffernan, N.: Scaffolding vs. hints in the assistment system. In: Ikeda, M., Ashley, K.D., Chan, T.-W. (eds.) ITS 2006. LNCS, vol. 4053, pp. 635–644. Springer, Heidelberg (2006)
7. Faria, L., Vale, Z., Ramos, C., Silva, A., Marques, A.: Training scenarios generation tools for an ITS to control center operators. In: Gauthier, G., VanLehn, K., Frasson, C. (eds.) ITS 2000. LNCS, vol. 1839, p. 652. Springer, Heidelberg (2000)
8. Sforna, M., Bertanza, V.: Restoration testing and training in italian iso. IEEE Transactions on Power Systems 17 (2002)

9. Jennings, N., Wooldridge, M.: Applying agent technology. Applied Artificial Intelligence: An International Journal 9, 351–361 (1995)

10. Ohlsson, S.: Constraint-based student modeling. In: Greer, McCalla (eds.) Student Modeling: the Key to Individualized Knowledge-based Instruction, pp. 167–189. Springer, Heidelberg (1993)

11. Silva, A., Vale, Z., Ramos, C.: Cooperative training of power systems restoration techniques. In: 13th International Conference on Intelligent Systems Applications to Power Systems, Washington (2005)

Adaptive State Space Abstraction Using Neuroevolution

Robert Wright and Nathaniel Gemelli

Air Force Research Laboratory Information Directorate
525 Brooks Rd. Rome, NY 13441
{Robert.Wright,Nathaniel.Gemelli}@rl.af.mil

Abstract. In this paper, we present a new machine learning algorithm, RL-SANE, which uses a combination of neuroevolution (NE) and traditional reinforcement learning (RL) techniques to improve learning performance. RL-SANE is an innovative combination of the neuroevolutionary algorithm NEAT[9] and the RL algorithm Sarsa(λ)[12]. It uses the special ability of NEAT to generate and train customized neural networks that provide a means for reducing the size of the state space through state aggregation. Reducing the size of the state space through aggregation enables Sarsa(λ) to be applied to much more difficult problems than standard tabular based approaches. Previous similar work in this area, such as in Whiteson and Stone [15] and Stanley and Miikkulainen [10], have shown positive results. This paper gives a brief overview of neuroevolutionary methods, introduces the RL-SANE algorithm, presents a comparative analysis of RL-SANE to other neuroevolutionary algorithms, and concludes with a discussion of enhancements that need to be made to RL-SANE.

Keywords: Reinforcement Learning, NeuroEvolution, Evolutionary Algorithms, State Abstraction.

1 Introduction

Recent progress in the field of neuroevolution has lead to algorithms that create neural networks to solve complex reinforcement learning problems [9]. Neuroevolution refers to technologies which build and train neural networks through an evolutionary process such as a genetic algorithm. Neuroevolutionary algorithms are attractive in that they are able to automatically generate neural networks. Manual engineering, domain expertise, and extensive training data are no longer necessary to create effective neural networks. One problem with these algorithms is that they rely heavily on the random chance of mutation operators to produce networks of sufficient complexity and train them with the correct weights to solve the problem at hand. As a consequence, neuroevolutionary methods can be slow or unable to converge to a good solution. Traditional reinforcement learning (RL) algorithms on the other hand take calculated measures to improve their policies and have been shown to converge very quickly. However, RL algorithms typically rely on costly Q-tables or predetermined function approximators to

J. Filipe, A. Fred, and B. Sharp (Eds.): ICAART 2009, CCIS 67, pp. 84–96, 2010.

enable them to work on complex problems. A hybrid of the two technologies has the potential to provide an algorithm with the advantages of both.

We present a new machine learning algorithm, which combines neuroevolution and traditional reinforcement learning techniques in a unique way. RL-SANE[1], Reinforcement Learning using State Aggregation via Neuroevolution, is an algorithm designed to take full advantage of neuroevolutionary techniques to abstract the state space into a more compact representation for a reinforcement learner that is designed to exploit its knowledge of that space. We have combined a neuroevolutionary algorithm developed by Stanley and Miikkulainen called NEAT[10] with the reinforcement learning algorithm Sarsa(λ)[12]. Neural networks serve as excellent function approximators to abstract knowledge while reinforcement learners are inherently good at exploring and exploiting knowledge. By utilizing the strengths of both of these methods we have created a robust and efficient machine learning approach in RL-SANE.

The rest of the paper is organized as follows. Section 2 will give an overview of the problem on which are working and other state aggregation approaches that have been done. Section 3 will provide a full description and discussion on the RL-SANE algorithm. In sections 4.3 and 4.4 we will provide experimental results for RL-SANE. Section 4.3 provides insight into how the β parameter affects performance and how it can be determined. Section 4.4 completes the analysis with a comparison of RL-SANE's performance versus a standard neuroevolutionary approach, NEAT, on two standard benchmark problems. Finally, section 5 will highlight our future work.

2   Background

The state aggregation problem for machine learning has started to gain more momentum over the past decade. Singh et al. used soft state aggregation methods in [8] and discussed how important compact representations are to learning methods. Stanley and Miikkulainen [11] discuss using neural networks for state aggregation and learning and demonstrated promising results for real-world applications. Further improvements to neuroevolutionary approaches were done in Siebel et al. in [7] and combinations of neuroevolution and reinforcement learning were studied in Whiteson and Stone in [15]. Neuroevolution has been shown to be one of the strongest methods for solving common benchmarking problems, such as pole-balancing [11]. This section will highlight neuroevolution and describe the NEAT algorithm which we use as a benchmark to compare our work against.

2.1   Neuroevolution

Neuroevolution (NE) [10] is a technology that encompasses techniques for the artificial evolution of neural networks. Traditional work in neural networks used

---

[1] It should be noted that RL-SANE is not related to the SANE algorithm[5], Symbiotic Adaptive NeuroEvolution. SANE is a competing neuroevolutionary algorithm with NEAT that uses a cooperative process to evolve NNs.

static neural networks that were designed by subject matter experts and engineers. This method showed the power of neural networks in being able to model and learn non-linear functions. However, the static nature of these structures limited their scope and applicability. More recent advances in the field of neuroevolution have shown that it is possible to build and configure these networks dynamically through the use of special mutation operators in a genetic algorithm (GA) [9]. These advances have made it possible to automatically generate and train special purpose neural networks for solving difficult multi-parameter problems. We recognize similar work in this area as in Siebel et al. [7], Stanley et al. [10], and Whiteston and Stone [15] exists and below we describe one such algorithm, NEAT.

2.2 NEAT

NeuroEvolution of Augmenting Topologies (NEAT) is a neuroevolutionary algorithm and framework which uses a genetic algorithm to evolve populations of neural networks. It is a neuroevolutionary algorithm that evolves both the network topology and weights of the connections between the nodes in the network[11]. The authors, Stanley et al., found that by evolving both the network topology and the connection weights, they solved typical RL benchmark problems several times faster than competing RL algorithms[10] with significantly less system resources. The algorithm starts with a population of simple perceptron networks and gradually, through the use of a genetic algorithm (GA), builds more complex networks with the proper structure to solve a problem. It is a bottom-up approach which takes advantage of incremental improvements to solve the problem. The results are networks that are automatically generated, not overly complicated in terms of structure, and custom tuned for the problem at hand.

Two of the most powerful aspects of NEAT which allow for such good benchmark performance and proper network structure are *complexification*[9] and *speciation*[11]. *Complexification*, in this context, is an evolutionary process of constructing neural networks through genetic mutation operators. These genetic mutation operators are: modify weight, add link, and add node [9]. *Speciation* is a method for protecting newly evolved neural network structures (specie) into the general population, and prevent them from having to compete with more optimized species [9]. This is important to preserve new (possibly helpful) structures with less optimized network weights.

3 RL-SANE

RL-SANE is the algorithm we have developed which is a combination of NEAT and a RL algorithm (Sarsa(λ)[12]). This combination was chosen because neural networks have been shown to be effective at reducing the complexities of state spaces for RL algorithms[13]. NEAT was chosen as a way for providing the neural networks because it is able to build and train networks automatically and effectively. Sarsa(λ) was chosen because of its performance characteristics. Any NE algorithm that automatically builds neural networks can be substituted for NEAT and any RL algorithm can be substituted in Sarsa(λ)'s place, and RL-SANE should still function.

RL-SANE uses NE to evolve neural networks which are bred specifically to interpret and aggregate state information. NEAT networks take the raw perception values and filter the information for the RL algorithm by aggregating inputs together which are similar with respect to solving the problem[2]. This reduces the state space representation and enables traditional RL algorithms, such as Sarsa(λ), to function on problems where the raw state space of the problem is too large to explore. NEAT is able to create networks which accomplish the filtering by evolving networks that improve the performance of Sarsa(λ) in solving the problem. Algorithm 1 provides the pseudo code for RL-SANE and describes the algorithm in greater detail.

What differentiates RL-SANE from other algorithms which use NE to do learning, such as in NEAT, is that RL-SANE separates the two mechanisms that perform the state aggregation and policy iteration. Previous NE algorithms[10] [15] use neural networks to learn a function which not only encompasses a solution to the state aggregation problem, but the policy iteration problem as well. We believe this overburdens the networks and increases the likelihood of "interference"[2]. In this context, "interference" is the problem of damaging one aspect of the solution while attempting to fix another. RL-SANE, instead, only uses the neural networks to perform the state aggregation and uses Sarsa(λ) to perform the policy iteration which reduces the potential impact of interference.

The purpose of the neural networks created by the NEAT component is to filter the state space for Sarsa(λ). They do this by determining if different combinations of perceptual inputs should be considered unique states with respect to the problem. This is done by mapping the raw multi-dimensional perceptual information onto a single dimensional continuum which is a real number line in the range [0..1]. The state bound parameter, β, divides the continuum into a specified number of discrete areas and provides an upper bound on the number of possible states there are in the problem. All points on the continuum within a single area are considered to be the same state by RL-SANE and are labeled with a unique state identifier. The output of the neural networks, given a set of perception values, is the state identifier of the area the perception values map on to the continuum. RL-SANE uses this state identifier as input to the Sarsa(λ) RL component as an index into its Q-table. The Q-table is used to keep track of values associated with taking different actions in specific states. The networks which are best at grouping perception values that are similar with respect to solving the problem will enable Sarsa(λ) to learn how to solve the problem more effectively.

---

[2] "With respect to the problem" in this paper refers to groupings made by NEAT which improve Sarsa(λ)'s ability to solve the problem, not necessarily the grouping of similar perceptual values.

Algorithm 1. RL-SANE(S,A,p,m_n,m_l,m_r,g,e,α,β,γ,λ,ϵ)

1: *//S: set of all states, A: set of all actions*
2: *//p: population size, m_n: set of all states, m_l: link mutation rate*
3: *//m_r: link removal mutation rate, g: number of generations*
4: *//e: number of episodes per generation, α: learning rate*
5: *//β: state bound, γ: discount factor, λ: eligibility decay rate*
6: *//ϵ: ϵ-Greedy probability*
7: $P[] \leftarrow$ INIT-POPULATION(S,p) *// create a new population P of random networks*
8: $Q_t[] \leftarrow$ new array size p *// initialize array of Q tables*
9: **for** $i \leftarrow 1$ to g **do**
10:    **for** $j \leftarrow 1$ to p **do**
11:       N,$Q[] \leftarrow P[j]$, $Q_t[j]$ *// select network and Q table*
12:       **if** $Q[] = null$ **then**
13:          $Q[] \leftarrow$new array size β *// create a new Q table*
14:       **end if**
15:       **for** $k \leftarrow 1$ to e **do**
16:          s,$s' \leftarrow null$, INIT-STATE(S) *// initialize the state*
17:          **repeat**
18:             $s'_{id} \leftarrow$INT-VAL(EVAL-NET(N,s')*β) *// determine the state id*
19:             **with-prob**(ϵ) $a' \leftarrow$RANDOM(A) *// ϵ-Greedy action selection*
20:             **else** $a' \leftarrow$argmax$_l Q[s'_{id}$,$l]$
21:             **if** $s \neq null$ **then**
22:                SARSA(r,a,a',λ,γ,α,$Q[]$,s_{id},s'_{id})*//update Q table*
23:             **end if**
24:             s_{id},s,$a \leftarrow s'_{id}$,s',a'
25:             r,$s' \leftarrow$TAKE-ACTION(a')
26:             $N.fitness \leftarrow N.fitness + r$ *// update the fitness of the network*
27:          **until** TERMINAL-STATE(s)
28:       **end for**
29:       $Q_t[j] \leftarrow Q[]$ *// update array of Q tables*
30:    **end for**
31:    $P'[]$,$Q'_t \leftarrow$ new arrays size p *// array for next generation*
32:    **for** $j \leftarrow 1$ to p **do**
33:       $P'[j]$,$Q'_t[j] \leftarrow$BREED-NET($P[]$,$Q_t[]$) *// make a new network and Q table based on parents*
34:          **with-probability** m_n: ADD-NODE-MUTATION($P'[j]$)
35:          **with-probability** m_l: ADD-LINK-MUTATION($P'[j]$)
36:          **with-probability** m_r: REMOVE-LINK-MUTATION($P'[j]$)
37:    **end for**
38:    $P[]$,$Q_t[] \leftarrow P'[]$,$Q'_t[]$
39: **end for**

Sarsa(λ) provides a fitness metric for each network to NEAT by reporting the aggregate reward it received in attempting to learn the problem. This symbiotic relationship provides NEAT with a fitness landscape for discovering better networks. This also provides Sarsa(λ) with a mechanism for coping with complex state spaces, via reducing complexity. Reducing the complexity of the state space makes the problem easier to solve by reducing the amount of exploration needed to calculate a good policy. It also needs to be stated that Q-tables persist with the neural networks they are trained on through out the evolution of the networks. When crossover occurs in NEAT'S GA, the Q-table of the primary parent gets passed on to the child neural network.

Selection of an appropriate or optimal state bound (β) is of great importance to the performance of the algorithm. If the selected state bound is too small, sets of perception values that are not alike with respect to the problem will be aggregated into the same state. This may hide relevant information from Sarsa(λ) and prevent it from finding the optimal, or even a good, policy. If the state bound is too large, then perception sets that should be grouped together may map to different states. This forces Sarsa(λ) to experience and learn more states than are necessary. Effectively, this slows the rate at which Sarsa(λ) is able to learn a good policy because it is experiencing and exploring actions for redundant states, causing bloat. Bloat makes the Q-tables larger than necessary, increasing the memory footprint of the algorithm. Improper state bound values have adverse effects on the performance of RL-SANE. Section 4.3 provides some insight into how varying β values affects RL-SANE's performance and gives guidance on how to set β.

We recognize that RL-SANE is not the first algorithm that attempts to pair NE techniques with a traditional RL algorithm. NEAT+Q by Whiteson and Stone [15] combines Q-learning [14] with NEAT in a different way than RL-SANE. In NEAT+Q the neural networks are meant to output literal Q-values for all the actions available in the given state. The networks are trained online to produce the correct Q-values via the Q-learning update function and the use of backpropagation [6]. We attempted to use NEAT+Q in our analysis of RL-SANE but we were unable to duplicate the results found in [15], so we are unable to present a comparison. Also, both RL-SANE and NEAT+Q are not easily adapted to work on problems with continuous action spaces, whereas NEAT is. RL-SANE and NEAT+Q both make discrete choices of actions.

4 Experiments

This section describes our experimental setup for analyzing the performance of the RL-SANE algorithm. In our experiments we compare RL-SANE's performance on two well known benchmark problems, mountain car [1] and double inverted pendulum [3], against itself with varying β values and against that of NEAT. NEAT was chosen as the benchmark algorithm of comparison because it is a standard for neuroevolutionary algorithms [15].

Both of the algorithms for our experiments were implemented using Another NEAT Java Implementation (ANJI)[4]. ANJI is a NEAT experimentation platform based on Kenneth Stanley's original work written in Java. ANJI includes a special mutation operator not included in the original NEAT algorithm that randomly prunes connections in the neural networks. This mutation operator is beneficial in that it attempts to remove redundant and excess structure from the neural networks. The prune operator was enabled and used by each algorithm in our experiments. For our experiments we used the default parameters for ANJI with the exceptions that we disallowed recurrent networks and we limited the range of the weight values to be [-1..1]. We disallowed recurrency to reduce their overall complexity of the networks and we bounded the link weights so the calculation of the state from the output signal is trivial.

4.1 Mountain Car

The mountain car problem [1] is a well known benchmark problem for reinforcement learning algorithms. In this problem a car begins somewhere in the basin of a valley, of which it must escape. See Figure 1 for an illustration of the problem. Unfortunately, the car's engine is not powerful enough to drive directly up the hill from a standing start at the bottom. To get out of the valley and reach the goal position the car must build up momentum from gravity by driving back and forth, up and down each side of the valley.

For this problem the learner has two perceptions: the position of the car, X, and the velocity of car, V. Time is discretized into time steps and the learner is allowed one of two actions, drive forward or backward, during each time step. The only reward the learner is given is a value of -1 for each time step of the simulation in which it has not reached the goal. Because the RL algorithms are seeking to maximize aggregate rewards, this negative reward gives the learner an incentive to learn a policy which will reach the goal in as few time steps as possible.

The mountain car problem is a challenging problem for RL algorithms because it has continuous inputs. The problem has an almost infinite number of states if each set of distinct set perceptual values is taken to be a unique state. It is up to the learner or the designer of the learning algorithm to determine under what conditions the driver of the car should consider a different course of action.

In these experiments the populations of neural networks for all of the algorithms have two input nodes, one for X and one for V, which are given the raw perceptions. NEAT networks have three output nodes (one for each direction the car can drive plus coasting) to specify the action the car should take. RL-SANE networks have a single output node which is used to identify the state of the problem.

Individual episodes are limited to 2500 time steps to ensure evaluations will end. Each algorithm was recored for 25 different runs using a unique random seed for the GA. The same set of 25 random seeds were used in evaluating all three algorithms. Runs were composed of 100 generations in which each member of the population was evaluated over 100 episodes per generation. The population size for the GA was set to 100 for every algorithm. Each episode challenged the learner with a unique instance of the problem from a fixed set that starts with the car in a different location or having a different velocity. By varying the instances over the episodes we helped ensure the learners were solving the problem and not just a specific instance.

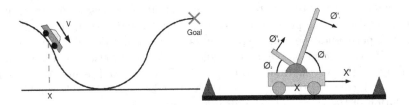

Fig. 1. These figures illustrate the two RL benchmark problems, (Above)mountain car and (Below) double inverted pendulum balancing

4.2   Double Inverted Pendulum Balancing

The double inverted pendulum balancing problem [3] is a very difficult RL benchmark problem. See figure 1 for an illustration of this problem. In this problem
the learner is tasked with learning to balance two beams of different mass and
length attached to a cart that the learner can move. The learner has to prevent
the beams from falling over without moving the cart past the barriers which
restrict the cart's movement. If the learner is able to prevent the beams from
falling over for a specified number of time steps the problem is considered solved.

The learner is given six perceptions as input: X is the position of the cart, X'
is the velocity of the cart, θ 1 and 2 are the angles of the beam, and θ' 1 and 2
are the angular velocities of the beams. At any given time step the learner can
do one of the following actions: push the cart left, push the cart right, or not
push the cart at all.

Like the mountain car problem, this problem is very difficult for RL algorithms because the perception inputs are continuous. This problem is much
more difficult in that it has three times as many perceptions giving it a dramatically larger state space. In these experiments each algorithm trains neural
networks that have six input nodes, one for each perception. NEAT has three
output nodes, one for each action. RL-SANE only has one output node for the
identification of the current state.

In this set of experiments the algorithms were tasked with creating solutions
that could balance the pendulum for 100000 time steps. Each algorithm was
evaluated over 25 different runs where the random seed was modified. Again, the
same set of random seeds was used in evaluating all three algorithms. Runs were
composed of 200 generations with a population size of 100. In each generation,
every member of the population was evaluated on 100 different instances of the
inverted pendulum problem. The average number of steps the potential solution
was able to balance the pendulum over the set of 100 determined its fitness.

4.3   Changing the Size of State Bounds

Figure 3 shows convergence graphs for RL-SANE using varying β values, the
state bound, on both benchmark problems. The graphs show the average number of steps the most fit members of the population were able to either solve
the mountain car problem in or balance the beams for. For the mountain car
experiments fewer steps taken is better and for the double inverted pendulum
problem the more steps the beams were balanced for the better. For these experiments we varied the β parameter from equal to the number of actions available
in the problems (3) up to 1000. The purpose of these experiments is to show how
varying the β parameter affects the performance of the RL-SANE algorithm.

When β is set equal to the number of actions available the neural networks do
all the work. They have to identify the state and determine the policy. All the
Sarsa(λ) component is doing is choosing which action best corresponds with the
output of the network. What is interesting is if you compare the corresponding
graphs from figure 3 and figure 4 when β equals 3 the performance of RL-
SANE is still better than that of NEAT. This is interesting because the networks

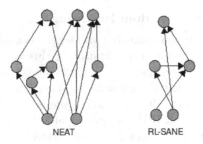

NEAT   RL-SANE

Fig. 2. This figure shows the structure of typical solution neural networks for the mountain car problem. The β value for RL-SANE was set to 50.

from both algorithms are both performing the same function and are effectively produced by the same algorithm, NEAT. The reason RL-SANE has an advantage over standard NEAT in this case is that Sarsa(λ) helps RL-SANE choose the best action corresponding with the network output whereas the actions in NEAT are fixed to specific outputs. This result shows that RL-SANE's performance on problems with discrete actions spaces is going to be as good or better than that of NEAT if β is set to a small value. Smaller β values preferred because they limit the size of the Q-tables RL-SANE requires and it improves the solution's generality.

As β is increased from 3 to larger values we see the performance of RL-SANE improve, but once the value of β exceeds a certain value the performance of RL-SANE begins to drop. The performance increase can be attributed to a burden shift from the NN to the RL component. Sarsa(λ) is much more efficient at performing the policy iteration than the NEAT component. Eventually the performance degrades because the NN(s) begin to identify redundant states as being different, which increases the amount of experience necessary for the RL component to learn. This behavior is what we expected and shows that there might convex fitness landscape for β values. A convex fitness landscape would mean that proper β values could be found by a simple hill climbing algorithm instead of having to specify them *a priori*.

It is also interesting that the double inverted pendulum problem runs are more sensitive to the value of β than the mountain car problem runs are. We are not certain of the reason for this behavior. This requires further investigation, and we hypothesize that more complex problems will be more sensitive to β value selection.

4.4   NEAT Comparison

Figure 4 show convergence graphs comparing the RL-SANE to NEAT on the benchmarks. The error bars on the graphs indicate 95% confidence intervals over the entire set of experiments to show statistical significance of the results. As can be seen from the charts the RL-SANE algorithm is able to converge to the final solution in far fewer generations and with a greater likelihood of success than

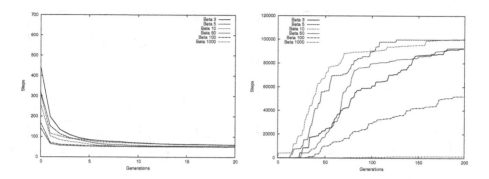

Fig. 3. Shows the performance of the RL-SANE algorithm using varying β values on the mountain car (Above) and the double inverted pendulum (Below)

NEAT in both sets of experiments. NEAT is only able to solve the entire double inverted pendulum problem set in just 2 of the 25 runs of experiments. The performance difference between RL-SANE and NEAT increases dramatically from the mountain car to double inverted pendulum problem which implies that RL-SANE may scale to more difficult problems better than NEAT.

Figures 2 and 5 show the structure of typical solution NN(s) for both RL-SANE and NEAT on both problems. In both figures the RL-SANE networks are much less complex than the NEAT networks. These results are not very surprising in that the RL-SANE networks are performing a less complicated function, state aggregation. NEAT networks need to perform their own version of state abstraction and policy iteration. The fact the RL-SANE networks do not need to be as complicated as NEAT networks explains why RL-SANE is able to perform better than NEAT on these benchmarks. If the NN's do not need to be as complex they have a greater likelihood of being produced by the GA earlier in the evolution. This result also supports our belief that RL-SANE will scale better to more complex problems than standard NEAT.

5   Future Work

5.1   State Bounding

RL-SANE, in its current form, depends on the state bound parameter, β, for the problem to be known *a priori*. For an algorithm that is intended to reduce or eliminate the need for manual engineering or domain expertise. In many cases, this *a prior* knowledge is not available. Much of our future research effort will be placed on developing a method for automatically deriving the state bound value. For reasons stated in section 3 and 4.3 we believe that the problem of finding good β values has a convex landscape that a hill climbing algorithm, such as a GA, can properly search. In our future research we will explore different methods

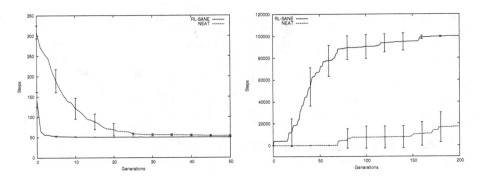

Fig. 4. Shows the performance of the RL-SANE algorithm compared to that of NEAT on the mountain car (Above) and the double inverted pendulum (Below) problems. The RL-SANE runs used a β value of 50 for the mountain car problem and 10 for the double inverted pendulum problem.

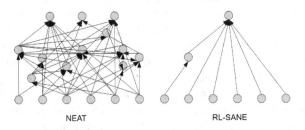

Fig. 5. This figure shows the structure of typical solution neural networks for the double inverted pendulum problem. The β value for RL-SANE was set to 10.

for calculating β automatically, even perhaps during the evolution of solving the problem.

5.2   RL-SANE Scaling

Our background research on RL and methods for handling large state spaces revealed a lack of work done to examine just how well these methods scale towards ever more complicated problems. This is surprising considering that there are so many algorithms designed to improve the scaling of RL algorithms towards larger state spaces. In the works that we have examined, the authors generally chose a single instance of a problem that was difficult or impossible for existing algorithms. They then showed how their algorithms could solve that instance. In our future work we intend to perform experiments that stress and examine the scalability of RL-SANE and the other neuroevolutionary based algorithms, such as NEAT, NEAT+Q, and EANT [7], to find out just how far these algorithms can be pushed.

6 Conclusion

In this paper, we have introduced the RL-SANE algorithm, explored its performance under varying β values, and provided a comparative analysis to other neuroevolutionary learning approaches. Our experimental results have show that RL-SANE is able to converge to good solutions over less iterations and with less computational expense than NEAT even with naively specified β values. The combination of neuroevolutionary methods to do state aggregation for traditional reinforcement learning algorithms appears to have real merit. RL-SANE is, however, dependent on the β parameter which must be calculated a priori. We have shown the importance of the derivation of proper β parameters and suggested finding methods for automating the derivation of β as a direction for future research.

Building off of what has been done by previous neuroevolutionary methods, we have found that proper decomposition of the problem into state aggregation and policy iteration is relevant. By providing this decomposition, RL-SANE should be more applicable to higher complexity problems than existing approaches.

References

1. Boyan, J.A., Moore, A.W.: Generalization in reinforcement learning: Safely approximating the value function. In: Tesauro, G., Touretzky, D.S., Leen, T.K. (eds.) Advances in Neural Information Processing Systems, vol. 7, pp. 369–376. The MIT Press, Cambridge (1995)
2. Carreras, M., Ridao, P., Batlle, J., Nicosebici, T., Ursulovici, Z.: Learning reactive robot behaviors with neural-q learning. In: IEEE-TTTC International Conference on Automation, Quality and Testing, Robotics. IEEE, Los Alamitos (2002)
3. Gomez, F.J., Miikkulainen, R.: Solving non-markovian control tasks with neuroevolution. In: IJCAI 1999: Proceedings of the Sixteenth International Joint Conference on Artificial Intelligence, pp. 1356–1361. Morgan Kaufmann Publishers Inc., San Francisco (1999)
4. James, D., Tucker, P.: A comparative analysis of simplification and complexification in the evolution of neural network topologies. In: Proceedings of the 2004 Conference on Genetic and Evolutionary Computation, GECCO 2004 (2004)
5. Moriarty, D.E., Miikkulainen, R.: Forming neural networks through efficient and adaptive coevolution. Evolutionary Computation 5, 373–399 (1997)
6. Rumelhart, D.E., Hinton, G.E., Williams, R.J.: Learning representations by back-propagating errors. Neurocomputing: foundations of research, 696–699 (1988)
7. Siebel, N.T., Krause, J., Sommer, G.: Efficient Learning of Neural Networks with Evolutionary Algorithms. In: Hamprecht, F.A., Schnörr, C., Jähne, B. (eds.) DAGM 2007. LNCS, vol. 4713, pp. 466–475. Springer, Heidelberg (2007)
8. Singh, S.P., Jaakkola, T., Jordan, M.I.: Reinforcement learning with soft state aggregation. In: Tesauro, G., Touretzky, D., Leen, T. (eds.) Advances in Neural Information Processing Systems, vol. 7, pp. 361–368. The MIT Press, Cambridge (1995)
9. Stanley, K.O.: Efficient evolution of neural networks through complexification. PhD thesis, The University of Texas at Austin. Supervisor-Risto P. Miikkulainen (2004)

10. Stanley, K.O., Miikkulainen, R.: Evolving neural networks through augmenting topologies. Tech. rep., University of Texas at Austin, Austin, TX, USA (2001)
11. Stanley, K.O., Miikkulainen, R.: Efficient reinforcement learning through evolving neural network topologies. In: GECCO 2002: Proceedings of the Genetic and Evolutionary Computation Conference, pp. 569–577. Morgan Kaufmann Publishers Inc., San Francisco (2002)
12. Sutton, R.S., Barto, A.G.: Reinforcement Learning: An Introduction (Adaptive Computation and Machine Learning). The MIT Press, Cambridge (1998)
13. Tesauro, G.: Temporal difference learning and td-gammon. Commun. ACM 38(3), 58–68 (1995)
14. Watkins, C.J.C.H., Dayan, P.: Q-learning. Machine Learning 8(3-4), 279–292 (1992)
15. Whiteson, S., Stone, P.: Evolutionary function approximation for reinforcement learning. Journal of Machine Learning Research 7, 877–917 (2006)

Goal-Based Game Tree Search for Complex Domains

Viliam Lisý, Branislav Bošanský, Michal Jakob, and Michal Pěchouček

Agent Technology Center, Dept. of Cybernetics, FEE, Czech Technical University
Technická 2, 16627 Prague 6, Czech Republic
{lisy,bosansky,jakob,pechoucek}@agents.felk.cvut.cz

Abstract. We present a novel approach to reducing adversarial search space by employing background knowledge represented in the form of higher-level goals that players tend to pursue in the game. The algorithm is derived from a simultaneous-move modification of the max^n algorithm by limiting the search to those branches of the game tree that are consistent with pursuing player's goals. The algorithm has been tested on a real-world-based scenario modelled as a large-scale asymmetric game. The experimental results obtained indicate the ability of the goal-based heuristic to reduce the search space to a manageable level even in complex domains while maintaining the high quality of resulting strategies.

1 Introduction

Recently, there has been a growing interest in studying complex systems, in which larger numbers of agents concurrently pursue their goals while engaging in complicated patterns of mutual interaction. Examples include real-world systems, such as various information and communication networks or social networking applications, as well as simulations, including models of societies, economies and/or warfare. Because in most such systems the agents are part of a single shared environment, situations arise in which their actions and strategies interact. Scenarios in which the outcome of agent's actions depends on actions chosen by others are often termed *games* and have been a topic of AI research from its very beginning. With the increasing complexity of environments in which the agents interact, however, classical game playing algorithms, such as minimax search, become unusable due to the huge branching factor, size of the state space, continuous time and space, and other factors.

In this paper, we present a novel game tree search algorithm adapted and extended for use in large-scale multi-player games with asymmetric objectives (non-zero-sum games). The basis of the proposed algorithm is the max^n algorithm [1] generalized to simultaneous moves. The main contribution, however, lies in a novel way in which background knowledge about possible player's goals and the conditions under which they are adopted is represented and utilized in order to reduce the extent of game tree search. The background knowledge contains:

- **Goals** corresponding to basic objectives in the game (goals represent elementary building blocks of player's strategies); each goal is associated with an algorithm which decomposes it into a sequence of actions leading to its fulfilment.

J. Filipe, A. Fred, and B. Sharp (Eds.): ICAART 2009, CCIS 67, pp. 97–109, 2010.
© Springer-Verlag Berlin Heidelberg 2010

- **Conditions** defining world states in which pursuing the goals is meaningful (optionally, representing conditions defining when individual players might pursue the goals).
- **Evaluation functions** assigning to each player and world state a numeric value representing desirability of the game state for the player (e.g. utility of the state for the player).

The overall background knowledge utilized in the search can be split into a player-independent part (also termed *domain knowledge*) and a player-specific part (further termed *opponent models*).

The proposed approach builds on the assumption that strategies of the players in the game are composed from higher-level goals rather than from arbitrary sequences of low-level actions. Adapted in game tree search, this assumption results in considerably smaller game trees, because it allows evaluating only those sequences of low-level actions that lead to reaching some higher-level goal. However, as with almost any kind of heuristics, the reduction in computational complexity can potentially decrease the quality of resulting strategies, and this fundamental trade-off is therefore an important part of algorithm's evaluation described further in the paper.

The next section introduces the challenges that complicate using game-tree search in complex domains. Section 3 describes the proposed algorithm designed to address them. Search space reduction, precision loss and scalability of the algorithm are experimentally examined in Section 4. Section 5 reviews the related work and the paper ends with conclusions and a discussion of future research.

2   Challenges

The complex domains of our interest include real-world domains like network security or military operations. We use the later for intuition in this section. The games appearing there are often n-player non-zero-sum games with several conceptual problems that generally prohibit using classic game-tree search algorithms, such as max^n.

- **huge branching factor (BF)** – In contrast to many classical games, in military operations the player assign actions simultaneously to all units it controls. Together with a higher number of actions (including parameters) which the units can perform, this results in branching factors several magnitudes bigger than in games such as Chess (BF $\approx$ 35) or Go (BF $\approx$ 361).
- **importance of long plans** – In many realistic scenarios, long sequence of atomic actions is needed before a significant change to the state of world/game is produced. A standard game tree in such scenarios needs to use a correspondingly high search depth, further aggravating the effect of huge branching factor mentioned earlier.
- **simultaneous moves** – The absence of alternation of different players' moves makes the original unmodified max^n algorithm inapplicable.

Let us emphasis also the advantages that using game tree search can bring. If we successfully overcome the mentioned problems, we can reuse the large amount of research in this area and further enhance the searching algorithm with many of existing extensions (such as use of various opponent models, probabilistic extensions, transposition tables, or other shown e.g. in [2]).

3 Goal-based Game-Tree Search

In this section, we present the Goal-based Game-tree Search algorithm (denoted as GB-GTS) developed for game playing in the complex scenarios and addressing the listed challenges. We describe the problems of simultaneous moves, present our definition of goals, and then follow with a description of the algorithm and how it can be employed in a game-playing agent.

3.1 Domain

The domains supported can be formalized as a tuple $(\mathcal{P}, \mathcal{U}, \mathcal{A}, \mathcal{W}, \mathcal{T})$, where $\mathcal{P}$ is the set of players, $\mathcal{U} = \bigcup_{p \in \mathcal{P}} \mathcal{U}_p$ is a set of units/resources capable of performing actions in the world, each belonging to one of the players. $\mathcal{A} = \mathsf{X}_{u \in \mathcal{U}} \mathcal{A}_u$ is a set of combinations of actions the units can perform, $\mathcal{W}$ is the set of possible world states and $\mathcal{T} : \mathcal{W} \times \mathcal{A} \to \mathcal{W}$ is the transition function realizing one move of the game where the game world is changed via actions of all units and world's own dynamics.

The game proceeds in moves. At the beginning of a move, each player assigns actions to all units it controls (forming the action of the player). Function $\mathcal{T}$ is subsequently invoked (taking the combination of assigned actions as an input) to modify the world state.

3.2 Simultaneous Moves

There are two ways simultaneous moves can be dealt with. The first one is to directly work with joint actions of all players in each move, compute their values and consider the game matrix (normal form game) they entail. The actions of individual players can then be chosen based on a game-theoretical equilibrium (e.g. Nash equilibrium in [3]). The second option is to fix the order of the players and let them choose their actions separately in the same way as in max^n, but using the unchanged world state from the end of the previous move for all of them and with the action execution delayed until all players have chosen their actions. This method is called *delayed execution* in [4]. In our experiments, we have used the approach with fixed player order, because of its easier implementation, allowing us to focus on core issue of utilizing the background knowledge.

3.3 Goals

For our algorithm, we define a goal as a pair (I_g, A_g), where $I_g(\mathcal{W}, \mathcal{U})$ is the initiation condition of the goal and A_g is an algorithm that, depending on its internal state and the current state of the world, *deterministically* outputs the next action that leads to fulfilling the goal.

A goal can be assigned to one unit and it is then pursued until it is successfully reached or dropped because its pursuit is no longer practical. Note that we do not specify any dropping or succeeding condition, as they are implicitly captured in the A_g algorithm. We allow the goal to be abandoned only if A_g is finished; furthermore, each unit can pursue only one goal at a time. There are no restrictions on the form of algorithm A_g, so it can represent any type of goal from the taxonomy of goals presented in [5] (e.g. maintain, achieve) and any kind of architecture (e.g. BDI, HTN) can be used to describe it.

The goals in GB-GTS serve as building blocks for more complex strategies that are created by combining different goals for different units and then explored via search. This contrasts with HTN-based approaches used for guiding the game tree search (see Section 5), where the whole strategies are encoded using decompositions from the highest levels of abstraction to the lower ones.

3.4   Algorithm Description

The main procedure of the algorithm (outlined in Figure 1 as procedure *GBSearch()*) recursively computes the value of a state for each of the players assuming that all the units will rationally optimize the utility of the players controlling them. The inputs to the procedure are the world state for which the value is to be computed, the depth to which to search from the world state and the goals the units are currently pursuing. The last parameter is empty when the function is called for the first time.

The algorithm is composed of two parts. The first is the simulation of the world changes based on the world dynamics and the goals that are assigned and pursued by the units, and the second is branching on all possible goals that a unit can pursue after it is finished with its previous goal.

The first part – *simulation* – consists of lines 1 to 15. If all units have a goal they actively pursue, the activity in the world is simulated without any need for branching. The simulation runs in moves and lines 3-10 describe the simulation of a single move. At first, for each unit, an action is generated based on the goal g assigned to this unit (line 5). If the goal-related algorithm A_g has finished, the goal is removed from the map of goals (lines 6-8) and the unit that was previously assigned to this goal becomes idle. The generated actions are then executed and the conflicting changes of the world are resolved in accordance with the game rules (line 10). After this step, one move of the simulation is finished. If the simulation has reached the required depth of search, the resulting state of the world is evaluated using the evaluation functions of all players (line 13).

The second part of the algorithm – *branching* – starts when the simulation reaches the point where at least one unit has finished pursuing its goal (lines 16-22). In order to ensure fixed order of players (see Section 3.2), the next processed unit is chosen from the idle units based on the ordering of the players that control the units (line 16). In the run of the algorithm, all idle units of one player are considered before moving to the units of the next one. The rest of the procedure deals with the selected unit. For this unit, the algorithm sequentially assigns each of the goals that are applicable for the unit in the current situation. The applicability is given by the goal's I_g condition. For each applicable goal, the algorithm assigns the goal to the unit and evaluates the value of the assignment by recursively calling the whole *GBSearch()* procedure (line 19). The current goals of the units are cloned because the state of the already started algorithms generating actions from goals for the rest of the units (A_g) must be the same for all the considered goals of the selected unit. After computing the value of each goal assignment, the one that maximizes the utility of the owner of the unit is chosen (line 21) and the values of this decision for all players are returned by the procedure.

Input: $W \in \mathcal{W}$: current world state, d: search depth, $G[\mathcal{U}]$: map from units to goals they pursue

Output: an array of values of the world state (one value for each player)

```
 1  curW = W
 2  while all units have goals in G do
 3      Actions = ∅
 4      foreach goal g in G do
 5          Actions = Actions ∪ NextAction(A_g)
 6          if A_g is finished then
 7              remove g from G
 8          end
 9      end
10      curW = T(curW, Actions)
11      d = d - 1
12      if d=0 then
13          return Evaluate(curW)
14      end
15  end
16  u = GetFirstUnitWithoutGoal(G)
17  foreach goal g with satisfied I_g(curW, u) do
18      G[u] = g
19      V[g] = GBSearch(curW, d, Copy(G))
20  end
21  g = arg max_g V[g][Owner(u)]
22  return V[g]
```

Fig. 1. GBSearch(W, d, G) – the main procedure of GB-GTS algorithm

3.5 Game Playing

The pseudo-code on Figure 1 shows only the computation of the values of the decisions; it does not deal with how the algorithm can be employed by a player to determine its next actions in the game. In order to do so, the player needs to extract a set of goals for its units from the searched game tree.

Each node in the search procedure execution tree is associated with a unit – the unit for which the goals are tried out. During the run of the algorithm, we store the maximizing goal choices from the top of the search tree representing the first move of the game. The stored goals for each idle unit of the searching player are the main output of the search.

In general, there are two ways the proposed goal-based search algorithm can be used in game-playing.

In the first approach, the algorithm in started in each move and with all units in the simulation set to idle. The resulting goals are extracted and the first actions generated for each of the goals are played in the game. Such an *eager* approach is better for coping with unexpected events should also be more robust in case the background knowledge does not exactly describe the activities in the game.

In the second approach the player using the algorithm maintains a list of current goals for all units it controls. If none of its units is idle (i.e. has no goal assigned), the player uses the goals to generate next actions for its units. Otherwise, the search algorithm is started with goals for the player's non-idle units pre-set and all the other units idle. The goals generated for the previously idle are assigned and pursued in following moves. This *lazy* approach is significantly less computationally intensive.

3.6   Opponent Models

The search algorithm introduced in this section is very suitable for the use of opponent models, that are proved to be useful in adversarial search in [6]. There are two types of opponent models in our approach. We now describe how they can be utilized in the algorithm. The first type, the evaluation function capturing preferences of each player is already an essential part of the max^n algorithm.

The other type of the opponent model can be used to reduce the set of all applicable goals (iterated in Figure 1 on lines 18-21) to the goals a particular player is *likely* to pursue. This can be done by adding player-specific constrains to conditions I_g defining when the respective goal is applicable. These constrains can be hand-coded by an expert or learned from experience; we call them *goal-restricting opponent models*.

The role of goal-restricting opponent model can be illustrated on a simple example of the goal representing loading a commodity to a truck. The domain restriction I_g could be that the truck must not be full. The additional player-specific constraint could be that the commodity must be produced locally at the location, because the particular opponent never uses temporary storage locations for the commodity and always transports it from the place where it is produced to the place where it is consumed.

Using a suitable goal-restricting opponent model can further reduce the size of the space that needs to be searched by the algorithm. A similar way of pruning is possible also in adversarial search without goals. We believe, however, that determining which goal a player will pursue in a given situation is more intuitive and easier to learn than to determine which low-level atomic action (e.g. going right or left on a crossroad) a player will execute.

4   Experiments

In order to practically examine the proposed goal-based (GB) adversarial search algorithm, we performed several experiments. Firstly, we compare it to the exhaustive search of the complete game tree performed by the simultaneous-move modification of max^n search (further called action-based (AB) search) in order to assess the ability to reduce the volume of search on one hand and to maintain the quality of resulting strategies on the other. Afterwards, we analyze the scalability of the GB algorithm in more complex scenarios.

Note that we use the eager, computationally more intensive version of the game playing algorithm in the experiments (see Section 3.5), in which all units are choosing their goals at the same moment in each move.

4.1   Example Game

The game we use as a test case is modelled after a humanitarian relief operation in an unstable environment, with three players - government, humanitarian organization, and separatists. Each of the players controls a number of units with different capabilities that are placed in the game world represented by a graph. Any number of units can be located in each vertex of the graph and change its position to an adjacent vertex in one game move. Some of the vertices of the graph contain cities, which can take in commodities the players use to construct buildings and produce other commodities.

The utilities (evaluation functions) representing the main objectives of the players are expressed as weighted sums of components, such as the number of cities with sufficient food supply, or the number of cities under the control of the government. The government control is derived from the state of the infrastructure, the difference between the number of units of individual players in the city and the state of the control of the city in the previous move. Detailed description of the game can be found in [7].

The goals, used in the algorithm, are generated by instantiation of fifteen goal types. Each goal type is represented as a Java class. Only four of the fifteen classes are unique and the rest nine classes are derived from four generic classes in a very simple way. The actions leading towards achieving a goal consist typically of path-finding to a specific vertex, waiting for a condition to hold, performing a specific action (e.g. loading/unloading commodities), or their concatenation. The most complex goal is escorting a truck by cop that consists of estimating a proper meeting point, path planning to that point, waiting for the truck, and accompanying it to its destination.

Simple Scenario In order to run the standard AB algorithm on a game of this complexity, the scenario has to be scaled down to a quite simple problem. We have created a simplified scenario as a subset of our game with the following main characteristics (see also the scheme in Figure 2):

- only two cities can be controlled (Vertices 3 and 6)
- a government's HQ is built in Vertex 3
- two "main" units - police (cop) and gangster (gng) are placed in Vertex 3
- a truck is transporting explosives from Vertex 5 to Vertex 7
- another two trucks are transporting food from Vertex 1 to the city with food shortage Vertex 3

There are several possible runs of this scenario. The police unit has to protect several possible threats. In order to make government to lose control in Vertex 3, the gangster can either destroy food from a truck to cause starving resulting in lowering the wellbeing and consequent destroying of the HQ (by riots), or it can steal explosives and build a suicide bomber that will destroy the HQ without reducing wellbeing in the city. Finally, it can also try to gain control in city in Vertex 6 just by outnumbering the police there. In order to explore all these options, the search depth necessary is six moves.

Even such a small scenario creates a game tree too big for the AB algorithm. Five units with around four applicable actions each (depending on the state of the world) considered in six consequent moves results in $(4^5)^6 \approx 10^{18}$ world states to examine. Hence, we further simplify it for the AB algorithm. Only the actions of two units (cop

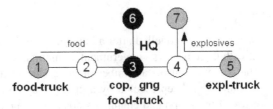

Fig. 2. A schema of the simple scenario. Black vertices represent cities that can be controlled by players, grey vertices represent cities that cannot be controlled, and white vertices do not contain cities.

and gng) are actually explored in the AB search and the actions of the trucks are considered to be a part of the environment (i.e. the trucks are scripted to act rationally in this scenario). Note that the GB algorithm does not need this simplification and actions of all units are explored in the GB algorithm.

4.2 Search Reduction

Using this simplified scenario, we first analyze how the main objective of the algorithm – *search space reduction* – is satisfied.

We run the GB and AB algorithms on a fixed set of 450 problems – world states samples extracted from 30 different traces of the game. On each configuration, we experiment with different values of the look-ahead parameter (1-6 for the AB algorithm and 1-19 for the GB algorithm). As we can see in Figure 3(a), the experimental results fulfilled our aim of substantial reduction of the search space. The number of nodes explored increases exponentially with the depth of the search. However, the base of the exponential is much lower for the GB algorithm. The size of the AB tree for six moves look-ahead is over 27 million, while GB search with the same look-ahead explores only

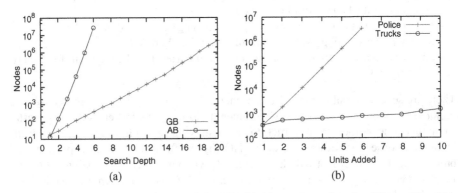

(a)   (b)

Fig. 3. (a) The search space reduction of the GB algorithm compared to the AB algorithm. An average number of the search tree nodes explored depending on the search depth is shown for both algorithms in a logarithmic scale. (b) Increase of the size of the searched tree when adding one to ten police units and explosives trucks to the simple scenario with a 6 move look-ahead.

385 nodes and even for the look-ahead of nineteen, the size of the tree was in average less than 5×10^6. These numbers indicate that using heuristic background knowledge can reduce the time needed to choose an action in the game from tens of minutes to a fraction of a second.

Our implementations of each of the algorithms processed approximately twenty five thousands nodes per second on our test hardware without any optimization techniques. According to [8], however, game trees with million nodes can be searched in real-time (about one second) when such optimization is applied and when efficient data structures are used.

4.3 Loss of Accuracy

With such substantial reduction of the set of possible courses of action explored in the game, some loss of quality of game-playing can be expected. Using the simplified scenario, we compared the actions resulting from the AB and the first action generated by the goal resulting from the GB algorithm. The action differed in 47% of cases. However, a different action does not necessarily mean that the GB search has found a sub-optimal move. Two different actions often have the same value in AB search. Because of the possibly different order in which actions are considered, the GB algorithm can output an action which is different from the AB output yet still has the same optimal value. The values of actions referred to in the next paragraph all come from the AB algorithm.

The value of the action resulting from the GB algorithm was in **88.1%** of cases exactly the same as the "optimal" value resulting from AB algorithm. If the action chosen by GB algorithm was different, it was still often close to the optimal value. We were measuring the difference between the values of GB and optimal actions, relative to the difference between the maximal and the minimal value resulting from the searching player's decisions in the first move in the AB search. The mean relative loss of the GB algorithm was 9.4% of the range. In some cases, the GB algorithm has chosen the action with minimal value, but it was only in situations, where the absolute difference between the utilities of the options was small.

4.4 Scalability

Previous sections show that the GB algorithm can be much faster than and almost as accurate as the AB algorithm with suitable goals. We continue with assessing the limits on the complexity of the scenario where GB algorithm is still usable. There are several possible expansions of the simple scenario. We explore the most relevant factor – number of units – separately and then we apply the GB algorithm on a bigger scenario. In all experiments, we ran the GB algorithm in the initial position of the extended simple scenario and we measured the size of the searched part of the game tree.

Adding Units. The increase of the size of the searched tree naturally depends on the average number of goals applicable for a unit when it becomes idle and the lengths of the plans that lead to their fulfilment. The explosives truck has usually only a couple of applicable goals. If it is empty, the goal is to load in one of the few cities where explosives are produced and if it is full, the goal is to unload somewhere where explosives can be consumed. On the other hand, a police unit has many possible goals. It can protect any transport from being robbed or it can try to outnumber the separatists in any city. We were adding these two unit types to the simple scenario and computed the size of the search tree with fixed six moves look-ahead.

When adding one to ten explosives trucks to the simple scenario, each of them has always only one goal to pursue at any moment. Due to our GB algorithm definition, where goals for each unit are evaluated in different search tree node, even adding a unit with only one possible goal increases the number of evaluated nodes slightly. In Figure 3(b) are the results for this experiment depicted as circles. The number of the evaluated nodes increases only linearly with increasing the number of the trucks.

Adding further police units with four goals each to the simple scenario increased the tree size exponentially. The results for this experiment are shown in Figure 3(b) as pluses.

Complex Scenario. In order to test the usability of the GB search in a more realistic setting, we implemented a larger scenario within our test domain. We used a graph with 2574 vertices and two sets of units. The first unit set was composed of nine units, including two police units with up to four possible goals in one moment, two gangster units with up to four possible goals, an engineer with three goals, stone truck with up to two goals and three trucks with only one commodity source and one meaningful destination resulting to one goal at any moment. The second unit set included seven units – one police, one gangster unit and the same amount of units of the other types. The lengths of the plans to reach these goals is approximately seven atomic actions. There are five cities, where the game is played.

A major difference of this scenario to the simple scenario is, besides the added units, a much bigger game location graph and hence higher length of the routes between cities. As a result, all plans of the units that need to arrive to a city and perform some actions there are proportionally prolonged. This is not a problem for the GB algorithm, because the move actions along the route are only simulated in the simulation phase and do not cause any branching.

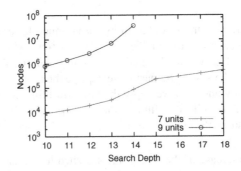

Fig. 4. Size of the trees searched by the GB algorithm in the complex scenario

In a simple experiment to prove this, we changed the time scale of the simulation, so that all the actions were split to two sub-actions collectively effecting the same change of the game world. After this modification, the GB algorithm explored exactly the same number of nodes and the time needed for the computation increased linearly, corresponding to more simulation steps needed.

If we assume that an optimized version of the algorithm can compute one million nodes in a reasonable time, then the look-ahead we can use in the complex scenario is 10 in the nine-unit case and 18 in the seven-unit case. Both values are higher than the average length of unit's typical plan, allowing it to find meaningful plans. If we wanted to apply the AB algorithm to the seven-unit case with the look-ahead of eighteen, considering only four possible move directions and waiting for each unit, it would mean searching through approximately $4^{7^{18}} \approx 10^{75}$ nodes of the game tree, clearly an impossible task.

5   Related Work

The idea to use domain knowledge in order to reduce the portion of the game tree that is searched during the play has already appeared in literature. The best-known example is probably [9], where authors used HTN formalism to define the set of runs of the game, that are consistent with some predefined hand-coded strategies. During game playing, they search only that part of the game tree.

A plan library represented as HTN is used to play GO in [10]. The searching player simulates HTN planning for both the players, without considering what the other one is trying to achieve. If one player achieves its goal, the opponent backtracks (the shared state of the world is returned to the previous state) and tries a different decomposition.

Both these works use quite detailed descriptions of the whole space of the meaningful strategies in HTN. Another approach for reducing the portion of the tree that is searched for scenarios with multiple units is introduced in [11]. The authors show successful experiments with searching just for one unit at a time, while simulating the movements of the other units using rule-based heuristic.

An alternative to using the game-tree search is summarized in [12]. They solve large scale problems with multiple units using, besides other methods, generation of the meaningful sequences of unit actions pruned according to various criteria. One of them is whether they can be intercepted (rendered useless or counterproductive) by traces of the opponent's units.

6   Conclusions

We have proposed a novel approach to introducing background knowledge heuristic to multi-player simultaneous-move adversarial search. The approach is particularly useful in domains where *long sequences* of actions are required to produce significant changes in the world state, each unit (or other resource) can only pursue a *few goals* at any time, and the decomposition of a each goal into low level actions is *uniquely defined* (e.g. using the shortest path to move between two locations).

We have compared the performance of the algorithm to a slightly modified exhaustive max^n search, showing that despite examining only a small fraction of the game tree (less than 0.002% for the look-ahead of six game moves), the goal-based search is still able to find the optimal solution in 88.1% cases; furthermore, even the suboptimal solutions produced are often close to the optimum. These results have been obtained with the background knowledge designed *before* implementing and evaluating the algorithm and without any further optimization to prevent over-fitting.

Furthermore, we have tested the scalability of the algorithm to larger scenarios where the modified max^n search cannot be applied at all. We have confirmed that although the algorithm's time complexity cannot escape exponential growth, this growth can be controlled by reducing the number of different goals considered for each unit and by making the action sequences generated by goals longer. Simulations on a real-world scenario modelled as a multi-player asymmetric game proved the approach viable, though further optimization and improved background knowledge would be necessary for the algorithm to discover more complex strategies.

An important advantage of the proposed approach is its compatibility with all existing extensions of general-sum game-tree search based on modified value back-up procedure and other optimizations. It is also insensitive to the granularity of space and time with which the game is modelled as long as the structure of the goals remains the same and their decomposition into low-level actions is scaled correspondingly.

In the future, we aim to implement additional technical improvements in order to make the goal-based search applicable to even larger problems. In addition, we would like to address the problem of the automatic extraction of goal-based background knowledge from game histories. First, we plan to learn goal initiation conditions for individual players and use them for additional search space pruning. Later, we plan to address a more challenging problem of learning the goal decomposition algorithms.

Acknowledgements. Effort sponsored by the Air Force Office of Scientific Research, USAF, under grant number FA8655-07-1-3083 and by the Research Programme No. MSM6840770038 by the Ministry of Education of the Czech Republic. The U.S. Government is authorized to reproduce and distribute reprints for Government purpose notwithstanding any copyright notation thereon.

References

1. Luckhardt, C., Irani, K.B.: An algorithmic solution of n-person games. In: Proc. of the National Conference on Artificial Intelligence (AAAI 1986), Philadelphia, Pa, August 1986, pp. 158–162 (1986)
2. Schaeffer, J.: The history heuristic and alpha-beta search enhancements in practice. IEEE Transactions on Pattern Analysis and Machine Intelligence 11(11), 1203–1212 (1989)
3. Sailer, F., Buro, M., Lanctot, M.: Adversarial planning through strategy simulation. In: IEEE Symposium on Computational Intelligence and Games (CIG), Honolulu, pp. 80–87 (2007)
4. Kovarsky, A., Buro, M.: Heuristic search applied to abstract combat games. In: Canadian Conference on AI, pp. 66–78 (2005)

5. van Riemsdijk, M.B., Dastani, M., Winikoff, M.: Goals in agent systems: A unifying framework. In: Padgham, Parkes, Müller, Parsons (eds.) Proc. of 7th Int. Conf. on Autonomous Agents and Multiagent Systems (AAMAS 2008), Estoril, Portugal, May 2008, pp. 713–720 (2008)
6. Carmel, D., Markovitch, S.: Learning and using opponent models in adversary search. Technical Report CIS9609, Technion (1996)
7. Semsch, E., Lisý, V., Bošanský, B., Jakob, M., Pavlíček, D., Doubek, J., Pechoucek, M.: Adversarial behavior testbed. Technical Report GLR 90/09, CTU, FEE, Gerstner Lab (2009), http://agents.felk.cvut.cz/publications
8. Billings, D., Davidson, A., Schauenberg, T., Burch, N., Bowling, M., Holte, R.C., Schaeffer, J., Szafron, D.: Game-tree search with adaptation in stochastic imperfect-information games. In: van den Herik, H.J., Björnsson, Y., Netanyahu, N.S. (eds.) CG 2004. LNCS, vol. 3846, pp. 21–34. Springer, Heidelberg (2006)
9. Smith, S.J.J., Nau, D.S., Throop, T.A.: Computer bridge - a big win for AI planning. AI Magazine 19(2), 93–106 (1998)
10. Willmott, S., Richardson, J., Bundy, A., Levine, J.: Applying adversarial planning techniques to Go. Theoretical Computer Science 252(1-2), 45–82 (2001)
11. Mock, K.J.: Hierarchical heuristic search techniques for empire-based games. In: ICAI, pp. 643–648 (2002)
12. Stilman, B., Yakhnis, V., Umanskiy, O.: 3.3. Strategies in Large Scale Problems. In: Adversarial Reasoning: Computational Approaches to Reading the Opponent's Mind, pp. 251–285. Chapman & Hall/CRC, Boca Raton (2007)

Generating Incomplete Data with DataZapper

Yingying Wen, Kevin B. Korb, and Ann E. Nicholson

Clayton School of Information Technology
Monash University, 3800, VIC, Australia
{ywen,kevin.korb,ann.nicholson}@infotech.monash.edu.au

Abstract. A nearly universal problem with real data is that they are incomplete, with some values missing. Furthermore, the *ways* in which values can go missing are quite varied, with arbitrary interdependencies between variables and their values leading to missing values. In order to test and compare data mining algorithms it is necessary to generate artificial data which have the same characteristics. We introduce DataZapper, a tool for uncreating data. Given a dataset containing joint samples over variables, DataZapper will make a specified percentage of observed values disappear, replaced by an indication that the measurement failed. DataZapper also supports any kind of dependence, and any degree of dependence, in its generation of missing values. We illustrate its use in a machine learning experiment and offer it to the data mining and machine learning communities.

Keywords: Machine learning, incomplete data, data generation, data analysis, missing data, data mining, machine learning evaluation.

1 Introduction

Machine learning (ML) research aims at finding the most effective algorithms for constructing models from data. Therefore, machine learning researchers need to find the means for assessing the performance of different ML algorithms applied to common datasets representing varying domains and degrees of difficulty. Although much work in machine learning has concentrated upon data without noise, real-world data always have noise, with the most extreme form being simply the absence of a measured value. In consequence, interest has grown in finding new methods to cope with incomplete datasets and in assessing those methods (e.g., [1,2,3]).

Absence of data values is ubiquitous in part because there are many ways in which measurements can fail. We illustrate with the simple causal Bayesian network of Figure 1. We shall assume that joint observations of these variables come from sample surveys, but similar failures to measure can arise from any measurement technique. First, some missing values may arise simply from survey takers entirely overlooking a question, independently of what the question is about or the values of any variables. Second, the failure to measure particular variables may depend upon the values of other variables; for example, it may turn out that lawyers as a class are less inclined to reveal their incomes than people of other occupations. Third, the failure to measure may be sensitive *additionally* to the unmeasured value of the variable at issue; for example, it may be that it is primarily the *wealthy* lawyers who are reluctant to reveal their incomes. Following Rubin [4], it has become common to refer to these three mechanisms for values to

J. Filipe, A. Fred, and B. Sharp (Eds.): ICAART 2009, CCIS 67, pp. 110–123, 2010.

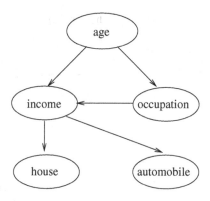

Fig. 1. An example model

be missed as, respectively, missing completely at random (MCAR), missing at random (MAR) and not missing at random (NMAR). These names are somewhat misleading, and we shall below present reasons for adopting a more descriptive nomenclature.

Given the prevalence of incompleteness in real data, and its variety, it is important for ML researchers to investigate how their various algorithms perform given these different types of incomplete data. However, the missing mechanism for real data is most likely unknown.

Of course, ML researchers *do* undertake these types of experiments with different missing data. For example, Ghahramani and Jordan [5] evaluated the performance of classification with missing data dealt by Expectation-Maximization (EM) and mean imputation (IM) (see Figure 2). Gill et al. [6] examined the performance of learning algorithms between artificial neural networks (ANNs) and support vector machines (SVMs) on data MAR.

Another example is Richman et al. [7], who compared different methods of handling missing value and presented in terms of mean absolute error (MAE) in Figure 3. They used real data with some values removed randomly, that is, MCAR.

However, it is difficult using only real data to compare the performance of algorithms for machine learning and methods for dealing with missing values, since the nature of the real system, including the mechanisms whereby data go missing, is at issue; it is difficult or impossible to determine which algorithm has produced a model closer to reality. For machine learning research, we want to test against artificial data generated from a known system with a known mechanism causing values to go missing. This provides more flexibility with the type of missing mechanisms, the type of datasets and the degree of dependence. Moreover, performances can then be evaluated against the true model.

Here we present DataZapper, a versatile software tool for generating artificial datasets with missing values. DataZapper renders some values in a dataset absent according to specified conditions based upon any variable and any value within that dataset; these conditions can be tuned precisely for degrees of dependence, allowing for systematic experimentation. We shall make this tool available to machine learning community via the Weka[1] machine learning platform. One of our motivations in

---

[1] http://www.cs.waikto.ac.nz/ml/weka/

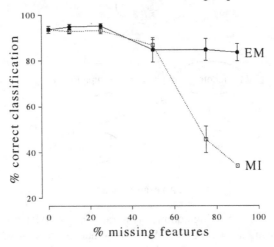

Fig. 2. Example 1 of ML research on varying missing values. Classification of the iris data from [5].

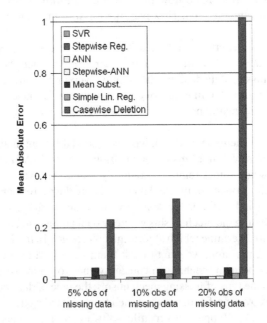

Fig. 3. Example 2 of ML research on varying missing values. A bar chart illustrating the difference of variance between the original and imputed data sets from [7].

producing this tool is to encourage the machine learning community to explore varieties of incompleteness beyond MCAR, which is the only kind assumed by many algorithms, such as the expectation maximization (EM) technique for replacing missing values in Weka. With a tool granting easy access to more realistic forms of incompleteness we expect more attention to them will be given.

The only previously reported tool we know of for generating incomplete data is that of Francois and Leray [8]. They employ Bayesian networks (BNs) as a useful way to generate artificial data with missing values. Unfortunately, their tool is limited to MCAR and limited forms of MAR incompleteness, with no ability to produce NMAR data. As Francois and Leray point out, all of these forms of generating missing data can be useful for generic software testing, beyond machine learning research.

The structure of our paper is as follows. Section 2 describes the three absent data mechanisms and introduces our nomenclature for them. In Section 3 we present a BNF (Backus-Naur Form [9]) grammar for scripting DataZapper. In Section 4 we present the details of DataZapper, including data formats in Section 4.1 and an overview of how it works in Section 4.2. Section 5 illustrates DataZapper's use in an experimental setting.

2 Absent Data Mechanisms

A dataset is a matrix in which rows represent the cases (joint samples) and columns represent variables measured for each case. Ideally, a dataset has all the cells filled—i.e., it is a complete data set. However, most real datasets have some values unobserved—i.e., they are incomplete.

As we mentioned, Rubin [4] introduced and named three types of missing data mechanisms. We shall now motivate the adoption of new names for these. First, we prefer to talk of "absent data" rather than "missing data", for the simple but sufficient reason that "absent" has a natural nominal form, "absence", while "missing" leads to the awkward neologism "missingness". More significantly, two of Rubin's labels are clearly inadequately descriptive of the mechanisms involved:

Missing Completely at Random (MCAR). as the absence of values is independent of all variable values, including the value for this particular cell, this label is actually appropriate. Therefore, we propose calling these cases *absent completely at random (ACAR)*.

Missing at Random (MAR). these missing cases have arbitrary dependencies upon the values of *other* variables. In consequence, they may not even be random at all, but functionally dependent upon the values of other variables in extreme cases. Hence, we prefer calling them *absent under dependence (AUD)*.

Not Missing at Random (NMAR). The natural way of interpreting this phrase is by negating the second kind of "missingness", which would be entirely wrong. This case is simply a generalization of AUD, allowing the absence of data to depend also upon the actual value which has failed to be measured. Hence, we have *absent under self-dependence (AUSD)*.

We submit that the most common case in real data is the case most commonly ignored, AUSD, where the values going unmeasured depend both on the values of some other variables and the absent values themselves, as in wealthy lawyers hiding their wealth.

3   Scripting DataZapper

<m-statement> :: = if <antecedent> then <consequent>
<antecedent> ::= <condition>*
<condition> ::= <variable> in <range>
<variable> ::= alpha alphanum*
<range> ::= [<value>, <value>]
<value> ::= alpha alphanum* | number | symbol
<consequent> ::= (<prob>) <variable>*
<prob> ::= number

The specifications for how the data should go missing are made in a simple scripting language, with the above BNF grammar. These rules are applied to a dataset file to generate a new dataset file with some observed values replaced by a token indicating absence. The basic form of a sentence is that of an "if... then..." production rule. The antecedent describes the dependencies that absence has on variables and values in the system, while the consequent lists the variables that take absent values on these conditions and with what probability. If the antecedent is empty, then the absent data generation is unconditional—i.e., the data are ACAR in so far as this production rule is concerned. If the consequent is empty, then the absence mechanism is applied to *all* variables in the dataset. When the data are AUD or AUSD, the antecedent grammar rule specifies which variable(s) the absence depends upon and for what values or value ranges. The effects of the script rules are cumulative. The result is a language in which

BNF:

1. if then (20)
2. if then (30) A C
3. if C in [?] then (40) E
4. if $Gender$ in $[F]$ Age in $[10, 20]$ then (40) $Income$
5. if $Gender$ in $[F]$ $Income$ in $[70000, 90000]$ then (40) $Income$
6. if A in $[A1]$ B in $[B1]$ then (60) A D

Explanation:

1. ACAR: every variable will have 20% of its values absent
2. ACAR: each of the variables "A" and "C" will have 30% of its values absent
3. AUD: variable "E" will have 40% of its values absent when variable "C" takes the value "?", namely variable "C" is already absent
4. AUD: variable "Income" will have 40% of its values absent when "Age" is between 10 and 20 (inclusive) and "Gender" is "F".
5. AUSD: variable "Income" will have 40% of its values absent when variable "Gender" has value "F" and "Income" is between "70000" and "90000"
6. AUSD: variable "A" and "D" will both have 60% of their values absent when variable "A" has value "A1" and "B" has value "B1"

Fig. 4. Examples of absent data specification in the DataZapper script language (above) with the corresponding English descriptions (below)

any strength of dependence upon any set of variables can be specified, and such dependencies may be combined arbitrarily. For example, "OR" can be represented by having two different conditions.

Figure 4 shows some examples of the absent data specifications, across the range of types, together with a corresponding English description. Note that example 6 is of a mixed type, producing AUD for variable D and AUSD for variable A.

4   Technical Details

4.1   Data Format

DataZapper accepts two data formats: a default format and Weka's [10] data format—Attribute-Relation File Format (ARFF).[2]

The default format is the data format used by the BN learning software CaMML [11], Tetrad [12] and BNT [13]. (We describe how we used DataZapper for the empirical comparison of some of these methods in Section 5.) An example of complete data in the default format is shown on the left side in Table 1. The first two lines are the number of variables and the number of observations, respectively. The next line lists the names of the variables in the dataset. Columns are separated by tab. Consider again Example 2 in Figure 5 above: "if then (30) A C", the corresponding corrupted data after applying dataZapper is given on the right side in Table 1, with the absent values represented by "?" in the default data format. (The token used to represent absence can be changed from this default using a runtime parameter.)

DataZapper supports the ARFF format in order to be compatible with the Weka machine learning platform, which has become a standard toolkit for ML studies (e.g. [10]). In Table 2 we reproduce the above example in an ARFF file. Note that an additional attribute for absent values must be indicated for those variables which are consequents of a DataZapper rule.

Table 1. Examples of complete and corrupted data in Datazapper's default format

Complete data					Corrupted data				
5					5				
10000					10000				
E	A	B	C	D	E	A	B	C	D
E0	A1	B1	C0	D1	E0	A1	B1	?	D1
E1	A0	B0	C1	D1	E1	A0	B0	C1	D1
E0	A1	B0	C1	D0	E0	A1	B0	C1	D0
E1	A1	B1	C0	D0	E1	?	B1	?	D0
E1	A0	B1	C1	D1	E1	A0	B1	C1	D1
...					...				

---

[2] http://www.cs.waikato.ac.nz/ml/weka/arff.html

Table 2. Examples of complete data and corrupted data in ARFF format

Complete data		Corrupted data	
5		5	
10000		10000	
@RELATION input		@RELATION input	
@ATTRIBUTE E	{E0,E1}	@ATTRIBUTE E	{E0,E1}
@ATTRIBUTE A	{A0,A1}	@ATTRIBUTE A	{A0,A1,?}
@ATTRIBUTE B	{B0,B1}	@ATTRIBUTE B	{B0,B1}
@ATTRIBUTE C	{C0,C1}	@ATTRIBUTE C	{C0,C1,?}
@ATTRIBUTE D	{D0,D1}	@ATTRIBUTE D	{D0,D1}
@DATA		@DATA	
E0,A1,B1,C0,D1,input		E0,A1,B1,?,D1,input	
E1,A0,B0,C1,D1,input		E1,A0,B0,C1,D1,input	
E0,A1,B0,C1,D0,input		E0,A1,B0,C1,D0,input	
E1,A1,B1,C0,D0,input		E1,?,B1,?,D0,input	
E1,A0,B1,C1,D1,input		E1,A0,B1,C1,D1,input	
...		...	

4.2 DataZapper Operation

DataZapper processes the absent data specifications one line at a time. For each script command, DataZapper first parses it, and then renders some values in the complete data absent, using a uniform random variate in comparison with the specified probability. DataZapper writes the resultant incomplete dataset to an intermediate file. DataZapper emulates parallelism by generating intermediate output files for each command line and, in the end, merging them into a final output file. In the merging process absent values dominate; that is, a value ends up missing if it is missing in *any* intermediate file. DataZapper finishes by generating a data report on the final dataset, comparing the proportions of absent values with the original dataset.

We will now look at in more detail at the DataZapper functions.

```
function main(script file, complete data)
   for each condition line in script file
      if parser(condition line, &absentInfo) > 0
         corruptedDataGenerator(complete data, corrupted data
                                 filename, &absentInfo, pass)
         dataReport(complete data, corrupted data, &absentInfo,
                    pass)
      endif
   endfor
   mergeData(corrupted data, corrupted data)
   dataReport(complete data, corrupted data,
              &absentInfo, pass)
end
```

The Parser. This function returns 0 for lines which are commented out, 1 for data ACAR, 2 for data AUD, 3 for data AUSD, and 4 for mix of data AUD and AUSD.

Input: a line from script file represent a condition
           address of *absentInfo* passed from main function
Output: value 0 to 4

function *parser*(*commandLine*, &*absentInfo*)
  if the first letter in a lilne is not
     a proper start of command
     return 0

  % Read tokens and save them in *tokens*[ ].
  tokenizer(commandLine, tokens)

  % If <condition> is empty, then the data is ACAR.
  if <condition> is empty
    if absent variable names are not given
       % The absence is for all variables
       return 1
    else
      if variable names are valid
         store the names
         return 1
      endif
    endif
  endif

  % If <condition> is not empty, then the data is AUD,
  %   AUSD or mixed.
  else
    for each dependent variable in <condition>
      if the name is valid
        if the value/range of the variable is valid
          store the names and value/range
        endif
      endif
    endfor

    if the proportion of absence is valid
      store the proportion
    endif

    % Check absent variable names
    for each absent variable

```
      if the name is valid
        store the names
      endif
    endfor

    % Check absent type and return different value.
    if absent type == AUD
      return 2
    elseif absent type == AUSD
      return 3
    % mix of AUD and AUSD
    else
      return 4
  endif
end
```

The Corrupted Data Generator. This is the key processing step that renders some values in the input data absent. The proportion of the absence is applied to each selected target variable, evenly distributed over all the relevant observations for that variable – that is, those observations which satisfy the dependency condition.

function *corruptedDataGenerator*(complete data, corrupted data
                              filename, *absentInfo*, *pass*)
```
  for each record in complete data
    get values

    % For data ACAR.
    if pass == 1
      for each value in the record
        if the value is for one of the absent variables
          generate a random value
          if random value < absent proportion
            change the value to ''?''
          else
            keep the value unchanged
          endif
        else
          keep the value unchanged
        endif
      endfor

    % For data AUD, AUSD or mix.
    else
      for each value in the record
        if the value is for one of the absent variables
          if the record satisfies the <condition>
```

```
      generate a random value
      if random value < absent proportion
        change the value to ''?''
      else
        keep the value unchanged
      endif
    else
      keep the value unchanged
    endif
  endif
  endfor
endif
  output the record to corrupted data file
  endfor
end
```

Merging Data Files. In this processing step, DataZapper merges multiple corrupted datasets with the same variables and the same number of observations. The datasets having a common source, the only differences between them are those required by processing distinct script file commands. The merged data is a kind of union of the corrupted datasets, with the absence of a value in any cell forcing its absence in the final output. If there are many script commands being executed, or if the initial input file itself contained incomplete data, then the final dataset may contain less information (more absent values) than anticipated.

Table 3. Examples of two corrupted datasets

corrupted data 1				corrupted data 2			
5				5			
10000				10000			
E	A	B C D		E	A	B C D	
E0	A1	B1 ? D1		E0	?	B1 C0 ?	
E1	A0	B0 C1 D1		E1	A0	B0 C1 D1	
E0	A1	B0 C1 D0		E0	A1	B0 C1 D0	
E1	?	B1 ? D0		E1	?	B1 C0 ?	
E1	A0	B1 C1 D1		E1	A0	B1 C1 D1	
...				...			

For example, consider again Examples 2 and 6 from Figure 5 for specifying absent data. Table 3 displays some examples from the two corrupted datasets respectively, while Table 4 shows the same examples in the final merged corrupted dataset.

Data Report. DataZapper presents a statistical summary of the incompleteness of the final dataset. Figure 5 gives an example data report. This report can be used to fine tune the scripting rules in the event that the overall sparseness of the data is unexpectedly high, possibly due to the cumulative effect of multiple rules on some variables.

Table 4. Merged data from the examples in Table 3

corrupted data			
5			
10000			
E	A	B C	D
E0	?	B1 ?	?
E1	A0	B0 C1	D1
E0	A1	B0 C1	D0
E1	?	B1 ?	?
E1	A0	B1 C1	D1
...			

Content of script file:
    if then (20) E A
    if (C) in [C0] D in [D1] then (20) D
    if A in [A1] then (30) B E
Percent of absent values:

	A	B	C	D	E
Final:	20.00%	12.64%	0.00%	5.14%	30.35%
Original:	0.00%	0.00%	0.00%	0.00%	0.00%

Overall:
    6813 values are absent, 13.63% of all values.
    5201 cases contain absent values, 52.01% of 10,000 total cases.

Fig. 5. Example of DataZapper's absent data report

5 Application

We now describe an application of DataZapper in generating incomplete data for use in some of our machine learning research. The specific application is an empirical comparison of the performance of causal discovery algorithms in finding the causal Bayesian network (a kind of directed acyclic graph, or DAG) which has generated some observational data. The algorithms under test were K2 [14], GES (Greedy Equivalence Search) [15], and the PC algorithm from Tetrad [12]. The first algorithm, K2, returns a single DAG which fits the data best.[3] The other two algorithms return an equivalence class of DAGs (a pattern); that is, a set of DAGs which all have equal maximum-likelihood scores based upon any given set of observational data [16]. In effect, these algorithms are asserting that all the DAGs within the pattern are equally likely to be the source of the observed data. In assessing such results, therefore, we use a pattern-to-DAG conversion algorithm [17] algorithm which returns two DAGs: that nearest to the original

---

[3] We have enhanced K2 by utilizing a Minimum Weighted Spanning Tree algorithm as a pre-processing step to produce the total ordering of variables that K2 demands.

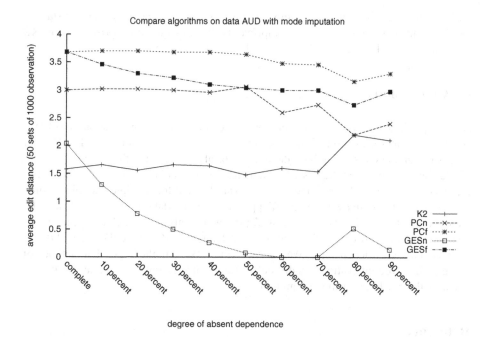

Fig. 6. Example experimental results using DataZapper: comparison of 3 causal discovery algorithms, on data generate with AUD absence mechanism, varying the degree of data completeness.

causal Bayesian network in structure (as measured in edit distance) and that farthest from the original network. This provides a range of performance for assessing such algorithms (assuming that the data are artificial, of course, since otherwise the original network is unknown).

The experiment we ran was three dimensional: we varied the algorithm, the proportion of absence and the absent data mechanism .

We used 50 sets of complete data generated from a known Bayesian network. We then applied DataZapper to produce 3×9 incomplete datasets for each complete dataset, given the three absence mechanisms and 9 steps of proportion of absence. We then designed comparison experiments for different combinations of these experimental parameters.

For example, one experiment involved selecting the absence mechanism and then comparing the performance of the causal discovery algorithms given varying proportion of absence. The results of this particular experiment are shown in Figure 6. Here the evaluation measure we used is the edit distance of the learned BN to the true model—Figure 1, averaged over the 50 datasets. For the PC and the GES algorithms we report two results, one based on the DAG within the pattern returned that is *nearest* to the true model (PCn and GESn), another for the DAG within the pattern that is *farthest* from the true model (PCf and GESf). In this experiment we used one of the simplest methods for handling absent values, namely modal imputation (i.e., replacing each absence token with the modal value for that variable). Results are available for all ACAR, AUD and

AUSD. Only AUD is used as an example. Figure 6 shows that under these circumstances the performances for PC and GES improve as the data quality improves, while K2 appears to be stuck. Overall, GESn shows the best performance.

6   Conclusions

DataZapper is a powerful and flexible tool for incomplete data generation, developed specifically for use in research comparing machine learning algorithms. DataZapper allows researchers to specify both the amount of absent data and the nature of the dependencies in generating the absent data, using simple conditional rules. Multiple conditions of absence can be described for each variable and for multiple variables, which will be applied cumulatively by DataZapper to the input dataset, which itself may be either complete or already corrupt. DataZapper is the only tool which can generate incomplete data for all types of absent data mechanisms (ACAR, AUD or AUSD) and with any degree of dependence. We offer it through Weka in the hopes that methods of coping with more interesting and difficult varieties of incomplete data may be investigated by the machine learning community.

References

1. Onisko, A., Druzdzel, M.J., Wasyluk, H.: An experimental comparison of methods for handling incomplete data in learning parameters of bayesian networks. In: Proceedings of the IIS 2002 Symposium on Intelligent Information Systems, pp. 351–360. Physica-Verlag (2002)
2. Twala, B., Cartwright, M., Shepperd, M.J.: Comparison of various methods for handling incomplete data in software engineering databases. In: 2005 International Symposium on Empirical Software Engineering, Noosa Heads, Australia, pp. 105–114 (2005)
3. Twala, B.E.T.H., Jones, M.C., Hand, D.J.: Good methods for coping with missing data in decision trees. Pattern Recogn. Lett. 29, 950–956 (2008)
4. Rubin, D.B.: Inference and missing data. Biometrika 63, 581–592 (1976)
5. Ghahramani, Z., Jordan, M.I.: Learning from incomplete data. Technical Report AIM-1509, Artificial Intelligence laboraory and Center for Biological and Computational Learning, Department of Brain and Cognitive Sciences, Massachusetts Institute of Technology (1994)
6. Gill, M.K., Asefa, T., Kaheil, Y., McKee, M.: Effect of missing data on performance of learning algorithms for hydrologic predictions: Implications to an imputation technique. Water Resources Research 43 (2007)
7. Richman, M.B., Trafalis, T.B., Adrianto, I.: Multiple imputation through machine learning algorithms. In: Artificial Intelligence and Climate Applications (Joint between 5th Conference on Applications of Artificial Intelligence in the Environmental Sciences and 19th Conference on Climate Variability and Change) (2007)
8. Francois, O., Leray, P.: Generation of incomplete test-data using bayesian networks. In: Proceedings of International Joint Conference on Neural Networks, Orlando, Florida, USA, pp. 12–17 (2007)
9. Backus, J., Naur, P.: Revised report on the algorithmic language algol 60. Communications of the ACM 3, 299–314 (1960)
10. Witten, I.H., Frank, E.: Data Mining: Practical Machine Learning Tools and Techniques with Java Implementations, 2nd edn. Morgan Kaufmann, San Francisco (2005)

11. Wallace, C., Korb, K.B., Dai, H.: Causal discovery via MML. In: Proceedings of the Thirteenth International Conference on Machine Learning, pp. 516–524. Morgan Kaufmann, San Francisco (1996)
12. Spirtes, P., Glymour, C., Scheines, R.: Causation, Prediction, and Search, 2nd edn. MIT Press, Cambridge (2000)
13. Leray, P., Francois, O.: BNT structure learning package: documentation and experiment s. Technical Report Laboratoire PSI - INSA Rouen-FRE CNRS 2645, Universitet INSA de Rouen (2004)
14. Cooper, G.F., Herskovits, E.: A Bayesian method for constructing Bayesian belief networks from databases. In: Proceedings of the Conference on Uncertainty in AI, pp. 86–94. Morgan Kaufmann, San Mateo (1991)
15. Meek, C.: Graphical Models: Selecting Causal and Statistical Models. PhD thesis, Carnegie Mellon University (1997)
16. Chickering, D.M.: A tranformational characterization of equivalent Bayesian network structures. In: Besnard, P., Hanks, S. (eds.) UAI 1995, San Francisco, pp. 87–98 (1995)
17. Wen, Y., Korb, K.B.: A heuristic algorithm for pattern-to-dag conversion. In: Proceedings of IASTED International Conference on Artificial Intelligence and Applications, pp. 428–433 (2007)

Extending Learning Vector Quantization for Classifying Data with Categorical Values

Ning Chen[1] and Nuno C. Marques[2]

[1] GECAD, Instituto Superior de Engenharia do Porto, Instituto Politecnico do Porto
Rua Dr. Antonio Bernardino de Almeida, 431 4200-072 Porto, Portugal
[2] CENTRIA/Departamento de Informatica
Faculdade de Ciencias e Tecnologia, Universidade Nova de Lisboa
Quinta da Torre, 2829-516 Caparica, Portugal
ningchen74@gmail.com, nmm@di.fct.unl.pt

Abstract. Learning vector quantization (LVQ) is a supervised neural network method applicable in non-linear separation problems and widely used for data classification. Existing LVQ algorithms are mostly focused on numerical data. This paper presents a batch type LVQ algorithm used for classifying data with categorical values. The batch learning rules make possible to construct the learning methodology for data in categorical nonvector spaces. Experiments on UCI data sets demonstrate the proposed algorithm is effective to improve the capability of standard LVQ to handle data with categorical values.

Keywords: Learning vector quantization, Self-organizing map, Categorical value, Batch learning.

1 Introduction

Classification is of fundamental importance to solve many practical problems in a wide range of fields such as credit scoring, customer management, image segmentation, pattern recognition, medical diagnostics, and control systems. It can be regarded as a two-stage process, i.e., model construction from a set of labeled data and class specification according to the retrieved model. Kohonen's learning vector quantization algorithm (LVQ) [1] is a supervised variant of the self-organizing map algorithm (SOM) that can be used for labeled input data. Both SOM and LVQ are based on neurons representing prototype vectors and use a nearest neighbor approach for clustering and classifying data. So, they are neural network approaches particularly useful for non-linear separation problems. In LVQ labels associated with input data are used for training. The learning process tends to perform the vector quantization starting with the definition of decision regions and repeatedly repositing the boundary to improve the quality of the classifier.

Existing LVQ algorithms are mostly focused on numerical data. However, the categorical values are commonly seen in data sets and it is worthwhile to study the LVQ algorithm for classifying data with categorical values. In this paper, the idea of the proposed algorithm originates from NCSOM [2], a batch SOM algorithm based on a new distance measurement and update rules in order to extend the usage of standard SOMs

J. Filipe, A. Fred, and B. Sharp (Eds.): ICAART 2009, CCIS 67, pp. 124–136, 2010.

to categorical data. In the present study, we advance the methodology of NCSOM to the batch type of learning vector quantization. We call this method BNCLVQ[1], an algorithm able to classify mixed numeric and categorical data. In one batch round, the Voronoi set of each map neuron is computed by projecting the input data to its best matching unit (BMU), then the prototype is updated according to incremental learning laws depending on class label and feature type. The main contribution of the proposed research is to extend the algorithms of LVQ family to handle data with categorical values for classification tasks. Experiments show that the algorithm outperforms standard LVQ with data preprocessing technique in terms of classification accuracy and achieves results as accurate as current state-of-the-art machine learning algorithms on various types of data sets.

The remaining of this paper is organized as follows. Section 2 reviews the related work. Section 3 presents a batch LVQ algorithm to handle numeric and categorical data during model training. In section 4, we evaluate the algorithm on some data sets from UCI repository. Lastly, contributions and future improvements are discussed in section 5.

2 Related Work

2.1 Data Type and Distance Measurement

Data could be described by features in numeric and categorical (nominal or ordinal) types [2]. Nominal features are categorical taking on values from a limited and predetermined set of categories without natural ordering. Ordinal features have particular order but unknown distance.

Let n be the number of input vectors, m the number of map units, and d the number of variables. Without loss of generality, we assume that the first p variables are numeric and the following $d - p$ variables are categorical, $\{\alpha_k^1, \alpha_k^2 \ldots \alpha_k^{n_k}\}$ is the list of variant values of the kth categorical variable (the natural order is preserved for ordinal variables). In the following description, $x_i = [x_{i1}, \ldots, x_{id}]$ denotes the ith input vector and $m_j = [m_{j1}, \ldots, m_{jd}]$ the prototype vector associated with the jth neuron. Data projection is based on the distance between input vectors and prototypes using squared Euclidean distance on numeric variables and simple mismatch measurement on categorical variables [3].

$$d(x_i, m_j) = \sum_{l=1}^{p} e(x_{il}, m_{jl}) + \sum_{l=p+1}^{d} \delta(x_{il}, m_{jl}) \tag{1}$$

where

$$e(x_{il}, m_{jl}) = (x_{il} - m_{jl})^2, \; \delta(x_{il}, m_{jl}) = \begin{cases} 0 & x_{il} = m_{jl} \\ 1 & x_{il} \neq m_{jl} \end{cases}$$

This distance simply conjoins the usual Euclidean distance between numeric values with the number of agreements between categorical categories. Complementing [2], this paper will also present how to handle ordinal data in BNCLVQ. However care must be taken so that measures are compatible with the Euclidean values.

---

[1] Batch Numeric and Categorical Learning Vector Quantization.

2.2 SOM Neural Networks

SOM is composed of a regular grid of neurons, usually in one or two dimensions for easy visualization. Each neuron is associated with a prototype or reference vector, representing the generalized model of input data. Due to its capabilities in data summarization and visualization, SOM is well suited for clustering, visualization and abstraction tasks [4]. Through a non-linear transformation, the data in high dimensional input space is projected to the low dimensional grid space while preserving the topology relations between input data. That is why resulting maps are sometimes called as topological maps.

Normally, standard SOMs are applicable to numeric features through arithmetic operations on distance calculation and map evolution. NCSOM [2] is an extension of standard SOM to handle categorical data. It is performed in batch manner based on the distance measure in Equation (1) and novel updating rules.

2.3 LVQ Neural Networks

LVQ is a variant of SOM, trained in a supervised way in the sense that the classification information is included in model learning. The prototypes define the class regions corresponding to Voronoi sets. LVQ starts from a trained map with class label assignment to neurons and attempts to adjust the class regions according to labeled data. Several versions of LVQ are presented, e.g., LVQ1, LVQ2, LVQ3, batch LVQ, differing in the process of searching for the optimum boundaries of classes [1]. In the past decades, LVQ has attracted much attention because of its simplicity and efficiency. The classic online LVQs are studied in literature [5]. In these algorithms, the map units are assigned class labels in the initialization and then updated at each training step. The online LVQs are sequential and sensitive to the order of presentation of input data to the network classifier [1]. Some algorithms have been proposed to perform LVQ in a batch way [5]. E.g., a batch network algorithm FKLVQ [6] fuses the batch learning, fuzzy membership functions and kernel-induced distance measures. The batch type learning vector quantization is employed for tissue differentiation from magnetic resonance images [7], efficient image compression [8] and bankruptcy prediction [9].

LVQ is also a viable way to tune the SOM results for better classification and therefore useful in data mining tasks. In classification problems, SOM is firstly used to concentrate the data into a small set of representative prototypes with respect to the topology of input data, then LVQ is used to fine tune the SOM prototypes for best class separation. It was reported that LVQ is able to improve the classification accuracy of a usual SOM rather than a standalone method [10,11].

It is well known that LVQ is designed for metric vector spaces in its original formulation. Some efforts were conducted to apply LVQ to nonvector representations. For this purpose, two difficulties are considered: distance measurement and incremental learning laws. The batch manner makes possible to construct the learning methodology for data in nonvector spaces such as categorical data. The SOM and LVQ algorithms expressed in batch version are proposed for symbol strings based on the so-called redundant hash addressing method and generalized means or medians over a list of symbol strings [12]. Also, a particular kind of LVQ is designed for variable-length and feature sequences to fine tune the prototype sequences for optimal class separation [13]. In BCNLVQ, differently from the conventional approaches of data conversion in prepro- cessing phase, the categorical values are handled in the model learning. Due to the use of batch learning method and the close relation between SOM and LVQ, the presence of categorical values can be handled using the same strategy as NCSOM.

3   BNCLVQ: A Batch LVQ Algorithm for Numeric and Categorical Data

In this section, a batch LVQ algorithm for mixed numeric and categorical data will be given. Similar to NCSOM, it adopts the distance measure introduced in previous section. Before presenting the algorithm, we first define incremental learning rules that will be used in the proposed BNCLVQ algorithm.

3.1   Incremental Learning Rules

Batch LVQ uses the entire data for incremental learning in one batch round. During the training process, an input vector is projected to the best-matching unit, i.e., winner neuron with the closest reference vector. Following [1] a Voronoi set can be generated for each neuron, i.e., $V_i = \{x_k \mid d(x_k, m_i) \leq d(x_k, m_j), 1 \leq k \leq n, 1 \leq j \leq m\}$ denotes the Voronoi set of m_i. As a result, the input space is separated into a number of disjointed sets: $\{V_i, 1 \leq i \leq m\}$. At one training epoch, the Voronoi set is calculated for each map neuron, composed of positive examples (V_i^P) and negative examples (V_i^N) indicating the correctness of classifying. In Voronoi set an element is positive if its class label agrees with the map neuron, and negative otherwise. Positive examples fall into the decision regions represented by the corresponding prototype and consequently make the prototype move towards the input. While, negative examples fall into other decision regions and consequently make the prototype move away from the input.

The map is updated by different strategies depending on the type of variables. The update rules combine the contribution of positive examples and suppress the influence of negative examples for each neuron in a batch round. Batch version is required for handling categorical values, since the number of values in neuron Voronoi set is used to define the weights. Assume $m_{pk}(t)$ is the value of the pth unit on the kth feature at epoch t. Let h_{ip} be the indicative function taking 1 as the value if p is the winner neuron of x_i, and 0 otherwise. Also, s_{ip} the denotation function whose value is 1 in case of positive example, and -1 otherwise.

$$h_{ip} = \begin{cases} 1 \text{ if } p = \underset{j}{\arg\min} \ d(x_i, m_j(t)) \\ 0 \text{ otherwise} \end{cases} \qquad (2)$$

$$s_{ip} = \begin{cases} 1 & \text{if } label(m_p) = label(x_i) \\ -1 & \text{otherwise} \end{cases}$$

The update rule of reference vectors on numeric features conducts in the similar way as NCSOM. Since LVQ is used to tune the SOM result, here the neighborhood is ignored and the class label is taken into consideration. If the denominator is 0 or negative for some m_{pk}, no updating is done. Thus, we have the learning rule on numeric variables:

$$m_{pk}(t+1) = \frac{\sum_{i=1}^{n} h_{ip} s_{ip} x_{ik}}{\sum_{i=1}^{n} h_{ip} s_{ip}} \tag{3}$$

As mentioned above, the arithmetic operations are not applicable to categorical values. Intuitively, for each categorical variable, the category occurring most frequently in the Voronoi set of a neuron should be chosen as the new value for the next epoch. For this purpose, a set of counters is used to store the frequencies of variant values for each categorical variable, in a similar way to what was done for NCSOM algorithm. However, we have now taken into account the categorical information, so the frequency of a particular category is calculated by counting the number of positive occurrences minus the number of negative occurrences in the Voronoi set.

$$F(\alpha_k^r, m_{pk}(t)) = \sum_{i=1}^{n} v(h_{ip} s_{ip} \mid x_{ik} = \alpha_k^r), r = 1, 2, \ldots, n_k \tag{4}$$

$F(\alpha_k^r, m_{pk}(t))$ represents the accumulated absolute votes of each α_k^r value[2]. As in standard LVQ algorithm, this change is made to better tune the original clusters acquired by SOM to the available supervised data.

For nominal features, the best category $c = \max_{r=1}^{n_k} F(\alpha_k^r, m_{pk}(t))$, i.e., the value having maximal frequency, is accepted if the frequency is positive. Otherwise, the value remains unchanged. As a result, the learning rule on nominal variables is:

$$m_{pk}(t+1) = \begin{cases} c & \text{if } F(c, m_{pk}(t)) > 0 \\ m_{pk}(t) & \text{otherwise} \end{cases} \tag{5}$$

Different from nominal variables, the ordinal variables have specific ordering of values (here represented by index r). Therefore, the updating depends not only on the frequency of values also on the ordering of values. The category closest to the weighted sum of relative frequencies on all possible categories is chosen as the new value concerning about the natural ordering of values. So, the learning rule on ordinal variables is:

$$m_{pk}(t+1) = round(\sum_{r=1}^{n_k} r * \frac{F(\alpha_k^r, m_{pk}(t))}{\sum_{i=1}^{n} h_{ip} s_{ip}}) \tag{6}$$

3.2   Algorithm Description

As mentioned, the BNCLVQ algorithm is performed in a batch mode. It starts from the trained map obtained in an unsupervised way, e.g., the NCSOM algorithm. Each map neuron is assigned by a class label with a labeling schema. In this paper the majority

---

[2] Function $v(y \mid COND)$ is y when $COND$ holds and zero otherwise.

class is used based on the distance between prototypes and input for acquiring the labeled map. Afterwards, one instance x_i is given as input and the distance between x_i and prototypes is calculated using Equation (1), consequently the input is projected to the closest prototype. After all inputs are processed, the Voronoi set is computed for each neuron, composed of positive examples and negative examples. Then the prototypes are updated according to Equation (3), Equation (5) and Equation (6), respectively. This training process is repeated iteratively enough iterations until the termination condition is satisfied. The termination condition could be the number of iterations or a given threshold denoting the maximum distance between prototypes in previous and current iteration. In summary, the algorithm is performed as follows:

1. Compute the trained and labeled map with prototypes: $m_i, i = 1, ..., m$;
2. For $i = 1..., n$, input instance x_i and project x_i to its BMU;
3. For $i = 1, ..., m$, compute V_i^P and V_i^N for $m_i(t)$;
4. For $i = 1, ..., m$, calculate the new prototype $m_i(t+1)$ for next epoch;
5. Repeat Step 2 to Step 4 until the termination condition is satisfied.

4 Empirical Analysis

The proposed BNCLVQ algorithm is implemented based on *somtoolbox* [14] in *matlab* running Windows XP operating system. In the empirical analysis, we mainly concern about the effectiveness of BNCLVQ in classification problems.

4.1 Data Sets

Eight UCI [15] data sets are chosen for the following reasons: missing data, class composition (binary class or multi-class), proportion of categorical values (pure categorical, pure numeric or mixed) and data size (from tens to thousands of instances). These data sets are described in Table 1, including the number of instances, the number of features (nu:numeric, no:nominal, or:ordinal), the number of categorical values (#val), the number of classes (#cla), percentage of instances in the most common class (mcc) and proportion of missing values (mv).

Table 1. Description of data sets

datasets	#ins	#features nu	no	or	#val	#cla	mcc	mv
soybean	47	0	35	0	74	4	36%	0
mushroom	8124	0	22	0	107	2	52%	1.4%
tictactoe	958	0	9	0	27	2	65%	0
credit	690	6	9	0	36	2	56%	1.6%
heart	303	5	2	6	20	2	55%	1%
horse	368	7	15	0	53	2	63%	30%
zoo	101	1	15	0	30	7	41%	0
iris	150	4	0	0	-	3	33%	0

- soybean small data: a well-known soybean disease diagnosis data with pure categorical variables and multiple classes;
- mushroom data: a large number of instances in pure categorical variables (some missing data);
- tictactoe data: a pure categorical data encoding the board configurations of tic-tac-toe games, irrelevant features with high amount of interaction;
- credit approval data: a good mixture of numeric features, nominal features with a small number of values and nominal features with a big number of values (some missing data);
- heart disease: mixed numeric and categorical values (some missing data);
- horse colic data: a high proportion of missing values, mixed numeric and categorical values;
- zoo data: multiple classes, mixed numeric and categorical values;
- iris data: pure numeric values.

To ensure all features have equal influence on distance, numeric features are normalized to unity range. The missing values contained in some data sets are neglected in the distance calculation and update operation of the corresponding features.

4.2   Experimental Results

In our experiments we set termination condition as 50 iterations. For each data set, the experiments are performed in the following way:

1. The data set is randomly divided into 10 folds. In each trail, 9 folds are used for model training and labeling, and the remaining is used for performance validation.
2. The map is trained with the training data set in an unsupervised manner by NCSOM algorithm, and then labeled by the majority class according to the known samples in a supervised manner.
3. BNCLVQ is applied to the resultant map in order to improve the classification quality.
4. For validation, each sample of the test data set is compared to map units and assigned by the label of BMU. In order to avoid the assignment of an empty class, unlabeled units are discarded from classification. Then the performance is measured by classification accuracy, i.e., the percent of the observations classified correctly.
5. Ten-fold cross-validation is applied by repeating Step 2 to Step 4, while considering the different folds as test and training data and computing final average accuracy and standard deviation results.

As other ANN models, LVQ is sensitive to some parameters in which map size, i.e., number of representative patterns, is an important one. In Table 2, we investigate the effect of map size to the resulted classification accuracy. Four kinds of maps are studied for comparison: 'middle' map is determined by the number of instances with the side lengths of grid as the ratio of two biggest eigenvalues [14]; 'small' map has one-quarter neurons of 'middle' one; 'tiny' map has half neurons of 'small' one; 'big' map has four times neurons of 'middle' one. As the map enlarges from 'tiny' to 'middle', both training accuracy and test accuracy improve significantly for most data sets (e.g., test

Table 2. Effect of map size in classification accuracy

datasets	tiny(%)		small(%)		middle(%)		big(%)	
	train	test	train	test	train	test	train	test
soybean	100	100	100	100	100	100	100	100
mushroom	92	91	96	96	99	98	99	98
tictactoe	73	75	83	75	92	85	95	84
credit	85	83	86	82	89	86	92	85
heart	86	80	86	78	87	82	88	79
horse	83	81	84	84	87	84	90	81
zoo	82	81	90	89	99	99	100	96
iris	95	97	97	96	98	95	99	95

accuracy increases from 81% to 99% for *zoo* and 75% to 85% for *tictactoe*, while *soybean* and *iris* are less sensitive to the change of map size). Further enlarging the map increases the accuracy of training data, however, in case of test data, the accuracy has a tendency of getting downward, indicating the map becomes overfitting. It is shown that the maps in middle size are best for generalization performance except on *iris* data, which achieves desirable accuracy using only a 'tiny' map of 2 by 2 units. For simplicity, in the following experiments, we only report the results of maps in middle size.

As summarized in Table 3, the results show the potential of BNCLQV compared with NCSOM in improving the accuracy on both training data and test data for not only data sets of pure categorical variables (e.g., soybean, mushroom and tictactoe) but also those of mixed variables (e.g., credit, heart, horse and zoo). Typically, BNCLVQ achieves increases up to 9% in classification accuracy, when comparing with using only NCSOM majority class for classification. The results where performance is almost equivalent are the *soybean* and *iris* data sets. In these datasets performance is already near reported maximum after running NCSOM[3]. It is observed that performance on BNCLVQ networks is always better than the one of NCSOM on all datasets without overfitting maps. This gives evidence in favor of the validity of the approach for refining hybrid SOM maps.

As mentioned previously, SOMs are very useful for data mining purposes. So, we have also analyzed our results from the visualization point of view. As an illustrative example we present the output map for credit data in Figure 1. Due to the topology preserving property of SOM, class regions are usually composed of neighboring prototypes of small u-distance (the average distance to its neighboring prototypes) values. The histogram of neurons contains the composition of patterns presented to the corresponding prototypes. Although neurons of zero-hit have no representative capability of patterns, they help to discover the boundary of class regions. From the visualization of u-distance and histogram, it becomes easy to obtain the visual insights into the cluster structures. Figure 1 shows the u-distance and histogram chart of trained map for credit data obtained by NCSOM (left graph) and BNCLVQ (right graph) respectively. Each

---

[3] Indeed, as it was just verified in Table 2, in BNCLVQ the simpler *iris* dataset is presenting overfitting with the 'middle' map size (used for all datasets in this experiment).

Table 3. Average values of accuracy and standard deviation

datasets	map size	train accuracy(%) NCSOM	BNCLVQ	test accuracy(%) NCSOM	BNCLVQ
soybean	[7 5]	100 ± 0	100 ± 2	98 ± 8	100 ± 0
mushroom	[14 8]	96 ± 1	99 ± 1	95 ± 4	98 ± 3
tictactoe	[13 11]	80 ± 2	92 ± 1	76 ± 3	85 ± 3
credit	[17 7]	85 ± 1	89 ± 1	80 ± 4	86 ± 3
heart	[10 8]	87 ± 2	87 ± 2	78 ± 9	82 ± 7
horse	[11 8]	83 ± 1	87 ± 1	79 ± 10	84 ± 7
zoo	[8 6]	99 ± 1	99 ± 1	99 ± 3	99 ± 3
iris	[16 3]	98 ± 1	97 ± 1	96 ± 3	95 ± 3

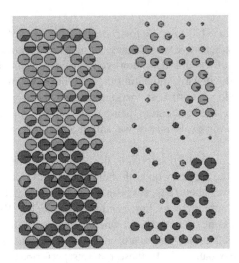

Fig. 1. Visualization of classification results on credit data

node has an individual size proportional to its u-distance value with slices denoting the percentage of two classes contained. It is observed (by the zero-hit neurons) that BN-CLVQ is able to separate two classes more clearly, showing the capability of BNCLVQ as a fine-tuning method of original NCSOM for better class discrimination.

We also study the convergence properties of BNCLVQ algorithm on the data sets mentioned above. In this experiment, the convergence was measured as the overall distance in prototypes between the current iteration and the previous iteration: $od(t) = \sum_{i=1}^{m} d(m_i(t-1), m_i(t))$. As observed in Figure 2, the evolution of prototypes reflects the significant tendency of convergence. The distance decreases rapidly in the beginning, and tends to be more stable after a number of iterations (less than 50 iterations for these data sets). The convergence speed depends on the size of data, number of variables and intrinsic properties of data distribution. For example, it takes only 3 iterations to converge for *soybean* data. However for *heart* data, the variation of prototypes is not stable after 30 iterations due to the local optimum solution. As mentioned previously, overfitting is a critical problem for BNCLVQ as other ANN architectures.

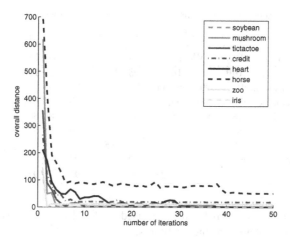

Fig. 2. Convergence study of BNCLVQ

Indeed, according to our results, overfitting is already detected in some tested datasets. So, small improvements could be achieved with smaller maps or an earlier stop criterion. A practical approach to prevent overfitting is to use an independent validation data set to select the best model trained with different parameters, namely for defining the earlier stop in the training process. However, in this paper for the easier comparison among the methods discussed in next section, we have decided to use the conservative value of 50 iterations for assuring proper convergence of all maps.

4.3 Comparative Studies

For better comparison with other advanced approaches, the performance of proposed algorithm is compared with some representative algorithms for supervised learning. Six representatives implemented by Waikato Environment for Knowledge Analysis (WEKA) [16] with default parameters are under consideration:

- Naive Bayes (NB): a well-known representative of statistical learning algorithm estimating the probability of each class based on the assumption of feature independence;
- Sequential minimal optimization (SMO): a simple implementation of support vector machines (SVM). Multiple binary classifiers are generated to solve multi-class classification problems;
- K-nearest neighbors (KNN): an instance-based learning algorithm classifying an unknown pattern to the its nearest neighbors in training data based on a distance metric (the value of k was determined between 1 and 5 by cross-validation evaluation);
- J4.8[4]: the decision tree algorithm to first infer a tree structure adapted well to training data then prune the tree to avoid overfitting;
- PART: a rule-based learning algorithm to infer rules from a partial decision tree;

---

[4] A Java implementation of popular C4.5 algorithm.

Table 4. Accuracy ratio comparison (the value of k is given for KNN)

datasets	Naive Bayes	SMO SVM	KNN	J4.8	PART	MLP	LVQ	BNC LVQ
soybean	**100**	**100**	**100**(1)	98	**100**	**100**	**100**	**100**
mushroom	93	**99**	**99**(1)	98	98	**99**	98	98
tictactoe	70	98	**99**(1)	85	95	97	96	85
credit	78	85	85(5)	86	85	84	80	**86**
heart	83	**84**	83(5)	77	80	79	78	82
horse	78	83	82(4)	**85**	**85**	80	79	84
zoo	95	93	95(1)	92	92	95	92	**99**
iris	95	96	95(2)	96	94	**97**	95	95

– Multi-layer perceptron (MLP): a supervised artificial neural network with back propagation to explore non-linear patterns.

We should stress that this comparison may be unfair to LVQ. Indeed, as discussed, LVQ is mainly a projection method that can also be used for classification purposes. So, for the sake of comparison, the standard LVQ is also tested. For doing so, categorical data are preprocessed by translating each categorical feature to multiple binary features (i.e., the standard approach for applying LVQ to categorical data).

The summary of results is given in Table 4. For each data set, the best accuracy is emphasized by bold. Although BNCLVQ is not the best one for all data sets, it produces desirable accuracy in most cases, and reported best values are always within standard deviation (deviation values are reported in Table 3). Results are especially relevant on data of mixed types. Also, BNCLVQ is always better than standard LVQ, the only exception on tictactoe data is probably caused by the presence of interaction between features [17]. Indeed in tictactoe, other tested algorithms with categorical features (e.g. J4.8 and PART) had systematically worst performances. Maybe the inclusion of extra variables is somehow helping generalization and showing the need for better feature encoding. E.g., in [18], the inclusion of domain knowledge and feature generalization in tictactoe improved classification accuracy. Similarly, author's research using tictactoe (namely in ongoing work and in [19]), also has confirmed the high relevance and need of domain data in this particular dataset.

5   Conclusions

Learning vector quantization is a promising and robust approach for classification modeling tasks. Although originally designed in metric vector spaces, LVQ could be performed on non-vector data in a batch way. In this paper, a batch type LVQ algorithm capable of dealing with categorical data is introduced. The analysis of predictive effectiveness is undertaken to demonstrate the capability of proposed method as a promising alternative to existing advanced classification models.

SOM topological maps are very effective tools for representing data, namely in data mining frameworks. LVQ is, by itself, a powerful method for classifying supervised data. Also, it is the most suitable method to tune SOM topological maps to supervised data. Unfortunately original LVQ can not handle categorical data in a proper way. In this paper we show that BNCLVQ is a feasible and effective alternative for extending previous NCSOM to supervised classification on both numeric and categorical data so that it servers as a promising complement to existing methods. Since BNCLVQ is performed on an organized map, only a limited number of known samples is needed for the fine-tuning and labeling of map units. Therefore, BNCLVQ is a suitable candidate for tasks in which scarce labeled data and abundant unlabeled data are available. BNCLVQ is also easy for parallelization [20] and can be applicable in frameworks with very large datasets.

Moreover, BNCLVQ is easily extended to fuzzy case to solve the prototype under-utilization problem [6], i.e., only the BMU is updated for each input, simply replacing the indicative function by the membership function [21]. The membership assignment of fuzzy projection implies the specification to classes, and can be used for the validity estimation of classification [22].

In the future plan, the benefit of fuzzy strategies in BNCLVQ will be investigated by cross-validation experiments on both UCI data sets and state-of-art real world problems. In a first real world case study, we are currently applying NCSOM topological maps to fine tune mixed numeric and categorical data in a natural language processing problem [23]. In this domain we have some pre-labeled data available and NCSOM is helping to investigate accurateness and consistency of manual data labeling. However, already known correct cases (and possible some previously available prototypes) should be included in the previously NCSOM trained topological map. For that we intend to use BNCLVQ as the appropriate tool. According to our results, BNCLVQ can achieve good accuracy in most domains. Moreover BNCLVQ is more than a classification algorithm. Indeed BNCLVQ is also a fine-tuning tool for topological features maps, and, consequently, a tool that will help the data mining process when some labeled data is available.

Acknowledgements. This work was supported by project C2007-FCT/442/2006-GEC-AD/ISEP (Knowledge Based, Cognitive and Learning Systems).

References

1. Kohonen, T.: Self-Organizing Maps, 3rd edn. Springer, Berlin (2001)
2. Chen, N., Marques, M.N.: An Extension of Self-Organizing Maps to Categorical Data. In: Bento, C., Cardoso, A., Dias, G. (eds.) EPIA 2005. LNCS (LNAI), vol. 3808, pp. 304–313. Springer, Heidelberg (2005)
3. Huang, Z.: Extensions to the K-means Algorithms for Clustering Large data Sets with Categorical Values. Data Mining and Knowledge Discovery 2, 283–304 (1998)
4. Kohonen, T.: The Self-Organizing Map. Proceedings of IEEE Special Issue on Neural Networks 78(9), 1464–1480 (1990)
5. Kohonen, T.: The Self-organizing Map. Neurocomputing 21, 1–6 (1998)
6. Zhang, D., Chen, S., Zhou, Z.: Fuzzy-kernel Learning Vector Quantization. In: Yin, F.-L., Wang, J., Guo, C. (eds.) ISNN 2004. LNCS, vol. 3173, pp. 180–185. Springer, Heidelberg (2004)

7. Yang, M., Lin, K., Liu, H., Lirng, J.: Magnetic Resonance Imaging Segmentation Techniques using Batch-type Learning Vector Quantization Algorithms. Magnetic Resonance Imaging 25(2), 265–277 (2007)
8. Tsekouras, G.E., Antoniosa, M., Anagnostopoulosa, C., Gavalasa, D., Economou, D.: Improved Batch Fuzzy Learning Vector Qantization for Image Compression. Information Sciences 178(20), 3895–3907 (2008)
9. Chen, N., Vieira, A.: Bankruptcy Prediction Based on Independent Component Analysis. In: Proceedings of International Conference on Agents and Artificial Intelligence (ICAART 2009), Portugal, pp. 150–155 (2009)
10. Visa, A., Valkealahti, K., Iivarinen, J., Simula, O.: Experiences from Operational Cloud Classifier based on Self-organising Map. In: Procedings of SPIE, Orlando, Florida. Applications of Artificial Neural Networks V, vol. 2243, pp. 484–495 (1994)
11. Priyono, A., Ridwan, M., Alias, A.J., Rahmat, R., Hassan, A., Ali, M.: Application of LVQ Neural Network in Real-Time Adaptive Traffic Signal Control. Journal Teknologi 42(B), 29–44 (2005)
12. Kohonen, T., Somervuo, P.: Self-Organizing Maps of Symbol Strings. Neurocomputing 21(10), 19–30 (1998)
13. Somervuo, T., Kohonen, T.: Self-Organizing Maps and Learning Vector Quantization for Feature Sequences. Neural Processing Letters 10(2), 151–159 (1999)
14. Laboratory of computer and information sciences & Neural networks research center, Helsinki University of Technology: SOM Toolbox 2.0
15. Asuncion, A., Newman, D.J.: UCI Machine Learning Repository. University of California, Department of Information and Computer Science, Irvine, CA, http://www.ics.uci.edu/~mlearn/MLRepository.html
16. Witten, H., Frank, E.: Data Mining: Practical Machine Learning Tools and Techniques, 2nd edn. Morgan Kaufmann, San Francisco (2005)
17. Bay, S.D.: Nearest Neighbor Classification from Multiple Feature Subsets. Intelligent Data Analysis 3(3), 191–209 (1999)
18. Matheus, C.: Adding Domain Knowledge to sbl through Feature Construction. In: Proceedings of the Eighth National Conference on Artificial Intelligence, pp. 803–808. AAAI Press, Boston (1990)
19. Bader, S., Holldobler, S., Marques, N.: Guiding Backprop by Inserting Rules. In: dAvila Garcez, A.S., Hitzler, P. (eds.) Proceeding of 18th European Conference on Artificial Intelligence, the 4th InternationalWorkshop on Neural-Symbolic Learning and Reasoning. ISSN, vol. 366, Patras, Greece, pp. 1613–0073 (2008)
20. Silva, B., Marques, N.C.: A Hybrid Parallel SOM Algorithm for Large Maps in Data-Mining. In:Proceedings of 13th Portuguese Conference on Artificial Intelligence (EPIA 2007), Workshop on Business Intelligence, Portugal. IEEE Guimaraes (2007)
21. Bezdek, J.C.: Pattern Recognition with Fuzzy Objective Function Algorithms. Plenum Press, New York (1981)
22. Chen, N.: Fuzzy Classification using Self-Organizing Map and Learning Vector Quantization. In: Shi, Y., Xu, W., Chen, Z. (eds.) CASDMKM 2004. LNCS (LNAI), vol. 3327, pp. 41–50. Springer, Heidelberg (2005)
23. Marques, N., Bader, S., Rocio, V., Holldobler, S.: Neuro-Symbolic Word Tagging. In: Proceedings of 13th Portuguese Conference on Artificial Intelligence (EPIA 2007), 2nd Workshop on Text Mining and Applications, Portugal, pp. 779–790. IEEE Guimaraes (2007)

Action Knowledge Acquisition with Opmaker2

T.L. McCluskey, S.N. Cresswell, N.E. Richardson, and M.M. West

School of Computing and Engineering,
The University of Huddersfield, Huddersfield HD1 3DH, U.K.
{T.L.McCluskey,S.N.Cresswell,
N.E.Richardson,M.M.West}@hud.ac.uk

Abstract. AI planning engines require detailed specifications of dynamic knowledge of the domain in which they are to operate, before they can function. Further, they require domain-specific heuristics before they can function efficiently. The problem of formulating domain models containing dynamic knowledge regarding actions is a barrier to the widespread uptake of AI planning, because of the difficulty in acquiring and maintaining them. Here we postulate a method which inputs a *partial* domain model (one without knowledge of domain actions) and training solution sequences to planning tasks, and outputs the full domain model, including heuristics that can be used to make plan generation more efficient.

To do this we extend GIPO's *Opmaker* system [1] so that it can induce representations of actions from training sequences without intermediate state information and without requiring large numbers of examples. This method shows the potential for considerably reducing the burden of knowledge engineering, in that it would be possible to embed the method into an autonomous program (agent) which is required to do planning. We illustrate the algorithm as part of an overall method to acquire a planning domain model, and detail results that show the efficacy of the induced model.

Keywords: Planning and Scheduling; Machine Learning.

1 Introduction

Applications of AI planning technology require persistent resources comprising of teams of highly skilled engineers to formulate and maintain a planner's knowledge base. The amount of effort needed to encode error free, accurate action specifications and planning heuristics, and to maintain them, is significant. *Actions* are real world operations that change the state of object(s) in the world in some way. These actions are invariably encoded in planning knowledge bases as generalised representations called *operator schema*. Additionally, heuristics are often hand coded in the form of *methods* which encapsulate the preferred solutions of a generalised subtask. Our work is aimed at automating the formulation of such operators and methods by employing a trainer to create training tasks and example solution sequences of these tasks. The solutions are fed to a knowledge acquisition tool, *Opmaker2*, as a sequence of action instances, where each action instance is identified by name plus the object instances that are affected by, or are necessarily present at, action execution. The sequences are produced by a trainer - a domain expert who may not be familiar with the languages and notations

J. Filipe, A. Fred, and B. Sharp (Eds.): ICAART 2009, CCIS 67, pp. 137–150, 2010.
© Springer-Verlag Berlin Heidelberg 2010

used by planners. *Opmaker*2 constructs operator schema and planning heuristics from training sessions which are composed of a handful of such action sequences. In other words, it outputs detailed specifications of operator schema from single action traces automatically, without requiring intermediate state information for each training example. The induced actions are detailed enough for use in planning engines and compare well with hand crafted operators.

This paper describes *Opmaker*2, an extension of the earlier *Opmaker* system [2], in that the latter is an interactive learning tool, whereas the former can be run in batch mode without the need for user assistance. *Opmaker* was implemented within the GIPO system [1], an experimental tools environment for use in the acquisition of AI planning knowledge, containing a wide range of engineering and validation tools. GIPO was based on the planning language of OCL [3]. To motivate the rest of the paper, we will describe in a little more depth the problem that we are aiming to solve, in terms of a learning, or more specifically a *knowledge acquisition* problem. Automated planning systems can be logically described as having three components.

(a) *The domain model* (sometimes referred to as a *domain description*) is the specification of the objects, structure, states, goals and dynamics of the domain of planning. The language family used for the communication of domain models is PDDL [4], although in this paper we use a higher level language called OCL[5] for domain modelling. Component (a) is further split into:

(i) knowledge of objects, object sorts, domain constraints, and possible states of objects - collectively called static knowledge.

(ii) knowledge of action and change - knowledge of dynamics. This knowledge is in both PDDL and OCL represented as a set of parameterised operator schema representing generic actions in the domain of interest.

(b) The *planning engine* is the software that reasons with the knowledge in (a) to solve planning goals. The development of fast planning engines which can deal with expressive variants of PDDL (e.g. modelling domains containing durative actions and metric resources) has been a primary goal of the AI Planning community.

(c) A set of *planning heuristics*. The general problem of AI Planning is well known to be intractable, and a set of heuristics for each domain is required to make the application of (b) to (a) tractable. Whereas the form and content of (a) and (b) are well understood, what form heuristics take is more contentious. Putting domain heuristics with the planning engine may limit its application (they anticipate the domain). Encoding heuristics into the domain model when constructing it is equally contentious - as the authors of PDDL claim it is for "physics and nothing else"[4].

The knowledge acquisition problem that this paper addresses is:

Given knowledge of (a)(i), can we design a simple process to enable a system to automatically acquire knowledge of type (a)(ii) and (c)?

The reason for setting up this knowledge acquisition problem is that hand crafting knowledge of dynamics (in particular operator schema), and planner and domain specific heuristics, is much harder than acquiring knowledge of type (a)(i). The difficulty in acquiring knowledge of actions is invariably pointed out in reports of AI planning applications (for example, in reports of Space applications [6]).

The general method that we are proposing is for a system to acquire knowledge from examples of solved tasks, represented as sequences of actions, given to it by a benevolent trainer. *Operator schema* (type (a)(ii) knowledge) are induced from each example action, whereas *heuristics* (type (c) knowledge) are induced from the whole sequence of actions the trainer uses to solve a task. The heuristics are in the form of HTN-type methods.

The rest of the paper is structured as follows: in section 2 we outline the *Opmaker2* system, starting with its inputs and outputs, and then detail the operation of its state-deriving component. We use a *tyre-change domain* to illustrate the algorithm which contains the knowledge acquisition process. Section 3 contains our experimental results, and Section 4 a brief survey of related work.

2   The Opmaker2 System

In this section we describe the *Opmaker2* system, and explain how it can form a solution to the knowledge acquisition problem introduced in the last section. We will use as a running example throughout the rest of the paper a domain which represents changing the tyre of a car wheel. This domain is an extended version of the simpler 'tyre world'[7]. It involves knowledge about such objects as tyre, wheel, nuts, wheel-trim, jack, wrench, and such actions as undo-nuts, put-on-wheel etc. In *Opmaker2*, components of type (a)(i) knowledge are referred to collectively as the *partial domain model* $\mathcal{PDM}$. For our running example, the partial domain of the tyre-change domain is provided in the appendix, in the native code of $\mathcal{OCL}$. There are two inputs to *Opmaker2*: the $\mathcal{PDM}$ and a set of hand crafted solution sequences to planning tasks. A $\mathcal{PDM}$ consists of:

object identifiers and sort names: denoted $Objs$ and $Sorts$ respectively; there are a number of sorts, each containing a set of objects where each object belongs to one set (called a sort). An example of an object is $hub1$ belonging to the hub sort. The behaviour of each object in a sort is assumed to be the same as all others in the sort.

predicate definitions: denoted $Prds$, where each object of each sort may be related to objects of other sorts, and have property - value relationships with sets of basic values (boolean or scalar). Examples are $on\_ground(hub)$, $jacked\_up(hub, jack)$, relating to whether an object of sort hub is on the ground or jacked up.

object state expressions: denoted $Exps$; these define all the possible values of an object's state. An object's state is defined by its relationship with other objects and/or the value of its properties. Sorts are engineered so that the object state space is defined by a small number of expressions. For example, the tyre-change $\mathcal{PDM}$ specifies that any object H of sort hub can occupy a state satisfying exactly one of the following object expressions:

```
[on_ground(H),fastened(H)],
[jacked_up(H,J),fastened(H)],
[free(H),jacked_up(H,J),unfastened(H)],
[unfastened(H),jacked_up(H,J)]
```

(as a convention we choose upper case variables as parameters - here J represents any object of sort $jack$).

domain invariants: denoted *Invs*; these are used to define domain constraints and are written in terms of the predicates given above. Informally, a set of invariants is adequate if it disqualifies states which are inconsistent. For example: "Only a single wheel can be on a hub".

$$\forall H{:}hub . \forall W_1{:}wheel . \forall W_2{:}wheel . \left[\begin{pmatrix} wheel\_on(W_1, H) \\ \wedge \\ wheel\_on(W_2, H) \end{pmatrix} \Rightarrow (W_1 = W_2) \right]$$

The second input is a set of *solution sequences* and the tasks that they solve. These are supplied by a trainer (a domain expert). For the purposes of training in $Opmaker2$, we define a task in terms of:

- an initial state comprising the initial states of objects in the domain,
- a set of desired goal states for a set of objects.

A solution sequence solves such a task and is written in terms of verbs (action names) and affected objects. The trainer is expected to include references to all objects that are needed for each action to be carried out, indicating whether or not the objects change as a result of the action. Typical tasks in the domain should be chosen that often form the basis of solutions to larger tasks. For example, in the sequence below a changed wheel is secured on the hub and the vehicle is made ready for use.

```
do_up       unchanged - wrench0, jack0, wheel1;
            changing - hub1, nuts1
jack_down   unchanged - null
            changing - hub1, jack0
tighten     unchanged - wrench0, hub1;
            changing - nuts1
apply_trim  unchanged - hub1;
            changing - trim1, wheel1
```

Objects preceded by **unchanged** remain unaffected by the action, but have to be present in the state during execution of the action. In the first element of the sequence, wrench0, jack0 and wheel1 all have to be in a certain state specified by initial state of the task (wrench0 is available, jack0 is jacking up the hub, and wheel1 is trimless to allow the nuts to be screwed). The **changing** objects *must* change state (hub1 becomes $fastened$ and nuts1 are $done\_up$).

The output of $Opmaker2$ is a full domain model, consisting of:

operator schema: they make up the knowledge of type (a)(ii), and represent actions or events that change objects' states. They are specified by a name, a list of parameters, and a set of object transitions. Transitions may be null (in which case they act as *prevail* conditions), necessary or conditional. The template of a schema is as follows:

head: name(list of parameters)
body: ≥ 0 prevail conditions;
        ≥ 1 necessary transitions;
        ≥ 0 conditional transitions

Prevails are represented by object state expressions, whereas necessary and conditional transitions are written in the form $LHS \Rightarrow RHS$, where LHS, RHS are object state expressions.

methods: each training sequence results in a parameterised method, similar in form to hierarchical (HTN) methods found in AI Planning. A method comprises of a name, prevail conditions, and a sequence whose members can comprise both operator schema and (other) methods. Methods can be used as a heuristic in planning engines as they encapsulate preferred ways to solve planning problems.

2.1   The Opmaker2 Algorithm

The main innovation of Opmaker2 is that it computes its own intermediate states using a combination of heuristics and inference from the $\mathcal{PDM}$ and the training tasks and solutions. This gives a fully automated solution to the knowledge acquisition problem described above - there is no need for user advice. In contrast, its predecessor $Opmaker$ is a *mixed initiative* knowledge acquisition tool which requires the same inputs as above (a $\mathcal{PDM}$ and a set of solution sequences to tasks) and, additionally, it requires user advice. As $Opmaker$ creates an operator schema from each action in a training solution sequence, it asks the user to input, if needed, the target state that each object would occupy after execution of the action. In order to build up transitions that form an operator schema, the LHS is taken as the current state of the object (object transitions are tracked as each action is processed). The RHS is taken from the user input, which indicates, where there is a choice, the state an object is left in (this becomes that object's current state). Having the start and end states for each object involved in the action, $Opmaker$ proceeds with a *generalisation* phase where object instances are replaced with sort parameters, which then form the parameter variables $X_1, \ldots, X_n$ of the resulting operator schema. In supplying the solution sequences, the trainer specifies what objects take part in what actions. As actions are executed, objects go through state transitions and occupy intermediate states en route to reaching their goal states. The space of states that an object may occupy are defined by the state expressions of the $\mathcal{PDM}$. To be able to automatically acquire operator schema, $Opmaker$ was able to resolve exactly what are the intermediate states of each object affected in the training sequence by asking for user advice.

In contrast, $Opmaker2$ uses a procedure called *DetermineStates*, which performs this function by tracking the changing states of each object referred to within a training example. It takes advantage of the static, object-state information and invariants within the domain model. The output from *DetermineStates* is, for each point in the training sequence, a map which associates each object with a unique state value. Uniqueness is not guaranteed, however, and depends on the information in the $\mathcal{PDM}$, hence sometimes this map may return a set of states rather than a unique one (we return to this problem below). Once the map determining intermediate states has been generated, the techniques of the original *Opmaker* algorithm are used to generalise object references and create parameterised operator schema.

A Description of the *DetermineStates* Procedure. To illustrate the workings of the Procedure, we will use the example tyre domain solution sequence to form the initial stage of an example walk-though. Let us consider $A(1) - A(4) =$ do_up, jack_down, tighten, apply_trim as given above. The algorithm is as follows:

Procedure **DetermineStates**
In:
    $\mathcal{PDM}$,
    I, F are maps of objects to their Initial, Final state, resp.
    $T = A(1)..A(N)$: training sequence of N actions
Out:
    maps $C_i, i = 1, \ldots, N + 1$, mapping from object names
        to object states such that action $A(i)$ of T
        represents a transition from C_i to C_{i+1}
Define $A.c$ to be the set of $A.obj$'s changing objects
1. $C_1 := I; C_{N+1} := F$;
2. for each $i \in 1, \ldots, N$
3.   for each object $O \notin A(i).c$
4. $C_{i+1}(O) := C_i(O)$;
5.   end for
6.   for each object $O \in A(i).c$
7.   if $O \notin A(i+1).c \cup \ldots \cup A(N).c$ then
8. $C_{i+1}(O) := F(O)$
9.   else
10.   **choose** $C_{i+1}(O) :=$ any legal state using $\mathcal{PDM}$
11.   with parameters bound to objects in $A(i)$ or $C_i(O)$
12.   test the choice using the following constraints
13.   – $C_{i+1}(O) \neq C_i(O)$
14.   – the transition $C_i(O) \Rightarrow C_{i+1}(O)$
15.   must be consistent with transitions at
16.   previous occurrences of $A(i).name$
17.   end if
18.   end for
19.   test that the conjunction of $C_{i+1}(O)$ for all O
20.   is consistent with $\mathcal{PDM}$'s invariants
21. end for

In **Line 1** the first and last components of the map C are initialised to be the same as the initial and final state respectively. The algorithm then iterates for all actions in the sequence. When $i = 1$, **Lines 3-5** define C_2 as the same as C_1 when applied to non-changing objects in the domain. **Lines 6-18** attempt to determine the rest of map C_2 where it is applied to objects that change as a result of the execution of $A(1)$. **Line 6** identifies the changing objects (hub1 and nuts1) - let us consider hub1. **Lines 7-8** look ahead to see if hub1 will not change again in a subsequent action and find that it does in the second action in the sequence. If we had chosen an example where the object does not change again after the first action then **Line 8** would set the object's state to be the final state. Considering **Line 10**, using the partial domain model there are four potential values for $C_2(\texttt{hub1})$:

a. `[on_ground(hub1),fastened(hub1)]`
b. `[free(hub1),jacked_up(hub1,jack0),unfastened(hub1)]`
c. `[jacked_up(hub1,jack0),fastened(hub1)]`
d. `[unfastened(hub1), jacked_up(hub1,jack0)]`

Lines 12-16 of the algorithm determine which of these states is appropriate. The constraint in **Line 12** makes sure the new object state is different from the last. `hub1`'s current state is `[unfastened(hub1), jacked_up(hub1,jack0)]`, so this eliminates d. **Lines 13-14** checks that an object state has no unreferenced parameters (if part of the state description references an object not taking part in the transition, then that state would be inappropriate). This does not eliminate any of the choices in the example. **Lines 15-16** check that the union of all the chosen states (in this case incorporating choices for `hub1` and `nuts1`) are consistent. Using these constraints, the states a. and b. are eliminated, leaving c. to be chosen as the value of $C_2(\texttt{hub1})$.

To complete the *Opmaker* process, once the state space map C has been determined, operator instances are constructed by creating *prevail* components for each unchanging object, and creating *necessary transitions* for each object that is changed by an action. These instances are generalised to schema on the basis that each object in a sort behaves the same, and can be replaced by a *sort parameter*. The systems stores the definition of the operator schema and checks them against any previous definition. Finally, a method is generated by combining the induced operator schema, using the original *Opmaker* code.

Non-deterministic choices in the selection of an object's state expression, and the binding of the variables in the object state expression (**Lines 10-11**) mean that sometimes the new state cannot be uniquely determined. However, we have found that this depends on the strength of the invariants that are supplied with the $\mathcal{PDM}$.

3   Experiments and Results

Opmaker2 has been implemented in Sicstus Prolog incorporating the algorithm detailed above. We use the same experimental approach that was used to test the original *Opmaker* system, which was to:

1. Compose training tasks and solution sequences from a range of domains that have already been captured within a hand-crafted model. The set of training tasks should contain at least one instance of each action in the domain, and each task is selected on the basis of whether it is likely to form building blocks for the solution of more complex tasks. The (initial) partial domain model input into *Opmaker2* is the hand crafted domain without its operator schema.
2. Use *Opmaker2* to induce operator schema and methods from the training tasks and solution sequences, and the partial domain model.

3. Use a planning engine to check that the automatically acquired operator schema can solve the same set of problems that the hand-crafted set has been applied to.

4. Use a planning engine to compare performance of the old hand-crafted action schema versus the induced schema and methods. In this case HyHTN [8], a HTN planner which can take advantage of the induced methods, was used. For a comparison with a planner which uses only operator schema (without methods), we use Hoffmann's FF planner [9]. [1]

Success is judged using the following kinds of criteria:

1. Uniqueness: is a set of unique operator schema acquired from the training tasks and the partial domain model that originated from the hand crafted domain model? Or, more subtly, can *Opmaker*2 induce unique schema without having to encode many invariants into the domain models?

2. Validity: Can a set of operator schema output from *Opmaker*2 be used by a planner to solve the same tasks that the original training sequences were aimed at?

3. Efficiency: Is it more efficient, in terms of planning time, to solve tasks using *Opmaker*2 defined operator schema and methods, rather than the original hand-crafted operators?

We detail the results for the extended tyre domain below, and describe other domains on which we have experimented. More details can be found in a recent doctoral thesis [10].

Results in the Extended Tyre Domain. The handcrafted version of the extended tyre domain has 26 objects in 9 sorts, with 22 operators. We engaged a researcher (who was not the author of *Opmaker*2 software) to create 7 sequences of tasks of between 2 and 5 actions in length, encapsulating useful subtasks such as taking a wheel off a hub, or bringing tools out of the car's boot. When input to *Opmaker*2 with the initial partial domain model, procedure *DetermineStates* did not have enough information to discover unique sequences of states for all objects in the training sequences. However, adding extra 'common sense' invariants to the partial domain model (shown in the appendix) was sufficient to allow *DetermineStates* to generate a unique set of state sequences, leading to a set of 22 operator schema generated [10]. An example follows:

```
operator(putaway_jack(Container1,Jack2),
    [(container,Container1,[open(Container1)])],
    [(jack,Jack2,[have_jack(Jack2)]
    =>[jack_in(Jack2,Container1)])],
    []
).
```

On inspection, these were identical in structure to the original hand crafted version. This was confirmed by running the full domain model with a planner and ensuring that all tasks were correctly solved. In addition to operators, the 7 sequences of training tasks

---

[1] We use the GIPO tool to translate the generated OCL domain models into PDDL (the strips version with typing, equality, conditional effects) so that they can be input to generally available planners.

input lead to 7 methods being output. For example, one of the 7 generated methods encapsulating solution heuristics is as follows:

```
method(ex_putaway_tools(Boot,Jack0,Wrench0),
    % Dynamic constraints
    [],
    % Necessary transitions
    [(container,Boot,[open(Boot)]=>[closed(Boot)]),
    (jack,Jack0,[have_jack(Jack0)]=>[jack_in(Jack0,Boot)]),
    (wrench,Wrench0,[have_wrench(Wrench0)]
    =>[wrench_in(Wrench0,Boot)])],
    % Temporal constraints
    [before(1,2),before(2,3)],
    % Static constraints
    [],
    % Decomposition
    [putaway_wrench(Boot,Wrench0),
    putaway_jack(Boot,Jack0),
    close_container(Boot)
    ]).
```

Generating plans up to 10-12 operations in length was possible with standard planning engines, but tasks demanding solutions of greater length were not possible with the planning engines at our disposal. However, when the induced operator schema and the methods were used together with HyHTN, plan times were significantly shorter. For example, a complex planning problem for this extended domain is paraphrased as: "A car has two flat tyres: one is intact and can be fixed by use of the pump, whilst the other is punctured and requires a full tyre change". No solution was found to this problem after 36 hours using FF or HyHTN without the induced methods. However, using the induced domain schema and methods a correct solution of length 24 was found by Hy-HTN after only 11 seconds. It is not surprising that HTN-type domain models are so efficient: this is supported by fielded planning applications. What is significant here is that both the operator schema and the HTN-type methods used in the domain model were generated by *Opmaker2*.

Experiments with other Domains. We experimented with an *OCL* encoding of a Blocks Domain, with 7 blocks stacked on a table. 6 action names were devised and one long training sequence that solved the following task was created: given a set of seven blocks stacked bottom to top block1 to block7, use a gripper to move one block at a time until the blocks are in two stacks. The order of the blocks in these stacks (bottom to top) is block6, block2, block4 form first stack; block5, block1, block7, block3 form the other stack. A 22 solution sequence was composed and fed into *Opmaker* in 6 separate batches, to enable methods and operator schema to be induced. With the original partial domain model enhanced with 4 invariants, 6 operator schema were output by *Opmaker2*. These operators were identical in structure to the hand-coded ones for this domain, and can be used operationally by planning engines. Table 1 shows that the overall task can be tackled in chunks (Tasks 1 - 6), as well as in one sequence (task 7). Each of the 7 tasks resulted in unique and accurate operator schema.

Table 1. Operator Testing in Blocks World Full Problem

Task No.	Actions	Operator Schema
1	4	2
2	2	2
3	2	2
4	4	4
5	2	2
6	2	2
7	22	6

The Hiking Domain was used to illustrate the original *Opmaker*, and models 'lazy' hikers (recreational walkers) who use two cars to carry their equipment around a long (several day) circular route. Automated planning is used to work out the logistics of where to leave their cars, put up their tent, transport their luggage etc. For *Opmaker*2 to produce an accurate, unique set of operator schema, the partial domain model required *one* extra invariant to strengthen it sufficiently. This compares well with the original use of the domain [2] which required a fairly laborious interactive session before outputting operator schema.

4   Related Work

Many machine learning systems are driven by the input of both positive and negative examples. Whilst it was thought to be advantageous to use both kinds of example, many systems like *Opmaker*2, use only positive examples. In particular Vere's [11] Maximal Unifying Generalisation (1987), and Wang's [12] OBSERVER system learn from just positive examples, whilst Grant's [13] POI system learns from positive examples and uses a default rule to provide negative information which boosts the positive training instances. *Opmaker*2 is similar: it uses positive examples in the solution sequence and it makes deductions from the partial domain model.

Learning expressive theories from examples is a central goal in the Inductive Logic Programming community. In his thesis [14], Benson describes an ILP method for learning more expressive operator schema than *Opmaker*2, using multiple examples. However, the focus of *Opmaker*2 is to learn from (ideally) one example sequence, and to learn heuristics as well as operator structure.

Perhaps closest to our work is ARMS [15], a system in which operators are learned without the need for user intervention. Further work by these authors [16] involves learning recursive HTN structures. The authors focus on matching sub-sequences to tasks assuming no knowledge of observed states achieved by low-level actions. The output consists of pairs of action sequences and the high-level tasks achieved by them. As with our system they begin with solution sequences of defined tasks, and compare learned methods to hand-crafted ones to judge success. Whilst ARMS does not require a partial domain model, it requires many training sets (about 40 training sets is quoted). Once learned they were fine tuned by domain experts by hand. By contrast our system does not require multiple examples, as we focus on an expert transferring heuristic knowledge encapsulated in a handful of well chosen examples solution sequences.

5 Conclusions

In this paper we have set up a knowledge acquisition problem which is very relevant to tackling the central problem of using AI planning engines - the acquisition of formulations of actions (in the form of operator schema), and acquisition of heuristics (in the form of HTN-type methods). Our work and the results reported here depend on a structured view of partial domain knowledge about objects being available. Whereas in propositional, classical planning (epitomised by the PDDL language [4]), states are fairly arbitrary sets of propositions, we assume that the space of states is restricted in that objects are pre-conceived to occupy a fixed set of plausible states. Within this framework, we have described a method for inducing operator schema that advances the state of the art in that it requires no intermediate state information, or large numbers of training examples, to induce a valid operator set. Further, our results give some evidence that the methods induced with the operator schema lead to more efficient domain models.

Opmaker2 is an improvement on *Opmaker* in that it eliminates the need for the user or trainer to give the system intermediate state information. After *Opmaker2* automatically infers this intermediate state information, it proceeds in the same fashion as *Opmaker* and induces the same operator schema. Our experimental results show, however, that partial domain models may have to be strengthened with extra invariants before a unique set of operator schema can be synthesised. Hence, we could summarise our work as arguing for the creation of planning domain models by the crafting of a strong partial domain model, and a set of training tasks, *rather* than crafting operator schema and planning heuristics manually.

There are several directions for future work:

1. Can our work be extended to capturing domains with durative or probabilistic actions, or other, more expressive formulations for action? What extra details would be required as input to the operator induction process?
2. Can the *Opmaker2* system be extended to deal with model maintenance (for instance by incremental learning), so that old operator schema can be refined in the presence of new example solution sequences?
3. What resilience does our approach offer in the face of errors in training tasks or in the partial domain model?

Finally, we believe that this line of research is essential if *intelligent agents* are to have general planning capabilities. If this is to be the case, it seems unlikely that intelligent agents will always rely on human experts to encode and maintain their knowledge. It seems reasonable that they would need the capability to acquire knowledge of actions themselves, perhaps by observing the actions of other agents, and using pre-existing static domain knowledge, to induce operator schema and domain heuristics.

References

1. Simpson, R.M., Kitchin, D.E., McCluskey, T.L.: Planning Domain Definition Using GIPO. Journal of Knowledge Engineering 1 (2007)
2. McCluskey, T.L., Richardson, N.E., Simpson, R.M.: An Interactive Method for Inducing Operator Descriptions. In: The Sixth International Conference on Artificial Intelligence Planning Systems (2002)
3. McCluskey, T.L., Porteous, J.M.: Engineering and Compiling Planning Domain Models to Promote Validity and Efficiency. Technical Report RR9606, School of Computing and Maths, University of Huddersfield (1996)
4. AIPS-98 Planning Competition Committee: PDDL - The Planning Domain Definition Language. Technical Report CVC TR-98-003/DCS TR-1165, Yale Center for Computational Vision and Control (1998)
5. Liu, D., McCluskey, T.L.: The OCL Language Manual, Version 1.2. Technical report, Department of Computing and Mathematical Sciences, University of Huddersfield (2000)
6. Chien, S.A. (ed.): 1st NASA Workshop on Planning and Scheduling in Space Applications. NASA, Oxnard, CA (1997)
7. Russell, S.J.: Execution architectures and compilation. In: Proc. IJCAI (1989)
8. McCluskey, T.L., Liu, D., Simpson, R.M.: GIPO II: HTN Planning in a Tool-supported Knowledge Engineering Environment. In: Proceedings of the Thirteenth International Conference on Automated Planning and Scheduling (2003)
9. Hoffmann, J.: A Heuristic for Domain Independent Planning and its Use in an Enforced Hill-climbing Algorithm. In: Proceedings of the 14th Workshop on Planning and Configuration - New Results in Planning, Scheduling and Design (2000)
10. Richardson, N.E.: An Operator Induction Tool Supporting Knowledge Engineering in Planning. PhD thesis, School of Computing and Engineering, University of Huddersfield, UK (2008)
11. Vere, S.: In Pattern Directed Inference Systems. Academic Press, New York (1978)
12. Wang, X.: Learning Planning Operators by Observation and Practice. PhD thesis, Computer Science Department, Carnegie Mellon University, 5000 Forbes Avenue, Pittsberg, PA 15213 (1996)
13. Grant, T.J.: Inductive Learning of Knowledge-Based Planning Operators. PhD thesis, de Rijksuniversiteit Limburg te Maastricht, Netherlands (1996)
14. Benson, S.S.: Learning Action Models for Reactive Autonomous Agents. PhD thesis, Dept. of Computer Science, Stanford University (1996)
15. Wu, K., Yang, Q., Jiang, Y.: Arms: Action-relation modelling system for learning acquisition models. In: Proceedings of the First International Competition on Knowledge Engineering for AI Planning, Monterey, California, USA (2005)
16. Yang, Q., Pan, R., Pan, S.J.: Learning recursive htn-method structures for planning. In: Proceedings of the ICAPS 2007 Workshop on Artificial Intelligence Planning and Learning (2007)

APPENDIX

```
% Sorts
sorts(primitive_sorts,[container,nuts,hub,
 pump,wheel, wrench,jack,wheel_trim,tyre]).

% Objects
objects(container,[boot]).
objects(nuts,[nuts1,nuts2,nuts3,nuts4]).
```

```
objects(hub,[hub1,hub2,hub3,hub4]).
objects(pump,[pump0]).
objects(wheel,[wheel1,wheel2,
               wheel3,wheel4,wheel5]).
objects(wrench,[wrench0]).
objects(jack,[jack0]).
objects(wheel_trim,[trim1,trim2,trim3,trim4]).
objects(tyre,[tyre1,tyre2,tyre3,tyre4,tyre5]).

% Predicates
predicates([ closed(container),open(container),
 tight(nuts,hub),loose(nuts,hub),have_nuts(nuts),
 on_ground(hub),fastened(hub),jacked_up(hub,jack),
 free(hub),unfastened(hub),have_pump(pump),
 pump_in(pump,container),have_wheel(wheel),
 wheel_in(wheel,container),wheel_on(wheel,hub),
 have_wrench(wrench),wrench_in(wrench,container),
 have_jack(jack),jack_in_use(jack,hub),
 jack_in(jack,container),trim_on(wheel_trim,wheel),
 trim_off(wheel_trim),fits_on(tyre,wheel),
 full(tyre),flat(tyre),punctured(tyre)]).

% Object State Expressions
substate_classes([
 container(C,[[closed(C)], [open(C)] ]),
 nuts(N,[[tight(N,H)],[loose(N,H)],[have_nuts(N)]]),
 hub(H, [[on_ground(H),fastened(H)],
    [jacked_up(H,J),fastened(H)],
    [free(H),jacked_up(H,J),unfastened(H)],
    [unfastened(H),jacked_up(H,J)] ]),
 pump(Pu, [[have_pump(Pu)],[pump_in(Pu,C)] ]),
 wheel(Wh, [[have_wheel(Wh)],[wheel_in(Wh,C)],[wheel_on(Wh,H)]]),
 wrench(Wr,[[have_wrench(Wr)],[wrench_in(Wr,C)]]),
 jack(J,[[have_jack(J)],[jack_in_use(J,H)],[jack_in(J,C)] ]),
 wheel_trim(WT,[[trim_on(WT,Wh)],[trim_off(WT)]]),
 tyre(Ty, [[full(Ty)],[flat(Ty)],[punctured(Ty)]]) ]).

% Invariants
atomic_invariants([ fits_on(tyre1,wheel1),
   fits_on(tyre2,wheel2), fits_on(tyre3,wheel3),
   fits_on(tyre4,wheel4), fits_on(tyre5,wheel5)]).
invariant( all(H:hub,fastened(H)<==>
   ex(N:nuts,tight(N,H)\/loose(N,H))) ).
invariant( all(H:hub,all(J:jack,jack_in_use(J,H)
   <==>jacked_up(H,J))) ).
invariant( all(H:hub,~free(H)<==>ex(W:wheel,wheel_on(W,H))) ).
invariant(
 all(T:wheel_trim,all(W:wheel,trim_on_wheel(T,W)
 <==>trim_on(W,T))) ).
%Hub may only have one set of nuts attached
```

```
invariant(all(H:hub,all(N1:nuts,all(N2:nuts,
 (tight(N1,H)\/loose(N1,H)) /\
 (tight(N2,H)\/loose(N2,H))==>(N1=N2) ))) ).
%Hub may only have one wheel attached.
invariant( all(H:hub,all(W1:wheel,all(W2:wheel,
 wheel_on(W1,H)/\wheel_on(W2,H)==>(W1=W2) ))) ).
%If nuts are tight then hub must be on the ground.
invariant( all(H:hub, ex(N:nuts,tight(N,H))==>on_ground(H))).
%If a trim is on a wheel, then the wheel is on
% a hub and the nuts are tight.
invariant(
 all(W:wheel,ex(T:wheel_trim,trim_on_wheel(T,W))
  ==>
 ex(H:hub,wheel_on(W,H)/\ex(N:nuts,tight(N,H))))).
```

Application of Hidden Topic Markov Models on Spoken Dialogue Systems

Hamid R. Chinaei, Brahim Chaib-draa, and Luc Lamontagne

Computer Science and Engineering Department, University of Laval
1065 rue de la Medecine, Quebec (QC), G1V 0A6, Canada
{hrchinaei,chaib}@damas.ift.ulaval.ca,
luc.lamontagne@ift.ulaval.ca
http://www.damas.ift.ulaval.ca

Abstract. A common problem in spoken dialogue systems is finding the intention of the user. This problem deals with obtaining one or several topics for each transcribed, possibly noisy, sentence of the user. In this work, we apply the recent unsupervised learning method, Hidden Topic Markov Models (HTMM), for finding the intention of the user in dialogues. This technique combines two methods of Latent Dirichlet Allocation (LDA) and Hidden Markov Model (HMM) in order to learn topics of documents. We show that HTMM can be also used for obtaining intentions for the noisy transcribed sentences of the user in spoken dialogue systems. We argue that in this way we can learn possible states in a speech domain which can be used in the design stage of its spoken dialogue system. Furthermore, we discuss that the learned model can be augmented and used in a POMDP (Partially Observable Markov Decision Process) dialogue manager of the spoken dialogue system.

Keywords: Learning, User intentions, Spoken dialogue systems.

1 Introduction

Spoken dialogue systems are systems which help users achieve their goals via speech communication. The dialogue manager of a spoken dialogue system should maintain an efficient and natural conversation with the user. The role of a dialogue manager is to interpret the user's dialogue accurately and decides what the best action is to effectively satisfy the user intention. So, the dialogue manager of a spoken dialogue system is an agent that may have a personality [17]. Examples of dialogue agents are a flight agent assisting the caller to book a flight ticket, a wheelchair directed by her patient, etc. [22,4]. However, these agents have some sources of uncertainly due to automatic speech recognition and natural language understanding.

Figure 1 shows the architecture of a Spoken Dialogue System (SDS). The Automatic Speech Recognition (ASR) component receives the user's utterance (which can be a sequence of sentences) in the form of speech signals, and converts it to a sequence of transcribed noisy words. The Natural Language Understanding (NLU) component receives the transcribed noisy words, and generates the possible intentions that the user could mean. The dialogue agent may receive the generated intentions with a confidence

J. Filipe, A. Fred, and B. Sharp (Eds.): ICAART 2009, CCIS 67, pp. 151–163, 2010.

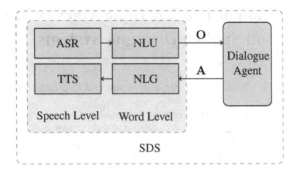

Fig. 1. The architecture of a spoken dialogue system

score as observation O since the output generated by Automatic Speech Recognition and Natural Language Understanding may consist of some uncertainty in the system. Based on observation O, the dialogue agent generates the action A, an input for Natural Language Generator (NLG) and Text-to-Speech (TTS) components.

Learning the intention of the user is crucial for design of a robust dialogue agent. Recent methods of design of dialogue agent rely on Markov Decision Process (MDP) framework. The basic assumption in MDPs is that the current state and action of the system determine the next state of the system (Markovian property). Partially Observable MDPs (or in short POMDPs) have been shown that are proper candidates for modeling dialogue agents [20,4]. POMDPs are used in the domains where in addition to the Markovian property, the environment is only partially observable for the agent; which it is the case in spoken dialogue systems.

Consider the following example taken from SACTI-II data set of dialogues [19]. SACTI stands for Simulated ASR-Channel: Tourist Information. Table 1 shows a sample dialogue in this corpus. The agent's observations are shown in braces. As the example shows, because of the speech recognition errors, each utterance of the user is corrupted. We assume that each user utterance contains one sentence. In POMDP framework, the user utterance can be seen as the agent's observation. And, one problem for the agent would be obtaining the user intention based on the user utterance, i.e. the agent partial observations. Without loss of generality, we can consider the user intention as the agent's state [4]. For instance, states could be: ask information about restaurants, hotels, bars, etc. The system observations could be the same as the states in the simplest case (the keywords restaurant, hotel, bar, etc.), and in more complex cases any word that can represent the states.

Thus, the problem would be estimating the intention of the user given the user utterance as the agent's observations. This can be seen as a typical problem in POMDPs, i.e. learning the observation model. In fact, capturing the intention of user is analogous to learning observation model in POMDPs and that the intention is analogues to the system's state in each turn of dialogue.

[2] used aspect Hidden Markov Models for learning topics in texts. Their experimental result shows that their method is also applicable to noisy transcribed spoken dialogues. However, they assumed that the sequence of utterances is drawn from one topic and there is no notion of mixture of topics. [6] introduced Hidden Topic Markov

Table 1. Sample dialogue from SACTI

U1 Is there a good restaurant we can go to tonight [Is there a good restaurant week an hour tonight] S1 Would you like an expensive restaurant U2 No I think we'd like a medium priced restaurant [No I think late like uh museum price restaurant] S2 Cheapest restaurant is eight pounds per person U3 Can you tell me the name [Can you tell me the name] S3 bochka S4 b o c h k a U4 Thank you can you show me on the map where it is [Thank you can you show me i'm there now where it is] S5 It's here U5 Thank you [Thank u] U6 I would like to go to the museum first [I would like a hour there museum first] ...

Model (HTMM), in order to be able to introduce mixture of topics similar to PLSA (Probabilistic Latent Semantic Analysis) model [7]. PLSA maps documents and words into a semantic space in which they can be compared even if they don't share any common words.

In this work, we observe that HTMM is a proper model for learning intentions behind user utterances at the word level (see Figure 1), which can be used in particular in POMDP framework. The rest of this paper is as follows. Section 2 describes the Hidden Topic Markov Models [6], an unsupervised method for learning topics in documents. We explain the model with a focus on dialogues for the purpose of learning user intentions. This section also describes Expectation Maximization and forward backward algorithm for HTMM. In Section 3, we describe our experiments on SACTI dialogue corpus. In Section 4, we discuss our observations followed by conclusion and future directions on the project, Robotic Assistant for Persons with Disabilities[1] in Section 5.

2   Hidden Topic Markov Models for Dialogues

Hidden Topic Markov Models (HTMM) is a method which combines Hidden Markov Model (HMM) and Latent Dirichlet Allocation (LDA) for obtaining some topics for documents [6]. HMM is a framework for obtaining the hidden states based on some observation in Markovian domains such as part-of-speech tagging [3]. In LDA, similar to PLSA, the observations are explained by groups of latent variables. For instance, if we consider observations as words in a document, then the document is considered as bag of words with mixture of some topics, where topics are represented by the words

---

[1] http://www.damas.ift.ulaval.ca/projet.php

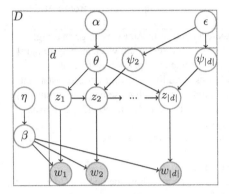

Fig. 2. The HTMM model adapted from [6], the shaded nodes are observations (w) used to capture intentions (z)

with higher probabilities. In LDA as opposed to PLSA, the mixture of topics are generated from a Dirichlet prior mutual to all documents in the corpus. Since HTMM adds the Markovian property inherited in HMM to LDA, in HTMM the dependency between successive words is regarded, and no longer the document is seen as bag of words.

In HTMM model, latent topics are found using Latent Dirichlet Allocation. The topics for a document are generated using a multinomial distribution, defined by a vector θ. The vector θ is generated using the Dirichlet prior α. Words for all documents in the corpus are generated based on multinomial distribution, defined by a vector β. The vector β is generated using the Dirichlet prior η. Figure 2 shows that the dialogue d in a dialogue set D can be seen as a sequence of words (w) which are observations for some hidden topics (z). Since hidden topics are equivalent to user intentions in our work, from now on, we call hidden topics as user intentions. The vector β is a global vector that ties all the dialogues in a dialogue set D, and retains the probability of words given user intentions. The vector θ is a local vector for each dialogue d, and retains the probability of intentions in a dialogue.

Algorithm 1 shows the process of generating and updating the parameters. First, for all possible user intentions β is drawn using the Dirichlet prior η. Then, for each dialogue, θ is drawn using the Dirichlet prior α.

The parameter ψ_i is for adding the Markovian property in dialogues since successive sentences are more likely to include the same user intention. The assumption here is that a sentence represents only one user intention, so all the words in a sentence are representative for the same user intention. To formalize that, the algorithm assigns $\psi_i = 1$ for the first word of a sentence, and $\psi_i = 0$ for the rest. Then, the intention transition is possible just when $\psi = 1$. This is represented in the algorithm between lines 6 and 18.

HTMM uses Expectation Maximization (EM) and forward backward algorithm [13], the standard method for approximating the parameters in HMMs. It is because of the fact that conditioned on θ and β, HTMM is a special case of HMMs. In HTMM, the latent variables are user intentions z_i and ψ_i which determines if the intention for the word w_i is drawn from w_{i-1}, or a new intention will be generated. In the expectation

Algorithm 1. The HTMM generative algorithm adapted from[6].

Input: Set of transcribed dialogues D, N number of intentions
Output: Finding intentions for D
1 **foreach** *intention z in the set of N intentions* **do**
2 | **Draw** $\beta_z \sim Dirichlet(\eta)$;
3 **end**
4 **foreach** *dialogue d in D* **do**
5 | **Draw** $\theta \sim Dirichlet(\alpha)$;

6 | **foreach** $i = 1 \ldots |d|$ **do**
7 | | **if** *beginning of a sentence* **then**
8 | | | $\psi_i = bernoli(\epsilon)$
9 | | **else**
10 | | | $\psi_i = 0$
11 | | **end**
12 | **end**
13 | **foreach** $i = 1 \ldots |d|$ **do**
14 | | **if** $\psi_i = 0$ **then**
15 | | | $z_i = z_{i-1}$
16 | | **else**
17 | | | $z_i = multinomial(\theta)$
18 | | **end**

19 | | **Draw** $w_i \sim multinomial(\beta_{z_i})$;
20 | **end**
21 **end**

step, for each user intention z, we need to find the expected count of intention transitions to intention z.

$$E(C_{d,z}) = \sum_{j=1}^{|d|} Pr(z_{d,j} = z, \psi_{d,j} = 1 | w_1, \ldots, w_{|d|})$$

where d is a dialogue in the corpus of dialogue D.
Moreover, we need to find expected number of co-occurrence of a word w with an intention z.

$$E(C^{z,w}) = \sum_{i=1}^{|D|} \sum_{j=1}^{|d_i|} Pr(z_{i,j} = z, w_{i,j} = w | w_1, \ldots, w_{|d|})$$

In the Maximization step, the MAP (Maximum A Posteriori) for θ and β is computed using Lagrange multipliers:

$$\theta_{d,z} \propto E(C_{d,z}) + \alpha - 1$$

$$\beta_{z,w} \propto E(C^{z,w}) + \eta - 1$$

Intention 0		Intention 1		Inteion 2		Inteion 3		Intention 4	
is	0.0599	is	0.0703	a	0.0620	the	0.0531	the	0.0653
the	0.0523	you	0.0403	i	0.0528	you	0.0446	you	0.0488
are	0.0498	where	0.0318	i'm	0.0330	can	0.0344	me	0.0443
where	0.0361	a	0.0315	for	0.0213	me	0.0311	a	0.0441
on	0.0275	there	0.0289	uh	0.0213	please	0.0268	is	0.0389
what	0.0189	e	0.0282	looking	0.0197	of	0.0235	of	0.0267
ah	0.0177	restaurant	0.0270	hotel	0.0196	is	0.0214	restaurant	0.0241
at	0.0175	me	0.0267	the	0.0177	hotel	0.0202	can	0.0238
tours	0.0167	uh	0.0264	to	0.0162	a	0.0192	could	0.0211
i	0.0166	can	0.0262	want	0.0147	and	0.0183	where	0.0174

Intention 5		Intention 6		Intention 7		Intention 8		Intention 9	
the	0.0612	i	0.0752	how	0.0626	the	0.0351	i	0.0534
are	0.0373	a	0.0463	the	0.0495	a	0.0271	thank	0.0407
to	0.0259	the	0.0310	i	0.0379	ok	0.0265	u	0.0370
i'm	0.0235	to	0.0269	to	0.0360	you	0.0239	no	0.0319
in	0.0219	are	0.0261	it	0.0320	much	0.0217	you	0.0254
and	0.0210	no	0.0220	from	0.0304	i	0.0217	to	0.0221
um	0.0202	um	0.0202	a	0.0300	me	0.0195	think	0.0198
museum	0.0191	is	0.0170	much	0.0262	fine	0.0192	like	0.0181
at	0.0183	bar	0.0162	does	0.0240	to	0.0189	a	0.0170
a	0.0177	in	0.0161	long	0.0209	is	0.0172	er	0.0165

Fig. 3. Captured Intentions by HTMM

The random variable $\beta_{z,w}$ gives the probability of an observation w given the intention z.

The parameter ϵ denotes the dependency of the sentences on each other, i.e. how likely it is that two successive uttered sentence of the user have the same intention.

$$\epsilon = \frac{\sum_{i=1}^{|D|} \sum_{j=1}^{|d|} Pr(\psi_{i,j} = 1 | w_1, \ldots, w_{|d|})}{\sum_{i=1}^{|D|} N_{i,sen}}$$

where $N_{i,sen}$ is the number of sentences in the dialogue i.

In this method, EM is used for finding MAP estimate in hieratical generative model similar to LDA. [5] argued that Gibbs sampling is preferable than EM since EM can be trapped in local minima. [9] also argued that EM suffer from local minima. However, they suggested methods for getting away from local minima. Furthermore, they also proposed that EM can be accelerated based on the type of the problem. In HTMM, the special form of the transition matrix reduce the time complexity of the algorithm to $O(|d|N^2)$, where $|d|$ is the length of the dialogue d, and N is the number of desired user intentions, given to the algorithm. The small time complexity of the algorithm enables the agent to apply it at any time to update the observation functions based on her recent observation.

3 Experiments

We evaluated the performance of HTMM on SACTI data set [19]. There are about 180 dialogues between 25 users and a wizard on this corpus. The user's sentences are first

Table 2. Sample dialogue from SACTI

[Is there a good restaurant week an hour tonight] [No I think late like uh museum price restaurant] [Can you tell me the name] [Thank you can you show me i'm there now where it is] [Thank u] [I would like a hour there museum first] ...

confused using a speech recognition error simulator [21,20], and then are sent to the wizard. However, the wizard's response to user is demonstrated on a screen in order to avoid speech confusion from wizard to the user. The dialogue is finished when the task is completed, or when the dialogue will last more than a limited time. This time is often more than 10 minutes. We assume that the intention transition is only possible from a sentence to the following one in a given utterance, which is more realistic than intention transition from a word to the following one within a sentence. We did our experiments on 95% dialogues with a vocabulary of 829 words, including some misspelled ones. On average, each dialogue contains of 13 sentences.

In our experiments, we removed the agent's response from the dialogues in order to test the algorithm only based on the noisy user utterances. Moreover, since HTTM is an unsupervised learning method, we did not have to annotate the dialogues, or any sort of preprocessing. Table 2 shows the sample dialogue in Section 1, after removing the agent's responds. As the table shows, this input data is quite corrupted. The results of our experiments show that the model is able to capture possible user intentions in the data set. Figure 3 shows 10 captured user intentions and their top 10 words. For each intention, we have highlighted the keywords which best distinguish the intention (the words which does not occur in many intentions). As Figure shows, intention 0, 1, 2, 5, and 6 represents the user asking information about tours, restaurants, hotels, museums, and bars, respectively. Intention 7, represents the user asking distance between two locations. Intentions 8 represents acknowledgement. Moreover, Intentions 3, 4, and 9 can represent hotels, restaurants, and acknowledgement, respectively. These three intentions have been previously recognized by the model; however, since the top words in each intention is slightly different, the agent assigns it in two different categories.

Table 3 shows highest obtained intentions for each sentence of the dialogue example in Section 1. As the table shows, the highest intention for U1 is ask information for restaurant, and with very small probability ask information for hotel. Interestingly, we can see that the obtained intention for U2 is I4, intention for restaurants, though the utterance consists of the word "museum" a strong observation for I4. This fact shows that the method is able to capture the Markovian property in U1 and U2. Another interesting observation is in U3, where the agent could estimate the user intention restaurants with 99% probability without receiving the word restaurant as observation. Yet another nice observation can be seen in the captured intentions for U4 and U5. The sentences in U4 and U5 contain "thank you" as observations. However, the captured intentions for U4 are I1 and I4, both of which represent restaurants. On the other hand, in utterance U5, the agent indeed is able to obtain intention I9, acknowledgement.

Table 3. Sample results of experiments on SACTI

U1 Is there a good restaurant we can go to tonight
[Is there a good restaurant week an hour tonight]
I4:0.9815 I2:0.0103 I1:0.0080
S1 Would you like an expensive restaurant
U2 No I think we'd like a medium priced restaurant
[No I think late like uh museum price restaurant]
I4:0.8930 I9:0.1005 I6:0.0041 I2:0.0015 I5:0.0005 I1:0.0001
S2 Cheapest restaurant is eight pounds per person
U3 Can you tell me the name
[Can you tell me the name]
I4:0.9956 I3:0.0034 I8:0.0008
S3 bochka
S4 b o c h k a
U4 Thank you can you show me on the map where it is
[Thank you can you show me i'm there now where it is]
I1:0.9970 I4:0.0029
S5 It's here
U5 Thank you
[Thank u]
I9:0.9854 I1:0.0114 I8:0.0013 I4:0.0007 I5:0.0006 I6:0.0003
U6 I would like to go to the museum first
[I would like a hour there museum first]
I9:0.9238 I6:0.0711 I5:0.0042 I2:0.0003 I7:0.0002
. . .

Moreover, we measured the performance of the model on the SACTI data set based on the definition of perplexity. For a learned language model on a train data set, perplexity can be considered as a measure of on average how many different equally most probable words can follow any given word, so the lower the perplexity the better the model. The perplexity of a test dialogue d after observing the first k words can be drawn using the following equation:

$$\text{Perplexity} = \exp(-\frac{\log\ Pr(w_{k+1}, \ldots, w_{|d|} | w_1, \ldots, w_k)}{|d| - k})$$

To calculate the perplexity, we have:

$$Pr(w_{k+1}, \ldots, w_{|d|} | w_1, \ldots, w_k) =$$
$$\sum_i^N Pr(w_{k+1}, \ldots, w_{|d|} | z_i) Pr(z_i | w_1, \ldots, w_k)$$

where z_i is a user intention in the set of N captured user intentions from the train set. Given a user intention z_i, probability of observing $w_{k+1}, \ldots, w_{|d|}$ are independent of each other, so we have:

$$Pr(w_{k+1}, \ldots, w_{|d|} | w_1, \ldots, w_k) =$$
$$\sum_i^N \prod_{j=k+1}^{|d|} Pr(w_j | z_i) Pr(z_i | w_1, \ldots, w_k)$$

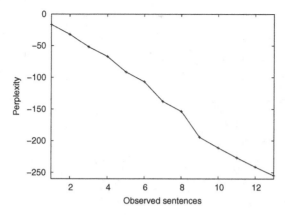

Fig. 4. Log likelihood for learning topics

To find out the perplexity, we learned the intentions for each test dialogue d based on the first k observed words in d, i.e. $\theta_{new} = Pr(z_i|w_1, \ldots, w_k)$ is calculated for each test dialogue. However, $Pr(w_j|z_i)$ is drawn using β, learned from the train dialogues.

We calculated the perplexity for 5% of the dialogues in data set, using the 95% rest for training. Figure 4 shows the average perplexity after observing the first k sentences of test dialogues (remember that each sentence of the dialogue consists of only one user intention). As the figure shows, the perplexity reduce significantly by observing new sentences.

4 Discussion

With the rise of spoken dialogue systems, the recent literature devoted on more robust methods of dialogue strategy design [16]. [10] evaluated the Markov assumption for spoken dialogue management. They argued that when there is not a proper estimate of reward in each state of dialogue, relaxing the Markovian assumption and estimating the total reward, using some features of the domain, could be more advantageous. Nevertheless, many researchers have found MDP and POMDP frameworks suitable for formulating a robust dialogue agent in spoken dialogue systems. In particular, [8] learned dialogue strategies within the Markov Decision Process framework.

[11] used MDPs to model a dialogue agent. They interpreted the observation mostly in the speech level and based on the definition of perplexity. [22] used POMDPs for modeling a dialogue agent and defined the observation function based on some features of the recognition system. However, these features are usually difficult to be determined and task dependent. We are particularly interested in the POMDP dialogue agent in [4]. The authors learned the observation function in a POMDP using Dirichlet distribution for the uncertainty in observation parameters. However, for each state they consider only one keyword as observation.

In this work, we learned the observation model based on the received noisy data in the word level, and abstract away the speech recognition features. The used method consider all the words in a sentence as observations which represent one state. This

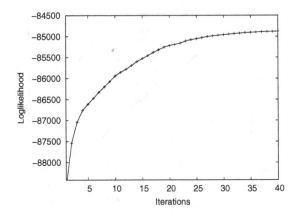

Fig. 5. Log likelihood for learning topics

is crucial for the frameworks such as POMDPs where the agent use an observation function to reason about the state of the system, and that the state of the system is the user intention. Based on our experiments on relatively small data set SACTI, we believe that this method can be used in the early design stage of practical dialogue agents, say in [4], in order to define the possible states of the domain (possible user intentions), as well as observation function. Moreover, the result shows that HTMM is able to capture a robust observation model which can be used in POMDPs with large number of observations such as [1]. Moreover, since HTMM use EM algorithm, this method is quite fast, and can be used by the agent at any time to learn new observations and update the observation function. Figure 5 shows the log likelihood of data for 50 iterations of the algorithm. For the given observations, the likelihood is computed by averaging over possible states:

$$\text{loglikelihood} = \sum_{i=1}^{|D|} \sum_{j=1}^{|d_i|} \log \sum_{t=1}^{N} Pr(w_{i,j} = w | z_{i,j} = z_t)$$

As the figure shows the algorithm converges after about 30 iterations which is an evidence for small time expense of the algorithm. This fact suggests use of the algorithm after finishing some tasks by agent to learn new states, observations, and hopefully a better policy.

The interesting property about HTMM includes in its combining LDA and HMM. On the one hand, LDA captures mixture of intentions for dialogues, and on the other hand, HMM adds the Markovian property. This makes the framework similar to POMDPs in terms of making a belief over possible states, besides the Markovian property. As Table 3 shows, the possible captured intentions for each sentence of the user can be seen as the agent's belief over possible states. Moreover, using this method, we learned the value of $\epsilon = 0.71$ on SACTI data set; which suggests it is likely that the user changes his intention in a dialogue in SACTI data data set; whereas for instance [4] assumes that the user may change his intention with a predefined low probability in wheelchair domain.

Fig. 6. Smart Wheeler Platform

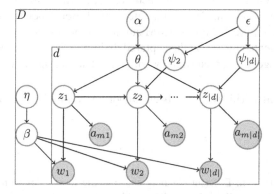

Fig. 7. The extended HTMM. The intention of user depends on both her words w_is and system actions a_{mi}s.

HTMM, however, assumes that the Dirichlet prior are known. During our experiments on SACTI, we observed that by feeding the Algorithm 1 with different α and η, the algorithm can derive slightly different intentions (the results presented here are for $\alpha = 1.1$ and $\eta = 1.1$). However, some of these intentions makes sense, for instance intentions for cost, dialogue initiation, etc. Moreover, the number of intentions (N in Algorithm 1) needs to be set. For instance, in our experiments we set $N = 10$ to be able to derive the maximum number of intentions, yet some intentions seem to be similar.

5   Conclusions and Future Works

We observed that HTMM can be used for capturing possible user intentions in dialogues. The captured intentions together with the learned observation function could be used in the design stage of a POMDP based dialogue agent. Moreover, the dialogue agent can use HTMM on the captured dialogues over time to update the states and observation function. Although, there is no notion of actions in HTMM, and it is a method which is used mostly on the static data, the similarity of HTMM and POMDP in terms of Markovian property and generating a belief over possible states suggest considering both these two models in practical applications, where the time complexity of POMDPs burden the problem. Moreover, our observation on SACTI data set suggests future works for use of HTMM for automatically annotating the corpus of dialogues, capturing the structure of dialogues, and dialogue agent evaluation [15,17,14,18].

In the future work we are going to use HTMM to learn the model for a POMDP dialogue agent. For instance, we would like to use HTMM for a a wheelchair robot similar to Figure 6, taken from [12]. This wheelchair is designed for patients with limited skills, say patients suffering from Multiple Sclerosis. The patients can direct the robot, with mentioning the goal, the path, and restrictions such as speed, instead of using a joystick. We are going to augment HTMM by considering actions of the system in the model. Since the actions performed by the agent carries much less noise comparing to the user utterance (agent's observations), agent's action can have more effect on the Markovian property of the environment. That is, the intention of user depends on both her words and agent actions. Figure 7 shows HTMM augmented with system actions. We are going to apply augmented model on the captured dialogues for a dialogue POMDP agent and compare the agent's learned strategy with that of similar models.

Acknowledgements. This work has been supported by a FQRNT grant.

References

1. Atrash, A., Pineau, J.: Efficient planning and tracking in pomdps with large observation spaces. In: AAAI 2006 Workshop on Empirical and Statistical Approaches for Spoken Dialogue Systems (2006)
2. Blei, D.M., Moreno, P.J.: Topic segmentation with an aspect hidden Markov model. In: Proceedings of the 24th annual international ACM SIGIR conference on Research and development in information retrieval (SIGIR 2001), pp. 343–348 (2001)
3. Church, K.W.: A stochastic parts program and noun phrase parser for unrestricted text. In: Proceedings of the second conference on Applied Natural Language Processing (ANLP 1988), Morristown, NJ, USA, pp. 136–143 (1988)
4. Doshi, F., Roy, N.: Efficient model learning for dialog management. In: Proceedings of the ACM/IEEE international conference on Human-Robot Interaction (HRI 2007), pp. 65–72 (2007)
5. Griffiths, T., Steyvers, J.: Finding scientific topics. Proceedings of the National Academy of Science 101, 5228–5235 (2004)
6. Gruber, A., Rosen-Zvi, M., Weiss, Y.: Hidden Topic Markov Models. In: Artificial Intelligence and Statistics (AISTATS 2007), San Juan, Puerto Rico (2007)
7. Hofmann, T.: Probabilistic latent semantic analysis. In: Proceedings of the fifteenth conference on Uncertainty in Artificial Intelligence (UAI 1999), pp. 289–296 (1999)
8. Levin, E., Pieraccini, R., Eckert, W.: Learning dialogue strategies within the Markov decision process framework. In: 1997 IEEE Workshop on Automatic Speech Recognition and Understanding, pp. 72–79 (1997)
9. Ortiz, L.E., Kaelbling, L.P.: Accelerating EM: An empirical study. In: Proceedings of the fifteenth conference on Uncertainty in Artificial Intelligence (UAI 1999), Stockholm, Sweden, pp. 512–521 (1999)

10. Paek, Tim, Chickering, David: Evaluating the Markov assumption in Markov Decision Processes for spoken dialogue management. Language Resources and Evaluation 40(1), 47–66 (2006)
11. Pietquin, O., Dutoit, T.: A probabilistic framework for dialog simulation and optimal strategy learning. IEEE Transactions on Audio, Speech, and Language Processing 14(2), 589–599 (2006)
12. Pineau, J., Atrash, A.: Smartwheeler: A robotic wheelchair test-bed for investigating new models of human-robot interaction. In: AAAI Spring Symposium on Multidisciplinary Collaboration for Socially Assistive Robotics (2007)
13. Rabiner, L.R.: A tutorial on hidden Markov models and selected applications in speech recognition. pp. 267–296 (1990)
14. Singh, S.P., Kearns, M.J., Litman, D.J., Walker, M.A.: Empirical evaluation of a reinforcement learning spoken dialogue system. In: Proceedings of the Seventeenth National Conference on Artificial Intelligence and Twelfth Conference on Innovative Applications of Artificial Intelligence, pp. 645–651. AAAI Press / The MIT Press (2000)
15. Walker, M., Passonneau, R.: DATE: a dialogue act tagging scheme for evaluation of spoken dialogue systems. In: Proceedings of the first international conference on Human Language Rechnology research (HLT 2001), Morristown, NJ, USA, pp. 1–8. Association for Computational Linguistics (2001)
16. Walker, M.A.: An application of reinforcement learning to dialogue strategy selection in a spoken dialogue system for email. Journal of Artificial Intelligence Research (JAIR) 12, 387–416 (2000)
17. Walker, M.A., Litman, D.J., Kamm, A.A., Abella, A.: PARADISE: A Framework for Evaluating Spoken Dialogue Agents. In: Proceedings of the Thirty-Fifth Annual Meeting of the Association for Computational Linguistics and Eighth Conference of the European Chapter of the Association for Computational Linguistics, Somerset, New Jersey, pp. 271–280. Association for Computational Linguistics (1997)
18. Walker, M.A., Passonneau, R.J., Boland, J.E.: Quantitative and qualitative evaluation of darpa communicator spoken dialogue systems. In: Meeting of the Association for Computational Linguistics, pp. 515–522 (2001)
19. Weilhammer, K., Williams, J.D., Young, S.: The SACTI-2 Corpus: Guide for Research Users, Cambridge University. Technical report (2004)
20. Williams, J.D., Poupart, P., Young, S.: Factored partially observable markov decision processes for dialogue management. In: The 4th IJCAI Workshop on Knowledge and Reasoning in Practical Dialogue Systems, Edinburgh, Scotland (2005)
21. Williams, J.D., Young, S.: Characterizing task-oriented dialog using a simulated asr channel. In: Proceedings of International Conference on Spoken Language Processing (ICSLP 2004), Jeju, South Korea (2004)
22. Williams, J.D., Young, S.: Partially observable markov decision processes for spoken dialog systems. Computer Speech and Language 21, 393–422 (2007)

Gossip Galore: An Embodied Conversational Agent for Collecting and Sharing Pop Trivia from the Web

Feiyu Xu, Peter Adolphs, Hans Uszkoreit, Xiwen Cheng, and Hong Li

DFKI GmbH, Language Technology Lab
Stuhlsatzenhausweg 3, D-66123 Saarbrücken, Germany
(feiyu,peter.adolphs,uszkoreit,xiwen.cheng,lihong)@dfki.de

Abstract. This paper presents a novel approach to a self-learning agent who collects and learns new knowledge from the web and exchanges her knowledge via dialogues with the users. The application domain is gossip about celebrities in the music world. The agent can inform herself and update the acquired knowledge by observing the web. Fans of musicians can ask for gossip information about stars, bands or people and groups related to them. This agent is built on top of information extraction, web mining, question answering and dialogue system technologies. The minimally supervised machine learning method for relation extraction gives the agent the capability to learn and update knowledge constantly from the web. The extracted relations are structured and linked with each other. Data mining is applied to the learned data to induce the social network among the artists and related people. The knowledge-intensive question answering technology enhanced by domain-specific inference and active memory allows the agent to have vivid and interactive conversations with users by utilizing natural language processing. Users can freely formulate their questions within the gossip data domain and access the answers in different ways: textual response, graph-based visualization of the related concepts and speech output.

1 Introduction

The development of information extraction and question answering in recent years opens new perspectives for simple but effective interactive dialogue systems [1,2,3]. Information extraction enables dialogue systems to access and understand natural language texts stored in semi- or unstructured formats, thus, allowing them to make use of the contents provided by the web, the world's largest information repository. Question answering technology gives a conversational agent the capability of understanding natural language questions and retrieving answers from a large knowledge or content pool. At the same time, question answering systems enhanced by some dialogue competence enable natural communication with the human users. The combination of information extraction, question answering and dialogue is a new approach to a conversational agent who is able to understand natural language questions and provide answers by extracting and mining information from a large amount of textual data in structured, semi- or unstructured form.

One of the hardest challenges in our information world is to constantly keep the information up to date and to prepare it in such a way that users can easily understand

J. Filipe, A. Fred, and B. Sharp (Eds.): ICAART 2009, CCIS 67, pp. 164–176, 2010.

and exploit it. We have developed a new architecture for conversational agent systems that can learn, update and interpret information from the web and make conversations with end users, provide answers to their questions and even help them to gain insights into the application domain. We selected gossip about celebrities in the music world as the domain for our experimental setup, because many of them exhibit interesting and dynamic aspects with respect to both their private and professional life. Furthermore, they are connected to each other in a variety of ways. Internet news and blogs report on them from different perspectives. Our task is to model this domain by covering relevant facts and trivia on the musicians and their communities and by discovering new properties and relations. The acquired information will be utilized as a knowledge resource for conversations with end users. Users can raise natural language questions about a special artist or ask for relationships between artists. Our system provides answers from its knowledge base or even hints at newly discovered information.

In comparison to existing systems, our conversational agent, called "Gossip Galore", is an active self-learning system. It starts with only a very small number of artists and bands and then gradually finds many more artists and bands. This is realized by the application of a minimally supervised relation extraction system (see section 3). Users can actively give comments on the answers provided by the agent, which is useful for self validation. Thus, "Gossip Galore" contains two major parts: one is the knowledge acquisition component and the other one is the component for communication and conversation. Both parts interact with each other and contribute to the self-learning process.

The paper is structured as follows: Section 2 describes the project and the general context in which "Gossip Galore" is embedded. Section 3 explains the web mining techniques for the knowledge acquisition, while section 4 presents the dialogue modelling and question answering component. Section 5 gives an overview of the related work. The conclusions and future steps are described in section 6.

2   RASCALLI

The research presented here is conducted within the project *Responsive Artificial Situated Cognitive Agents Living and Learning on the Internet* (RASCALLI). RASCALLI is supported by the Sixth Framework Programme of the European Commission in the area of Cognitive Systems (IST-27596-2004). Its goal is to develop and implement cognitively enhanced artificial companions by combining natural language processing, question answering, web-based information extraction, semantic web technology and interaction-driven profiling with cognitive modelling [4]. This work is further supported by the project KomParse, which is devoted to equipping non-player characters in computer games with dialogue capacities.

In the realized system, the RASCALLI agents assist users in extracting information from the web and other resources. Users can own their own RASCALLI agents, which are 3D modelled virtually embodied conversational agents. The perception and action components of the RASCALLI agents are modelled by a combination of information extraction, question answering and dialogue capabilities. Within the project some major strands of research are devoted to the investigation and modelling of architectures that combine all major components of cognitive systems. This is an ambitious and demanding task, and as a step on the way, the results reported here are a pragmatic compromise that combines state-of-the-art and novel methods from information extraction, question answering, semantic technologies and visual animation with insights from cognitive modelling into a robust fun application.

3   Web Mining for Knowledge Acquisition

One of the major competences of the RASCALLI agents is that they can learn and acquire knowledge constantly from the web according to user interests. The minimally supervised machine learning methods for relation extraction provided by the system DARE can be easily utilized for realizing this competence [5,6]. DARE can be initialized with several examples of relations about artists or bands as seed provided by the users and then learn rules which map the linguistic structures to these semantic relations. The rules can be applied to texts to discover new relation instances, which can be reused as seed again for new rule discovery.

The experimental domain selected for RASCALLI is gossip about celebrities in the pop world. We start with domain modelling to define the potentially relevant concepts and relations that will serve as a framework for the musician profiles and the associated gossip information to be acquired. Given the relevant concepts and their relations, we apply DARE to acquire instances of the relations from the web.

3.1   Domain Modelling

The aim of the domain modelling is to identify and structure the relevant concepts and relations within the gossip domain. The current domain contains properties of a musician such as personal profiles, social contexts, achievements, gossip topics and career relevant issues. The gossip content is modeled as an ontology, utilizing the formal language OWL [7]. The concepts and properties centering on musicians are depicted in Fig. 1.

3.2   Knowledge Acquisition

Many resources on the web report on celebrities, e.g., online news sites, Wikipedia, music portals, fan blogs and forums. The information mentioned above is stored in different formats: unstructured (free text), semi-structured (e.g., Wikipedia) or almost structured (e.g. NNDB). Therefore, we propose a hybrid information discovery strategy to detect as much information as possible, as shown in Fig. 2.

We apply information wrapping, information extraction and information merging techniques to acquire new knowledge. The whole discovery is embedded in a bootstrapping framework, namely, starting with some examples and then learning more and more information after several iterations.

Relation Extraction with DARE. DARE provides a general framework for the extraction of relations and events with various complexities [5,6]. This method is minimally supervised since the system works with a collection of free natural language texts without any annotation of domain information. The only domain knowledge for the whole

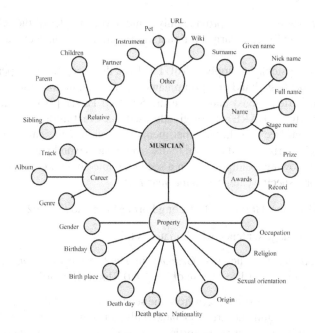

Fig. 1. Domain ontology (simplified)

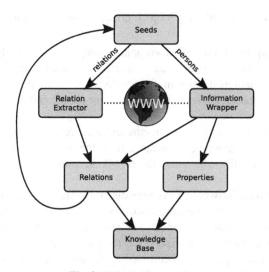

Fig. 2. Webmining workflow

process is the seed. DARE can use linguistic knowledge as it is provided, for example, by named entity recognizers and linguistic parsers. The complexity of the seed determines the complexity of the extracted relations. The seed helps us to identify the explicit linguistic expressions containing mentions of n-ary relation instances or instances of their k-ary projections where $1 \le k < n$. Therefore, DARE can be easily adapted to user interests. Users provide only some new examples of the relations they are interested in; DARE can learn additional information from the web based on these examples.

In the current system, we apply SProUT [8] for the recognition of person names and other concepts (e.g., band and group names, date time, nationalities, instruments, religions, sexual orientations) and utilize the Stanford Parser [9] to detect linguistic dependency structures. DARE was originally used to extract information about Nobel Prize winners from free text. Later experiments showed how to adapt learned DARE rules for the Nobel Prize award domain to discover awards won by musicians [10]. Let us look at the following example. Given a seed example about a *Grammy* award won by *Madonna* for a specific category in the year *1992*:

Example 1. ⟨Madonna, Grammy, Best Long Form Music Video, 1992⟩

The natural language sentence which matches this seed is:

Example 2. Madonna won her first *Grammy* in *1992* in the *Best Long Form Music Video category* for the laserdisc release of her 1990 Blond Ambition Tour.

DARE can extract a linguistic pattern from the seed example and the matched sentence where the linguistic arguments are associated with their semantic roles in the semantic relation, after applying linguistic analysis to the sentence. The simplified DARE rule looks as follows:

Example 3. ⟨subject: recipient⟩ <u>win</u> ⟨object: prize⟩ ⟨mod: year⟩ ⟨mod: category⟩

Information Wrapping. Information wrapping is responsible for collecting structured data from structured or semi-structured web sites. It discovers the HTML structures which indicate the relations defined in our ontology. We apply this technology to web sites such as Wikipedia and the special web portal for people and their profiles, namely, the NNDB.

The method starts with a set of musicians and their relation instances as seed. Our system sends a query containing an instance from the seed set as a query to the web sites and discovers the rules which map the HTML structures to the relation structures.

Induction of the Social Network. Given the discovered relations among the musicians themselves and other people, we developed a special system which can construct a social network from the relation instances. For example, Fig. 3 shows the social network of Madonna. The social network also serves as the basis for the active dialogue memory of the agent. Whenever a person is mentioned by the user, this person gets activated in the agent's memory, making related people also accessible.

4 Conversational Agents

In RASCALLI, the central method for users to access the acquired knowledge is to communicate with the user's personal embodied conversational agent (ECA). The core functionality of the agent is question answering, wrapped in a smooth natural language

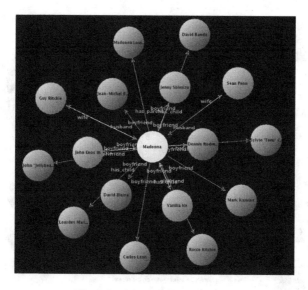

Fig. 3. Social network of Madonna

dialogue. One main design criterion is to create and enhance an immersive effect on the user when interacting with the system. The agent should be physically embodied, she should be situated in a consistent physical environment, and she should act naturally.

The interaction between users and RASCALLI can be described as follows. After logging in to the platform, a three-dimensional visualization of the user's agent is displayed (see Fig. 4). Just as in an instant messaging program, the user communicates with the agent by typing messages into a text field. The agent, on the other hand, responds with natural language utterances which are presented in their spoken form (by the use of the open source speech synthesis system *OpenMary* [11]), along with their written form. But the agent's means of communication are not restricted to verbal actions. Complex answers such as the social network of a star can be visualized on a TV screen, which is embedded into the agent's environment. Where it is appropriate, the agent also emphasizes her responses by facial and body gestures such as shrugging the shoulders, nodding or shaking her head and pointing to the screen.

4.1   Architecture

The RASCALLI system is realized as a server-client architecture. Users are connected to the server via the 3D client, which displays our ECA and manages the interaction with the user. The actual control logic of the agent is executed on the server. The server's function is to accept new connections, to manage users and their logins, and to route messages between the 3D client and the conversational agent.

Figure 5 shows the component hierarchy of our conversational agents. Some of the components are responsible for processing various linguistic aspects of the dialogue, whereas others are concerned with knowledge representation, management and retrieval as well as behavioural procedures. Details on the interplay between these components when processing dialogue turns are presented in the following subsection.

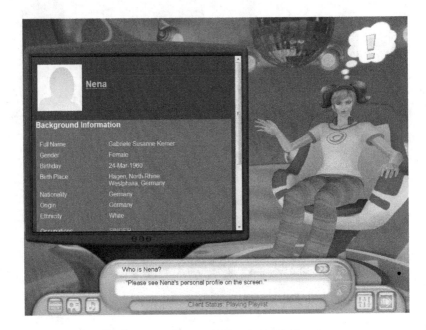

Fig. 4. The Gossip Galore conversational agent

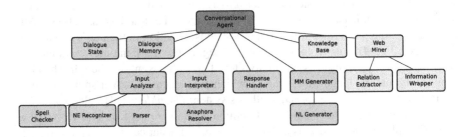

Fig. 5. Components of a conversational agent

4.2 Dialogue Processing

When the conversational agent receives an utterance from the user, its task is to compute a suitable dialogue turn in response. We follow a pipeline architecture for realizing this: the user's input string is first linguistically analyzed, then it is interpreted in the current dialogue context and turned into a suitable plan for a response action that is executed in the third stage, leading to an abstract representation for the answer, which is realized with verbal and non-verbal means in the fourth and final stage. This basic data flow when processing dialogue turns as a response is depicted in Fig. 6. In the following, the four main components are presented in greater detail.

The idea of having two separate components for input processing, namely, *input analyzer* and *input interpreter*, one for the analysis and one for the interpretation of the user's input, serves the purpose of drawing a clear boundary between the general

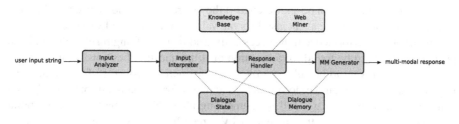

Fig. 6. Processing dialogue turns

and reusable and the domain-specific parts of the system. The input analysis component relies on standard domain-independent linguistic tools, namely a spell checker, a named entity recognizer, and a parser producing a linguistic analysis of the input, for which we currently employ a fuzzy paraphrase matcher to approximate the output of a deeper syntactic/semantic parser.

Each utterance is associated with a meaning representation as well as with the answer focus and the expected answer type in case of questions. Note that by mapping utterances of quite different sentence types such as plain questions ("Who is Madonna?"), statements with embedded questions ("I wonder who Madonna is."), statements about the user's interests without embedded questions ("I'm interested in Madonna.") to the same semantic representation, we can conflate sets of user utterances with the same intended meaning.

In the second stage, the input is interpreted in the current dialogue context, considering previously mentioned entities for resolving anaphora as well as the current dialogue state for modelling the system's expectations about the user's turn. If, for instance, a substring can only be resolved as a named entity with the help of the spell checker, the system poses a clarification question to the user and sets the dialogue state accordingly. This allows the system to interpret a following utterance by the user such as "yes" or "no", which would otherwise not be understood. The result of the input interpretation stage is an abstract plan to perform a certain action. For example, factoid in-domain questions result in a plan to look up the data in our knowledge base, general information requests about an in-domain person result in the plan to show a profile page of that person, out-of-domain questions about a known person result in the plan to present a suitable web link, and so on.

In the third stage, the Response Handler component executes the planned action. For factoid questions, this means that the corresponding query is looked up and submitted to the knowledge base. If the user asks for general information about a person, the URL of the corresponding profile page is constructed. The user may also have asked whether there are any new information about a musician he is interested in. In this case, an online search for new connections between people is performed.

The agent follows certain pragmatic principles of relevance when giving answers to questions. By assigning the same semantics to indirect speech acts ("I wonder who the boyfriends of Madonna are.") as to the corresponding direct speech act ("Who are the boyfriends of Madonna?"), we are able to return a relevant response to the user's request. Similarly to the previous example, certain yes/no-questions ("Does Madonna have any boyfriends?") can be answered as if they were wh-questions. Instead of giving a simple "yes" answer, the agent also lists the values for the queried variable.

Not all of the performed actions necessarily lead to a satisfying result, though. If no positive answer can be found for a question or if the question lies outside the covered domain, we still want to be able to provide a constructive answer. If, for instance, the user asks about a person we do not have detailed information about but for whom a Wikipedia entry exists (e.g. "Tell me something about Nicolas Sarkozy!"), we point the user to this page, using the embedded TV screen for displaying the page. If, on the other hand, the user asks an in-domain question for which the system does not have any results, we direct the user to a Google search page with appropriate query parameters in order to help him find relevant information.

The outcome of the performed action is always an abstract representation of the agent's response. This might be a simple boolean value for yes/no-questions, a list of entities for factoid questions, a URL for the system's own information services or to external sources, and so on. This information is finally realized as a communicative act in the fourth stage, the multi-modal generation. We currently employ template-based generators for both producing the natural language utterance as well as for the multi-modal message with gestures and TV screen commands. When generating natural language answers to questions, care has to be taken how these answers are provided. Since we have to expect that our knowledge base is incomplete and that the acquired information could partly be inaccurate (particularly in the gossip domain), special relativizing expressions such as "according to my sources" are produced as part of the answer. The introduction of pronouns for entities mentioned before helps making the utterance less static and the conversation more natural.

4.3   Multimodal Communication

Gossip Galore uses several modalities for communicating with the user. First of all, all the agent's utterances are spoken, with the help of a speech synthesis system. The verbal part of the answer is additionally supported by gestures. To make maximal use of the available means for communication, we also use a TV screen embedded in the agent's environment, which is able to display arbitrary web sites, to present illustrations of the current answer (see Fig. 7) or even to provide the very content of the answer where the answer is not a single fact or a small set of facts but would require a complex explanation involving heterogenous kinds of information (see Fig. 4) or would lead to a rather longish and tiring answer if it were realized verbally (see Fig. 8).

5   Related Work

Web-based question answering systems typically proceed in several stages: i) the question is turned into a query for a standard search engine, ii) a set of relevant web sites is retrieved, iii) text passages are selected as an answer from that document set [12]. A variation of this idea is applied in HITIQA [2], an interactive open-domain question answering system for complex exploratory questions, where the answer is retrieved based on complete semantic event frames which are matched against the frame of the

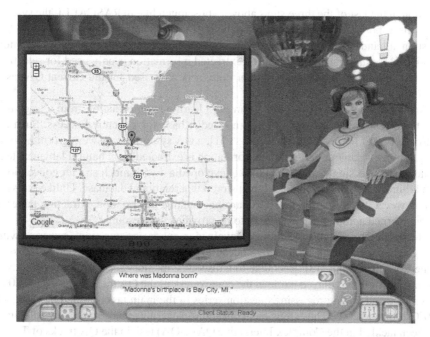

Fig. 7. Multimodal answers – illustrating locations using the Google Maps service

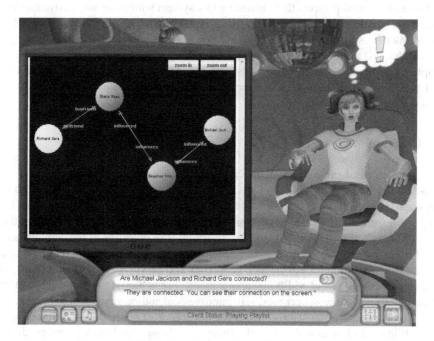

Fig. 8. Social network visualization

question in the last of the three steps above. Thus, much like in RASCALLI, the system performs a more structured semantic analysis of the original data (with respect to the question at hand). In contrast to our system, and more in accordance with the common web-based QA approach, however, the extracted information is only used for selecting the most relevant text passages. Furthermore, it is only used for the current QA task, not for building up world knowledge.

The approach followed in our system is to learn new information from a large unstructured text pool and store it in a knowledge base for structured access. Such an approach is also followed in BIRDQUEST [1], a QA system for answering ornithological questions. As in our system, the information is extracted from natural language text (although from a single source – a bird encyclopedia – with much stricter conventions than arbitrary web documents) and then stored in a relational database. Unlike our system, however, the system is not self-learning; suitable information extraction patterns are not learned automatically but have to be provided as a resource.

The potential benefits and sub-tasks involved in enhancing question answering with dialogue capacities to get interactive question answering have been briefly discussed in the Q&A Roadmap [13]. Recourse to a discourse memory for tracking entities over several questions has played a role in the context task of the QA track at the TREC 2001 conference and when processing question series in the main tasks in the QA tracks of TREC 2004 and 2005. The extension of question answering to more interactive dialogue has been tackled in the Complex Interactive QA (ciQA) task in the QA tracks of TREC 2006 and 2007[1].

There are several projects that enhance a QA system with more interactive capabilities, namely BIRDQUEST [1], HITIQA [2], RITEL [14], the IMIX demonstrator [3], and SMARTWEB [15]. RASCALLI has a different focus compared to all of these systems in that it i) aims to create a personal relationship with the user by the use of user-adaptive knowledge acquisition methods, and ii) conducts a vivid conversation with the user that mimics human-to-human communication, creating the immersive effect of a living entity with its own personality.

6 Conclusions

We have described the overall architecture and main components of a new class of web-based virtual agents. Although the design of the agents is strongly influenced by empirical observations and theoretical models of natural cognitive agents, our goal has not been a simulation of biological cognition. This aim is partially targeted by other research strands within the RASCALLI consortium. The objective of the demonstrated architecture and implementation has been a rather pragmatic and simplified agent model that exhibits the desired performance properties and serves as the starting point for a range of extensions and additional applications. The achieved relevant performance properties are: robustness, accuracy, self-improvement and nearly real-time behavior.

The planned future extensions include the integration of deeper language processing methods instead of or in addition to the fuzzy paraphrase matcher. A prime candidate

---

[1] Please refer to the TREC homepage at http://trec.nist.gov/ for further information and references.

for this extension is our own deep syntactic/semantic parser. Another plan concerns the required temporal aspects of relations. It is the dynamics of the domain that provide the basis for the gossip. Properties and relationships change quite often. By detecting and relating the utterance and report times of the various information sources, a multitude of answers may be temporally sorted. Once in a while, contradicting information is harvested. In some cases, these contradictions result from an unresolved temporal succession, i.e. the contradicting facts were true at different times. In other cases, one of the contradicting facts is simply false. In order to deal with such situations, we need to enrich the information extraction by methods for credibility checking, which will be adopted from IE/IR research.

Finally, we plan to exploit the dialogue memory for moving more of the dialogue initiative to the agent. In cases of missing or negative answers or in cases of pauses on the user side, the agent can use the active parts of the dialogue memory to propose additional relevant information or to guide the user to fruitful requests within the range of user's interests. However, the hardest test for the agent architecture will be the extension to other domains and tasks that may be less error forgiving than the colorful world of pop trivia.

Acknowledgements. The work presented here was supported by the international project RASCALLI funded by the Sixth Framework Programme of the European Commission in the area of Cognitive Systems (IST-27596-2004), and partially funded through a grant to the project KomParse by the ProFIT programme of the Federal State of Berlin and the EFRE programme of the European Union. We are also grateful to the cooperation with the HyLap project funded by the German Ministry for Education and Research (BMBF, FKZ: 01 IW F02). Many thanks go to our RASCALLI project partners, in particular, Radon Labs team, led by Nicolaas Bongaerts, for the development of the 3D client, and Brigitte Krenn and her team from OFAI and SAT as well as Rebecca Dridan from the Department of Computational Linguistics at the University Saarbrücken for their suggestions and comments.

References

1. Jönsson, A., Andén, F., Degerstedt, L., Flycht-Eriksson, A., Merkel, M., Norberg, S.: Experiences from combining dialogue system development with information extraction techniques. In: Maybury, M.T. (ed.) New Directions in Question Answering, pp. 153–168. MIT Press, Cambridge (2004)
2. Strzalkowski, T., Small, S., Hardy, H., Yamrom, B., Liu, T., Kantor, P., Ng, K., Wacholder, N.: HITIQA: A question answering analytical tool. In: Proceedings of the International Conference on Intelligence Analysis (IA 2005), Lean, VA (2005)
3. Theune, M., Krahmer, E., van Schooten, B., op den Akker, R., van Hooijdonk, C., Marsi, E., Bosma, W., Hofs, D., Nijholt, A.: Questions, pictures, answers: Introducing pictures in question-answering systems. In: Tenth international symposium on social communication, Cuba, pp. 450–463 (2007)
4. Krenn, B.: Responsive artificial situated cognitive agents living and learning on the internet. Poster presented at the International Conference on Cognitive Systems, CogSys 2008 (2008)

5. Xu, F., Uszkoreit, H., Li, H.: A seed-driven bottom-up machine learning framework for extracting relations of various complexity. In: Proceedings of the 45th Annual Meeting of the Association of Computational Linguistics, pp. 584–591 (2007)
6. Xu, F., Uszkoreit, H., Li, H.: Task driven coreference resolution for relation extraction. In: Proceedings of the European Conference for Artificial Inteligence ECAI 2008, Patras, Greece (2008)
7. Mcguinness, D.L., van Harmelen, F.: OWL Web Ontology Language Overview (2004), http://www.w3.org/TR/owl-features/ (accessed March 31, 2009)
8. Drozdzynski, W., Krieger, H.U., Piskorski, J., Schäfer, U., Xu, F.: Shallow processing with unification and typed feature structures – foundations and applications. Künstliche Intelligenz 1, 17–23 (2004)
9. Klein, D., Manning, C.D.: Accurate unlexicalized parsing. In: Proceedings of the 41st Meeting of the Association for Computational Linguistics (ACL 2003), pp. 423–430 (2003)
10. Xu, F., Uszkoreit, H., Li, H., Felger, N.: Adaptation of relation extraction rules to new domains. In: Proceedings of the Sixth International Conference on Language Resources and Evaluation, LREC 2008 (2008)
11. Schröder, M., Hunecke, A.: Mary tts participation in the Blizzard Challenge 2007. In: Proceedings of the Blizzard Challenge 2007, Bonn, Germany (2007)
12. Neumann, G.: Strategien zur Webbasierten Multilingualen Fragebeantwortung: Wie Suchmaschinen zu Antwortmaschinen werden. Computer Science - Research and Development 22, 71–84 (2008)
13. Burger, J., Cardie, C., Chaudhri, V., Gaizauskas, R., Harabagiu, S., Israel, D., Jacquemin, C., Lin, C.Y., Maiorano, S., Miller, G., Moldovan, D., Ogden, B., Prager, J., Riloff, E., Singhal, A., Shrihari, R., Strzalkowski, T., Voorhees, E., Weishedel, R.: Issues, tasks and program structures to roadmap research in Question & Answering, Q&A (2000)
14. Rosset, S., Galibert, O., Illouz, G., Aurélien, M.: Integrating spoken dialog and question answering: the Ritel project. In: Proceedings of INTERSPEECH 2006 (2006)
15. Reithinger, N., Herzog, G., Blocher, A.: SmartWeb - mobile broadband access to the semantic web. KI - Künstliche Intelligenz 2 (2007)

Biosignal Based Discrimination
between Slight and Strong Driver Hypovigilance
by Support-Vector Machines

David Sommer[1], Martin Golz[1,2], Udo Trutschel[2,3], and Dave Edwards[4]

[1] University of Applied Sciences Schmalkalden
Faculty of Computer Science, Schmalkalden, Germany
[2] Circadian, Stoneham, Massachusetts U.S.A.
[3] Institute for System Analysis and Applied Numerics, Tabarz, Germany
[4] Caterpillar Inc., Machine Research, Peoria, Illinois U.S.A.

Abstract. In the area of transportation research, there is a growing need for robust and reliable measures of hypovigilance, particularly due to the current volume of research in the development and validation of Fatigue Monitoring Technologies (FMT). Most of the currently emerging FMT is vision based. The parameter Percentage of Eyelid Closure (PERCLOS) is used for the fatigue detection. The development and validation of PERCLOS based FMT require an independent reference standard of drivers' hypovigilance. Most approaches utilized electrooculography (EOG) and electroencephalography (EEG) combined with descriptive statistics of a few time or spectral domain features. Typically, the power spectral densities (PSD) averaged in four to six spectral bands is used for fatigue characterization. This constricted approach led to sometimes contradicting results and questioned the validity of the EEG and EOG as gold standard for driver fatigue, wrongly as we will show. Here we present a more general approach using generalized EEG and EOG PSD features in combination with data fusion and advanced computational intelligence methods, such as Support-Vector Machines (SVM). Biosignal based discrimination of driver hypovigilance was performed by independent class labels which were derived from Karolinska Sleepiness Scale (KSS) and from variation of lane deviation (VLD). The first is a measure of subjectively self-experienced hypovigilance, whereas the second is an objective measure of performance decrements. For simplicity, two label classes were discriminated: slight and strong hypovigilance. The discrimination results of PERCLOS were compared with results from single and combined EEG and EOG channels. We conclude that EEG and EOG biosignals are substantially more suited to assess driver's hypovigilance than the PERCLOS biosignals. In addition, computational intelligence performed better when objective class labels were used instead of subjective class labels.

1 Introduction

Both distracted and fatigued driving accidents are thought to be underreported. Unlike alcohol related accidents where assessing the driver's state is relatively straight forward, there are no similarly objective means of ascertaining the driver's state of vigilance following a distraction or fatigue related accident. Unless the driver admits distraction or

J. Filipe, A. Fred, and B. Sharp (Eds.): ICAART 2009, CCIS 67, pp. 177–187, 2010.
© Springer-Verlag Berlin Heidelberg 2010

fatigue as a cause, one can only infer the driver's state from the physical evidence at the accident scene (Sirois et al 2007). Most drivers are reluctant to admit distraction or fatigue because they may fear being assigned blame for the incident. Therefore, the objective determination of driver's hypovigilance and distraction through the use of FMT systems could greatly reduce the occurrence of fatigue related incidents by informing drivers of their own level of vigilance. Vigilance describes the ability to sustain attention that is required for people to perceive and interpret random, relevant changes in the environment, in order to make effective decisions and perform precise motor actions. Hypovigilance is a deficit of vigilance. Two major causes of hypovigilance are central fatigue and task monotony. But, it is well known that several other factors influence driver's hypovigilance. It is a complex issue with several facets (Leproult et al 2002, Trutschel et al 2006).

Driver's hypovigilance depends for example on time-of-day due to the circadian rhythm, on time-since-sleep (long duration of wakefulness), on time-on-task (prolonged work), inadequate sleep, and accumulated lack of sleep. The last two factors may be caused by pathological sleepiness due to diseases, like sleep apnea or narcolepsy, or may be caused by intentionally sleep loss due to prolonged time awake. Moreover, there are also psychological factors influencing the actual level of vigilance, e.g. motivation, stress, and monotony. The last is believed to play a major role in driving, because it is mostly a simple lane-tracking task with a low event rate. Therefore, hypovigilance is considered as a psychophysiological variable not always decreasing monotonically during driving. It shows slow waxing and waning patterns, which can be observed in driving performance and repeatedly self-reported sleepiness.

There are many biosignals which contain more or less information on hypovigilance. Among them, EEG is a relatively direct, functional reflection of mainly cortical and to some low degree also sub-cortical activities. EOG is a measure of eye and eyelid movements and reflects activation / deactivation as well as regulation of the autonomous nervous system.

Until recently, for the assessment of driver's hypovigilance the analysis of EEG and EOG was based on a variety of definitions involving PSD summation in a few spectral bands which proved in clinical practice. The same applies to the location of EEG electrodes. Separate analysis of EEG of different electrodes and of alternative definitions of spectral bands led to inconsistent and sometimes contradicting results. Large inter-individual differences turned out to be another problematic issue.

Therefore, adaptive methods with less predefined assumptions are needed for comprehensive hypovigilance assessment. Here we propose a combination of different brain (EEG) and oculomotoric (EOG) signals whereby parameters of pre-processing and summation in spectral bands were optimized empirically. Moreover, modern concepts of discriminant analysis such as computational intelligence and concepts of data fusion were utilized. Using this general approach ensures optimal information gain even if unimodal data distributions are existent (Golz et al. 2007).

As a first step solution, we utilized SVM in order to map feature vectors extracted from EEG / EOG of variable segment lengths to two, independent types of class labels. For their generation a subjective as well as an objective measure was applied.

Both reflect different facets of hypovigilance: sleepiness and performance decrements, respectively.

For the first type of labels, an orally spoken self-report of sleepiness on a continuous scale, the so-called Karolinska Sleepiness Scale (KSS), was recorded every two minutes during driving. The second type of labels was determined through analyzing driving performance. In previous studies it was found that especially the variation of lane deviation (VLD) correlates well with hypovigilance and attention state of drivers (Pilutti et al. 1999). For the discrimination task, a total of 10891 biosignal segments with the corresponding labels were selected. In 3611 cases the labels indicated low KSS and low VLD (class 1—class 1), in 3746 cases the labels indicated high KSS and high VLD (class 2—class 2). Here the labels of two selected facets of hypovigilance were in agreement. But for 1922 cases KSS was high and VLD was low (class 2—class 1) and for 1611 cases KSS was low and VLD was high (class1—class 2) showing that the two facets of hypovigilance are pointing in opposite directions. The disagreements between the subjective and objective labels are caused mainly by inter-individual differences of drivers regarding the ability to tolerate extreme fatigue and still keeping a safe performance level.

2   Methods

2.1   Experiments

16 participants drove two nights (11:30 p.m. - 8:30 a.m.) in our real car driving simulation lab. One overnight experiment comprised of 8 x 40 min of driving. EEG (FP1, FP2, C3, Cz, C4, O1, O2, A1, A2) and EOG (vertical, horizontal) were recorded at a sampling rate of 256 Hz. PERCLOS as another oculomotoric measure was recorded utilizing an established eye tracking system at a sampling rate of 60 Hz. Also several variables of driving simulation, like e. g. steering angle and lane deviation, were recorded at a sampling rate of 50 Hz. Especially, variation of lane deviation (VLD) is a good measure of driving performance and is used here as an objective and independent measure of hypovigilance as described below. VLD is defined as the difference between two subsequent samples of lane deviation normalized to the width of lane. For example, moving the car from the left most to the right most position of the lane results in VLD = 100 %. The KSS was mentioned above and is a standardized, subjective, and independent measure of hypovigilance on a numeric scale between 1 and 10. KSS was asked at the beginning and after finishing driving. During driving only relative changes in percent of the full range were asked because subjects are more aware of relative than on absolute changes.

2.2   Feature Extraction

To allow a comparison of the selected biosignals regarding hypovigilance, pre-possessing and feature extraction were performed due to the same concept for all biosignals (Golz et al. 2007). First, non-overlapping segmentation with variable segment length was carried out, followed by linear trend removal and estimation of power spectral densities (PSD) utilizing the modified periodogram method. Other estimation techniques,

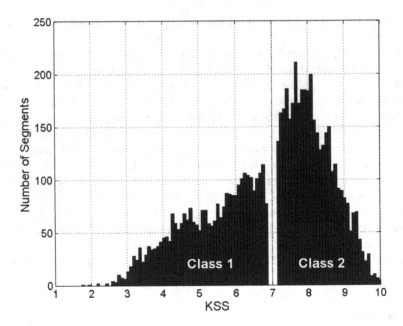

Fig. 1. Histogram of subjective ratings of sleepiness (KSS). Binarization leads to two classes: slight (class 1) and strong hypovigilance (class 2). Values in the immediate threshold region (around KSS=7) were eliminated.

such as Welch's method, the Multi-Taper method, and a parametric estimation (Burg method), were also applied, but resulted in slightly higher discrimination errors. It seems that these three methods failed due to reduced variance of PSD estimation at the expense of bias. In contradiction to explorative analysis, machine learning algorithms are not such sensitive to higher variances. Second, PSD values of all three types of signals were averaged in spectral bands. In case of EEG and EOG signals 1.0 Hz wide bands and a range of 1 to 23 Hz turned out to be optimal, whereas in case of PER-CLOS signals 0.2 Hz wide bands and a range of 0 to 4 Hz were optimal. All parameters were found empirically at lowest discrimination errors in the test set. Further improvements were achieved, but only in case of electrophysiological features, by applying a monotonic, continuous transform log(x).

2.3   Classification

KSS and VLD values were divided into categories 'slight hypovigilance' (class 1) and 'strong hypovigilance' (class 2). This was necessary to get labels for discriminant analysis (classification). For the subjective measure the threshold parameter was selected at KSS = 7 (Fig. 1). For a better visualization of separation between class 1 and class 2 samples in the range of KSS = 6.9 ... 7.1 were eliminated from data set. This step

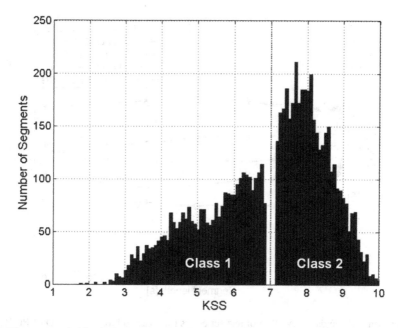

Fig. 2. Histogram of objectively measured performance (VLD). Binarization leads to two classes: slight (class 1) and strong hypovigilance (class 2). Values in the immediate threshold region were eliminated.

turned out to be not crucial. Results of classification (test set errors) showed not much of a difference.

The same binarization was applied also to the objective measure. Threshold was determined at VLD = 13.5 % and all samples in the range of VLD = 13.0 % ... 14.0 % were eliminated (Fig. 2). This data elimination also turned out to be not crucial.

Segment length was always optimized (see below) in order to get minimal test errors. Test errors were estimated by multiple, random cross validation (80 % training / 20 % test set). Due to the relatively high dimensionality of the feature space a powerful machine learning method, the Support-Vector Machine (SVM), was applied. SVM adapts an optimal separating hyperplane without any presumptions on data distribution. To achieve nonlinear discriminant functions special kernel functions have to be applied. Among several others, kernel functions such as radial basis function $k(\mathbf{x}_1, \mathbf{x}_2) = \exp\left(-\gamma \|\mathbf{x}_1 - \mathbf{x}_2\|^2\right)$ and the Coulomb function $k(\mathbf{x}_1, \mathbf{x}_2) = \left(1 + \gamma \|\mathbf{x}_1 - \mathbf{x}_2\|^2\right)^{-d}$ performed best in our application. Three SVM parameters (slack variable, two kernel parameters) were optimized carefully which requested high computational load (Golz et al. 2007). For each of the selected biosignals the segment length was varied in the range of 10 to 300 seconds to find an empirical optimum of the discrimination test error utilizing multiple hold-out cross validation. In general, small segment lengths lead to a high number of input vectors following to higher complexity presented to the discrimination algorithms and therefore to higher error rates for all signals.

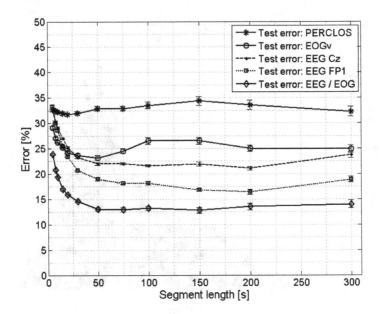

Fig. 3. Mean and standard deviation of test set errors for selected biosignals. PSD of PERCLOS had the lowest discrimination ability (largest errors), whereas PSD feature fusion of EEG and EOG performed best (lowest errors). Class labels were subjective (KSS).

3 Results

Discriminant analysis (classification) of different biosignals resulted in different errors for KSS labels (Fig. 3) and for VLD labels (Fig. 4). For the KSS labels, the PERC-LOS signal and the vertical component of EOG (EOGv) showed relatively high errors and depend in similar manner on segment length. EEG at location 'Fp1' showed lower errors for all segments length compared to EEG at location 'Cz'. The feature fusion of EEG at all 7 locations and of both EOG components resulted in lowest errors. This confirms our previous finding (Golz et al. 2007) that feature fusion of EEG and EOG lead to significant improvements in the discrimination between two classes utilizing SVM. Mean errors of about 13 % yielded in a relatively broad range of optimal segment lengths between 50 and 150 seconds. Similar results for EEG / EOG signals were found in a previous study (Golz et al. 2005). In this study, which was based on different data sets, the optimal segment length was as well between 50 to 150 seconds. Learning Vector Quantization was used instead of SVM as classification method. PERCLOS features resulted considerably worse (Fig. 3). Mean errors varied between 32 and 34 % in the whole range of segment lengths.

Slightly better, but basically comparable results yielded if the objective measure (VLD) was used as class labels. Lowest errors resulted if features of EEG and EOG were fused together (Fig. 4). Mean errors of about 10 % yielded at optimal segment lengths of about 150 seconds. PERCLOS results were considerably worse. Mean errors varied between 26 and 30 % if segment lengths were larger than 50 seconds. The

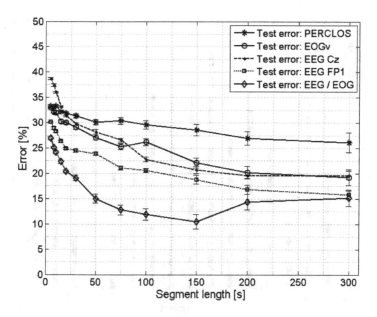

Fig. 4. Mean and standard deviation of test set errors for selected biosignals. PSD of PERCLOS had the lowest discrimination ability (highest errors), whereas PSD feature fusion of EEG and EOG performed best (lowest errors). Class labels were objective (VLD).

characteristics of the other signals EOG (vertical), EEG (Cz) and EEG (Fp1) as function of segment length is clearly more complex for the VLD labels than for KSS labels. In terms of mean errors, the achieved improvements due to feature fusion were considerable. The results (Fig. 3, 4) disclose two things: First, the driver hypovigilance detection is best for fused EEG / EOG biosignals when PSD features were utilized. This measure is suited as a benchmark to evaluate FMT devices. Second, the PERCLOS biosignal is able to detect driver hypovigilance with medium reliability.

The question arises if machine learning algorithms are able to find properties of driver hypovigilance, generally valid for all subjects under investigation. This was checked out by cross validation on the subject level. Learning algorithms were tested on all data of only one subject after they were trained on all data of all other subjects. This was repeated for every subject.

Results show high inter-individual variability (Fig. 5 , 6) not only between subjects, but also between the PERCLOS and the fused EEG/EOG biosignals, indicating that common characteristics were rarely found. Overall the inter-individual variability is larger for subjective labels (KSS) than for objective labels (VLD). This can be explained in that the subjects in our lab study were not professional drivers and could have difficulties to assess their own subjective sleepiness level using KSS. The classification errors between slight and strong hypovigilance are clearly biosignal and subject specific. Overall, the discrimination ability between the two classes is close to the optimal results only for subject '3' for EEG / EOG and for subjects '8, 15' for PERCLOS in case of subjective measures (KSS) as labels (Fig. 5). The picture differs completely when

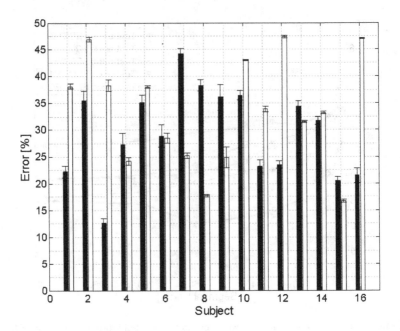

Fig. 5. Inter-individual differences of test set errors for the feature fusion of EEG and EOG (black bars) and for PERCLOS (white bars). Class labels were subjective (KSS).

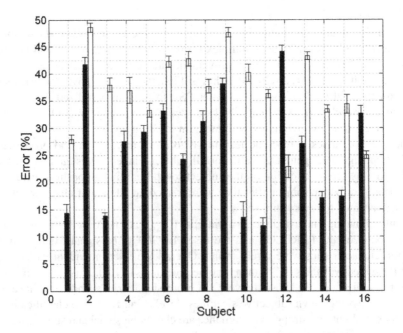

Fig. 6. Inter-individual differences of test set errors for the feature fusion of EEG and EOG (black bars) and for PERCLOS (white bars). Class labels were objective (VLD).

objective measures (VLD) were utilized as labels (Fig. 6). In this case driver hypovigilance detection is acceptable for subjects '1, 3, 10, 11' if EEG / EOG were processed. But there is no subject where PERCLOS reaches an acceptable low error level. This is somehow unexpected and questioned the validity of the application of stand alone vision based for FMT systems. In general, for most of the subjects the discrimination is limited. Test set errors are typically ranging between 15 % and 45 %, which is too high for practical applications. At the moment, it is necessary to have data of each subject also in the training set. In consequence, signal processing has to be specialised to any individual in order to get better results. Figures 5 and 6 also depict that in 9 out of 16 subjects for KSS labels and in 14 out of 16 subjects for VLD labels, the EEG / EOG feature fusion performed better than PERCLOS features.

4 Conclusions

Model free approaches are used in many different fields. Hence, it would be appropriate for the fatigue and performance research community to reach out and explore alternative algorithms beyond rule based statistical analysis of biosignals. This could help to advance the complex issue of driver hypovigilance which has eluded researchers for a long time.

Results of experimental investigations and subsequent adaptive data analysis yielded substantial differences in the usefulness of electrophysiological signals (EEG, EOG) compared to an oculomotoric signal (PERCLOS) which is at the moment the most often utilized measure of driver's hypovigilance in fatigue monitoring technologies, such as infrared video camera systems. This main result is regardless of the definition of hypovigilance, considering that subjective (KSS) as well as objective (VLD) labels has been utilized. Results were robust to different variations in parameters such as segment length which controls temporal resolution and amount of information to be involved. Mean test errors of 13 % and 10 % for subjective and objective labels, respectively, show that feature fused EEG and EOG has the potential to account for a reference standard (gold standard) to evaluate fatigue monitoring technologies (FMT). Mean test errors between 26 % and 32 % for subjective and objective labels, respectively, show that the PERCLOS signals seems to carry less information on driver's hypovigilance than fused EEG and EOG.

Our results contradict results of other authors (depicted in table 1 in Dinges et al 1998), where PERCLOS was found to be most reliable and valid for determination of driver's hypovigilance level. There, based on complete other data analyses, different measures of hypovigilance were compared. EEG resulted worse than PERCLOS, whereas measures of head position and of eye blink behaviour led to contradictory results between subjects. As a reference standard of hypovigilance they utilized measures of the well-known psychomotor vigilance task (PVT). Results are based on the fact that PERCLOS varies simultaneously with attention lapses in PVT which was repeated during 42 hours of sustained wakefulness. However, some doubts were raised (Johns 2003). It was pointed out that contradictions are possible, e. g. under demands of sustained attention some sleep-deprived subjects fall asleep while their eyes remain open. Unfortunately, PERCLOS does not include any assessment of eye and eye lid movements.

Important dynamic characteristics which are widely accepted, such as slow roving eye movements, reductions in maximal saccadic speed, or in velocity of eye lid re-opening, are ignored. Their spectral characteristics were picked up in our study through EOG and may account for the far better results of EEG / EOG data fusion presented here. Note, that highly dynamical alterations are better reflected by EOG than by PERCLOS. Our results support doubts stated in (Johns 2003) and clearly show limitations of PERCLOS. Some serious cautions should be considered when driver's hypovigilance is estimated relying solely on PERCLOS. In general, the aim of many researchers on driver's hypovigilance in the 90's to reduce such complex issue to a simple threshold parameter (Dinges et al 1998) was presumably misguiding. Fortunately, this has been corrected in recent projects. Different approaches were investigated Schleicher et al. 2007, among them also data fusion concepts (AWAKE 2004).

In addition, our previous findings (Trutschel et al. 2006, Golz et al. 2005, Golz et al. 2007) have shown that results on the assessment of driver states differ from subject to subject, as well as to some limited extent also from driving session to driving session. This was confirmed in the current investigations as well. This is a problematic issue for FMT systems. Individualization will be needed for reliable detection of driver's hypovigilance. To find practical solutions in order to address intra-individual differences in discrimination of slight and strong hypovigilance future investigations are required. For example, it could be futile to master group-average model predictions before exploring means of predicting individual hypovigilance. Due to large inter-subject variability in subjective alertness (KSS) and driving performance (VLD), it may turn out to be easier to develop reliable and accurate models of individualized measures of hypovigilance on the basis of an individual's data fusion concept than group-average vigilance models based on a single data stream.

References

1. AWAKE, System for Effective Assessment of Driver Vigilance and Warning According to Traffic Risk Estimation. In: Road Safety Workshop, Balocco, Italy (2004),
   http://www.awake-eu.org/index.html
2. Dinges, D., Grace, R.: PERCLOS: A Valid Psychophysiological Measure of Alertness As Assessed by Psychomotor Vigilance. TechBrief NHTSA, Publication No. FHWA-MCRT-98-006 (1998)
3. Golz, M., Sommer, D.: Detection of Strong Fatigue During Overnight Driving. In: 39th Annual Congress of the German Society for Biomedical Engineering (BMT 2005), Part 1, Nürnberg, Germany, pp. 479–480 (2005)
4. Golz, M., Sommer, D., Chen, M., Trutschel, U., Mandic, D.: Feature Fusion for the Detection of Microsleep Events. The Journal of VLSI Signal Processing 49, 329–342 (2007)
5. Johns, M.: The Amplitude-Velocity Ratio of Blinks: A new Method for Monitoring Drowsiness. Sleep 26, A51–A52 (2003)
6. Leproult, R., Coleccia, E., Berardi, A., Stickgold, R., Kosslyn, S.M., Van Cauter, E.: Individual differences in subjective and objective alertness during sleep deprivation are stable and unrelated. Am. J. Physiol. 284, R280–R290 (2002)
7. Pilutti, T., Ulsoy, G.: Identification of Driver State for Lane-Keeping Tasks. IEEE Transactions on Systems, Man, and Cybernetic, Part A: System and Humans 29, 486–502 (1999)

8. Schleicher, R., Galley, N., Briest, S., Galley, L.: Looking Tired? Blinks and Saccades as Indicators of Fatigue. Ergonomics 51, 982–1010 (2007)
9. Sirois, W., Dawson, T., Moore-Ede, M., Aguirre, A., Trutschel, U.: Assessing Driver Fatigue as s Factor in Road Accidents. In: Proceedings of the Fourth International Driving Symposium on Human Factors in Driver Assessment, Training and Vehicle Design, pp. 527–533 (2007)
10. Trutschel, U., Sommer, D., Aguirre, A., Dawson, T., Sirois, B.: Alertness Assessment Using Data Fusion and Discrimination Ability of LVQ-Networks. In: 10th International Conference on Knowledge-Based and Intelligent Information and Engineering Systems (KES 2006), Bournemouth, UK, pp. 1264–1271 (2006)

PART II

Agents

Tiered Logic for Agents in Contexts

Rosalito Perez Cruz and John Newsome Crossley

Faculty of Information Technology
Monash University, Australia, 3800
lcruz@sleekersoft.com, John.Crossley@infotech.monash.edu.au
http://www.csse.monash.edu.au/~jnc, http://www.sleekersoft.com

Abstract. We introduce a new kind of logic for agents in different localities, which works in tiers or layers. At the base are local worlds with their own logic. Above them is a global logic that takes statements from the local worlds and combines them. This allows communications between the different localities.

We give a basic example using first order logic as the local logic and propositional calculus at the global level. As a more sophisticated example we use the algebraic specification language CASL and take the locations as specifications. Moreover we then permit the combination of such specifications according to the architectural specifications of CASL.

Although we only consider two layers in the present paper, we see no reason why the approach should not be extended to any finite number of tiers. We prove soundness and completeness proofs for our logics.

Keywords: Agents, Logics for agent systems, Ontologies and agent systems.

1 Introduction

It is well established that the work of agents in a multi-agent system is enhanced by the presence of ontologies. For an ontology to be useful, people will have to agree to its terms and usage in the spirit of sharing. However, human nature ensures that people will not agree nor use something like an ontology consistently. Thus the idea of arriving at a *global ontology* for a domain of application appears to be wishful thinking. So it seems more appropriate to conceive of pockets of communities sharing their ontologies and coping with any differences. It is more realistic to think of communities adopting a number of ontologies, each created within their *local* community.

We shall adopt an approach which contextualizes the logics that support these ontologies, and thereby point a way for agent systems to deal with heterogenous ontologies. We shall describe two logics:[1] a first order logic (FOL) of localities, **Tiered FOL**, which we use as a basis, then we extend this technique to a language **Tiered CASL**, where the localities are architectural specifications in the Common Algebraic Specification Language, CASL, see [6,2]. We prove completeness results for both these logics.

In the field of AI and, by association, Logic, there are two major styles of embedding localities[2] in a logical system. The first is in the Propositional Logic of Context (PLC)

---

[1] We use natural deduction systems throughout.
[2] We use "locality" rather than "context" because the latter is so ambiguous.

J. Filipe, A. Fred, and B. Sharp (Eds.): ICAART 2009, CCIS 67, pp. 191–204, 2010.

of Buvač-Mason [5] and their extension of this to FOL. The second is the Local Models Sematics/MultiContext Systems (LMS/MCS) of [11,12]. By no means do we imply that these are the only two possible styles: there are others such as in [1].

One example of an LMS approach in the field of Description Logic (DL) is that taken by Borgida and Serafin, who describe a Distributed DL in [3]. A major problem has been the transfer of knowledge between localities. Bridge rules (see Section 3) were introduced in [11], but the form of the rules was very limited and only allowed the (partial) identification of one concept as a subset of another in a different locality. The idea is to align ontologies (or knowledge bases) by expressing the connections between them. The intent is that the logical system should allow the relationship of concepts to be stated in the said ontologies, for example subsumption of concepts between ontologies. To do this, Borgida and Serafini extend the usual DL formulation, taking their cue from the Distributed First Order Logic (DFOL) of [11]. In their formulation, a DL statement is preceded by a label that stands for the ontology. Then they state *bridge rules*, which relate a concept in one ontology to another one in a different ontology (see [3]). Thus they have semantic mappings in the system.

Serafini, Borgida and Ghidini take their technique from Giunchiglia's LMS which they call the *compose-and-conquer* way of dealing with differences of languages in contexts. PLC uses a divide-and-conquer technique and since we take our cue from PLC, Tiered Logic is a divide-and-conquer technique, though the terminology may not be entirely appropriate as there are similarities to both.

Gabbay's Fibring of Logics, see e.g. [10], is clearly related to our work but does not permit the up and down interaction between local and global systems that we have.

In global (natural language) discourse one often sees or hears statements in a foreign language used in the middle of something in the local language, for example in a television broadcast where the spoken foreign language is accompanied by subtitles. References may then need to be changed or at least clarified. Consider the following two assertions:

"Le président à dit qu'il n'y a aucune arme de destruction de masse en Irak."[3]
"The President said that there are weapons of mass destruction in Iraq."

Here the references are to the same country, however the reference to the president refers, in the first case, to the French one, and in the second, to the US President.[4] There is no contradiction between the quotations, but there is between the two men.

In the media there would be an indication of the locality, i.e. country. Thus we might have found in the USA: "The President *of France* said that there are no weapons of mass destruction in Iraq," and in France: "Aux États Unis, le Président a dit qu'il y a armes de destruction de masse en Irak". Finally, in a third country: "In the USA, the President said there are weapons of mass destruction in Iraq, but in France, its President said there are no weapons of mass destruction." Semantically we understand these utterances because we tag each utterance with its context or, as we shall say, "locality", in these cases, France and the USA, respectively. Then we interpret them in that locality.

---

[3] "The President said that there are no weapons of mass destruction in Iraq."

[4] The reference to weapons of mass destruction was more problematic because we did not know whether there were any in Iraq!

For agents in localities we again have the problem of them communicating across different languages. This paper provides a basic method of formalizing such situations by allowing the inclusion of powerful *bridge rules and axioms*.

We give the first presentation of what we call "Tiered Logic". In our logic, statements made in a local language are tagged with that local locality and then become "atomic" statements or basic propositions in a higher tier of what we call the *global* logic. With bridge rules or axioms any statement in one locality can have consequences in another. So information can be conveyed, or simply translated, from one locality to another.

We provide soundness and completeness proofs for two varieties of our underlying idea of tiered logic. For simplicity we assume that all our localities have the same underlying logic, but different languages. This restriction is not essential but a completely general approach would be notationally horrendous. The complications in our presentation come from the interactions between the tiers: when a sentence from one locality is used in a different locality, one has to refer back to the previous locality in order to determine the semantics.

Additionally we use Saša Buvač's notion of *flatness* (see e.g. [5] and Section 2). This entails that once a statement has been made (and its semantics determined for its own locality) then the truth or falsehood of the statement is unaffected by reporting it in another locality. Thus in the example above, a US newspaper reporting what had been said in the USA might include the statement that it had been reported in France that the (US) President had said there were weapons of mass destruction in Iraq. The semantics here would only depend on what was said in the US, not what was reported in France (assuming that the media tell the truth).

2   Tiered FOL

First we consider the informal semantics. We have a number of localities, think of France, the USA, etc., each with its own local theory. In our first example we simply use first order logic at each locality. These comprise Tier 0. At each locality we have a traditional model of the local theory, that is to say, a first order model. We collect these together to form a model for the global (tier 1) language. The underlying semantics at tier 1 is the standard semantics of propositional calculus except that traditional propositional letters are replaced by what we call "basic" global formulae.

However, we also have interaction between the global scene and the localities. So we have to specify how the semantics (the models) interact between tier 0 and tier 1. From an intuitive point of view the interaction is relatively simple and reflects our earlier informal example. Intuitively: a formula is interpreted in its local locality, so that a tier 0 formula is interpreted in a traditional first order logic model (in tier 0 at a locality, l, say). On the other hand a global, or tier 1, formula is interpreted using the values from the tier 0 model (or models) according to the usual rules for propositional calculus. When we go back down from tier 1 to tier 0, the semantic value is unchanged. (This depends on the fact that our formulae at tier 1 have no free variables and are therefore true or false.) The formal definitions follow the usual pattern.

Syntax. Because of going up and down between tiers the syntax looks a little complicated, however the actual formulae should be easily readable. We let $\mathbb{L}$ be a set of

$$\frac{\Gamma \vdash_l A}{\Gamma \vdash_\gamma A^l} \text{ (Exit)} \qquad \frac{\Gamma \vdash_\gamma A^l}{\Gamma \vdash_l A} \text{ (Enter)} \qquad \frac{\Gamma \vdash_\gamma (A \to B)^l}{\Gamma \vdash_\gamma (A^l \to B^l)} \text{ (K)}$$

$$\frac{}{\Gamma \vdash_\gamma (\neg A)^l \leftrightarrow \neg (A^l)} \text{ (DT)} \qquad \frac{}{\Gamma \vdash_\gamma (A^l)^k \leftrightarrow A^l} \text{ (Flat)} \qquad \frac{}{\Gamma \vdash_l A \leftrightarrow A^l} \text{ (Flat-0)}$$

Fig. 1. The transfer rules for Tiered FOL. Note that A and B must be local formulae of l for the (Exit) and (Enter) rules. (Of course this includes global formulae).

localities. At each locality $l \in \mathbb{L}$ we have a first order logic with a language $\mathcal{L}^l$ as usual. These generate the *strictly local formulae*, which we denote by φ, ψ, etc. Going up to the global level (tier 1) we define the basic global formulae as strictly local sentences tagged by their locality, e.g. φ^l. These are combined as in an ordinary propositional calculus and we denote global formulae by Φ, Ψ, etc. But now we can take these back down to the local level, where they interact with formulae already there (including *strictly lo-cal formulae*). We then take the inductive closure in the usual way, to get the set of local formulae at that locality.

Thus local formulae and global formulae are inductively defined using a pair of in-teracting inductive definitions. Notice that although global formulae are local formulae (for any locality) the reverse is definitely not the case. For example, a strictly local formula of locality l is not a global formula.

Examples. We assume that the language of locality l has *only* the predicate letter P, and that the locality k has *only* the predicate letters P_1 and P_2.

Strictly local formulae: $\forall x P(x)$ in the locality l; $(P_1(x) \to P_2(x))$ in the locality k; and $\exists y P_2(y)$ in the locality k.

Global formulae: $\forall x P(x)^l$, $((\forall x P(x))^l \to (\exists x P_1(x))^k)$, $(\exists y P_2(y))^k$. Notice that the localities are superscripts in the global formulae. Each global example is either a superscripted local *sentence* or a propositional combination of such sentences.

Local formulae for the locality k: $(\forall x P(x))^l$, $(P_1(x) \to P_2(x))$, and $\exists y((\forall x P(x))^l \to P_2(y))$. The first formula, $(\forall x P(x))^l$, is local (even in the local-ity k) because it is a global formula; the second is local in k because it is a strictly local formula of k; and the third is local in k, because it is a first order logic combination of a strictly local (and therefore also local) formula, $P_2(y)$, of k and a global (therefore also local) formula, $(\forall x P(x))^l$.

Our axiom system is designed from reflecting on the semantics. The (strictly) local syntax is simply first order logic in the language $\mathcal{L}^l$ for tier 0 and propositional calculus for tier 1. In addition to these we have the rules in Figure 1 which are essentially due to Buvač [5]. We read $\Gamma \vdash_\gamma A$ as "Γ globally proves A" and $\Gamma \vdash_l A$ as "Γ proves A in the locality l".

The (Exit) and (Enter) rules allow us to move up and down between the tiers, pro-vided we appropriately tag or untag the formula. The rules (K) and (DT), in which we follow Buvač, when used together with the (Exit) and (Enter) rules, ensure that the propositional connectives commute with moving between the tiers.[5] The rule (Flat),

---

[5] If we did not have the (Exit) and (Enter) rules we would be able to have, say, the apparent inconsistency of having $\neg A$ at the local level and yet A^l at the global level.

see [5], ensures that once a statement has been made in one locality its truth-value is unchanged when it is taken into another locality. (Flat-0), which is our addition to the ideas of Buvač, ensures consistency between local and global versions of a statement. cf. footnote 5 above.

Remark 1. If Ξ is the strictly local theory in the locality l, then we define the *lifting* of Ξ to the global tier to be $\Xi^l = \{\varphi^l : \Xi \vdash_l \varphi\}$.

Lemma 1. *1. If Φ is a global formula, then $\Gamma \vdash_\gamma (\Phi \leftrightarrow \Phi^l)$ for any locality l.*
2. $\Xi \vdash_l \varphi$ is equivalent to $\Xi^l \vdash_\gamma \varphi^l$.
3. If Φ and all formulae in Γ are global formulae, and $\Gamma \vdash_l \Phi$, then $\Gamma \vdash_\gamma \Phi$.

The proofs of these and all other results may be found in [9].

Theorem 1 (CNF for Global Formulae). *Every global formula is provably equivalent to a conjunction of disjunctions of basic global formulae.*

Proof. First show that every global formula is globally provably equivalent to a propositional combination of basic global formulae, and then, as usual, put this into conjunctive normal form. $\square$

Formal Semantics. We first define a strictly local model for a locality l as a model in the usual first order logic sense, and we denote such models as $\mathfrak{m}_l$. These are the tier 0 models. Then a model for the *global* system, or *tier 1 model*, is a set of such models: $\mathfrak{M} = \{\mathfrak{m}_l : l \in \mathbb{L}\}$.

Remark 2 (Overlap requirements). It is possible to have overlaps in the languages at the different localities. Then we impose the requirement that if two atomic *sentences* from different localities, are syntactically identical, then they are semantically identical also. This will then carry over to more complicated formulae in the usual way.

In order to define global satisfaction we need simultaneously to define local satisfaction, so we have a double inductive definition. The reader should be warned that the formal definitions, which may be found in [9] look much more forbidding than they are in practice. He or she should refer back to the beginning of this section, and here we shall only give an intuitive picture.

Given a *basic global sentence* φ^l (which means φ is a strictly local *sentence* of locality l), then φ^l is *(globally) true in* $\mathfrak{M} = \{\mathfrak{m}_l : l \in \mathbb{L}\}$, written $\mathfrak{M} \models_\gamma \varphi^l$, if, and only if, $\mathfrak{m}_l \models_l \varphi$. In this case we also say φ^l is *locally satisfied at l*, and we write this as $\mathfrak{M} \models_l \varphi^l$.[6]

If Φ is a *global sentence*, then we use the usual rules of propositional calculus to compute its truth value. This also covers local satisfaction.

This only partly defines global satisfaction, for it only defines it for propositional combinations of *basic* global sentences.

It remains to define local satisfaction for local formulae that are not global formulae. Such formulae may contain free variables from a particular locality. We simply do this in the obvious way, except that, because global formulae are sentences and have no free

---

[6] There will be no ambiguity, because *strictly* local satisfaction is not defined for such formulae.

variables, we can simply use the truth values of any global sentences contained in such a formula. Thus a local sentence A is *locally satisfied in* l if, and only if, $\mathfrak{m}_l \models_l A$. We also use the locutions "A is (strictly locally) true in $\mathfrak{m}_l$ (at l)", and "$\mathfrak{m}_l$ is a model of (the sentence) A".

To determine global satisfaction of a global formula put the formula into conjunctive normal form by Theorem 1, then determine the truth value of each basic global *sub-formula* φ^l by determining the local truth value of φ in l. Finally compute the global truth value from these truth values.

Consistency and Soundness. There are many varieties of consistency: strictly local, global and local. Happily, because of our rule system they are all essentially equivalent. For example, we say that a set of global formulae Γ is *globally consistent* if $\Gamma \not\vdash_\gamma \perp$ and that a set, Γ_l, of formulae local in l is *locally consistent in* l if $\Gamma_l \not\vdash_l \perp$. It then follows that if Σ is a set of strictly local formulae then Σ is strictly locally consistent if, and only if, it is locally consistent; if Σ is a locally consistent set of local formulae in a locality l, then $\Sigma^l = \{A^l : A \in \Sigma\}$ is globally consistent; and that if Σ is a set of global formulae, then Σ is globally consistent if, and only if it is locally consistent at some locality l if, and only if, it is locally consistent for every locality.

We define *soundness* in the obvious way: A rule $\Gamma, A, B \vdash_x C$ is *sound* (where x is γ or l) if, whenever Γ, A and B are satisfied (globally, or locally at l), then so is C, respectively.

Theorem 2. *1. The axioms and rules for Tiered FOL are both globally sound, and locally sound for any locality l.*

2. The rules and axioms for Tiered FOL are consistent.   □

3 Bridge Axioms and Rules

Having discussed the syntax and semantics of our system, we turn to the concept of *lifting axioms* which was introduced by McCarthy and Buvăc found in [16]. Lifting axioms relate truth in one context to truth in another. McCarthy and Buvăc use this term to mean "inferring" or "lifting" truth in one context to truth in another context. An example of this is when two global formulas are related to each other, e.g.: $\exists x P(x)^k \rightarrow \exists x Q(x)^l$, but the connection may also be as complicated as: $(\forall x P_1(x) \rightarrow \forall y P_2(y))^k \rightarrow ((\exists z Q_1(z) \rightarrow (\exists w Q_2(w))^j \wedge \exists v Q_3(v))^l$.

Bouquet and Serafini are advocates of LMS (see Section 1), and in [19] they opt for the concept of *bridge rules*. Bridge rules say that an assertion in one context leads to a conclusion in another context. For example:

(br) From φ^i infer ψ^j.

They construct these bridge rules as inference rules that lie outside the theories that operate as the contexts. In LMS's presentation of its system and in [4] (see their Definition 7), it is depicted as: $MS = \langle \{C_i = \langle L_i, \Omega_i, \Delta_i \rangle\}, \Delta_{br} \rangle$, where C_i is a theory having L_i as a language with axioms in Ω_i and its inference rules are in Δ_i. Finally Δ_{br} is the set of bridge rules. As can be observed, Δ_{br} is not part of the theory and stands outside it.

In the present work, we prefer to use the terms *Bridge Axioms* and *Bridge Rules*, because they are "bridges" relating truth in one context to truth in another context. We are using this phrase to recall that Tiered Logic was specifically designed to support navigation among heterogenous knowledge bases, especially ontologies, which may be viewed as contexts or localities. They are axioms because they are part of the system: they do not lie outside it.

Our idea of bridging is a cross between worlds–PLC and LMS. Lifting axioms and bridge rules are easily seen to be special cases of our bridge axioms and rules, respectively. Both may be embedded, and not just simulated, in Tiered Logic. To see this, take for example PLC: then Γ may be seen as the trivial context i.e., the context where there are no strictly local formulas at all, only global formulas, i.e. only superscripted formulas. Take now LMS; then the bridge rules (br) can simply be added to the global system as new rules of inference. Indeed, looking at the composition of MS, we may see TLM as $\langle C_i, \Gamma \rangle$ except unlike MS, our Γ and C_i interact with one another and are integrated in the reasoning machinery. Of course, the issue of consistency arises. However, this issue is not unique to Tiered Logic alone, but it is common to any system, be it PLC or LMS, etc. Consistency checking is a standard process necessary in the use of any logic system so this is expected and should not surprise us.

When, as in LMS, bridge rules are outside the system, independently existing reasoning engines will usually have to be modified and expanded to accommodate these rules. With Tiered Logic, the global system employs a labelling mechanism and so minimal changes to existing reasoning engines are required for the Tiered Logic system to work.

When we turn to description logic, suppose we have concepts, C and D, in localities k and l, respectively, then, our version of the rules in [3] would mean we would write $C^k \sqsubseteq D^l$ which corresponds to the informal sentence $\forall x(C^k(x) \to D^l(x))$. However, we cannot model this directly in our system.[7] Nevertheless we can certainly imitate the intent of Borgida and Serafini by adding rules of the form: For all constants c common to localities k and l

$$\frac{\Gamma \vdash_\gamma D^l(c)}{\Gamma \vdash_\gamma C^k(c)}$$

However, our system admits very powerful rules. For example, we can have rules that depend on not just one locality influencing another, but more than one. We can have bridge axioms of the form $\varphi^k \wedge \psi^l \to \chi^m$ or bridge rules of the form

$$\frac{\Gamma \vdash_\gamma \varphi^l \qquad \Gamma \vdash_\gamma \psi^k}{\Gamma \vdash_\gamma \chi^m}$$

or with even more premises. Further examples of bridge axioms involving quantification are: $\forall x P(x)^k \to \exists x Q(x)^l$, and
$(\forall x P_1(x) \to \forall y P_2(y))^k \to ((\exists z Q_1(z) \to \forall w Q_2(w) \wedge \exists v Q_3(v)))^l$.

---

[7] For an implementation of our scheme using description logic see the first author's forthcoming thesis [8].

Completeness and Decidability. In order to prove the completeness of our system under the tier scheme, we follow the technique of Leon Henkin [13]. Given a set, Γ, of consistent global formulae, we extend this to a maximal consistent set, Γ^∞, and show this has a model.[8] The main difference from the classical scheme is that we make maximal consistent sets of sentences both at the global level, Γ^∞, and at each locality.[9]

Now consider the strictly local sentences in $(\Gamma^\infty)_l = \{\varphi : \varphi^l \in \Gamma^\infty\}$. These include the atomic (strictly) local sentences and it is just these that are used, in the standard Henkin way, to build a *local* model, $\mathfrak{m}_l$. Then we collect these into $\mathfrak{M} = \{\mathfrak{m}_l : l \in \mathbb{L}\}$ as a global model for Γ^∞.

The only unusual part is to show that each $(\Gamma^\infty)_l$ is maximal consistent. Consistency has been treated above, so suppose A is a local sentence of l and A is *not* in $(\Gamma^\infty)_l$. Then we cannot have A^l in Γ^∞ by (Enter) and (Exit). Hence $\neg(A^l)$ is in Γ^∞, and by rule (DT), $(\neg A)^l$ is in in Γ^∞. Finally by (Enter), $\neg A$ is in $(\Gamma^\infty)_l$. The rest of the Henkin style completeness proof follows as before.

Theorem 3 (Completeness). *The system of rules and axioms for Tiered FOL is complete (both locally and globally).*

For decidablility we restrict ourselves to systems in which the first order logics in every locality are decidable and there is only a finite number of localities in our system.

Theorem 4. *If 1. the global system has only a finite number of localities and the strictly local theories at each locality are decidable, and 2. there is a finite number of bridge axioms and rules, them the global system is decidable.*

Proof. To decide whether

$$\Gamma \vdash_\gamma \bigwedge\{\Xi^l : l \text{ is a locality}\} \rightarrow \Phi$$

express the sentence as a propositional combination of basic global sentences.[10] Now use the truth values of these basic global *sentences* to compute the value of the sentence.

4   CASL

In the previous part of the paper there was no direct interaction between localities except in the presence of bridge axioms or rules, or overlapping languages (cf. Remark 2). There are other possibilities dealing with structured localities [10]. Here we consider algebraic specifications as the localities and build new specifications from old ones.

Each locality l will now be a specification described in a language such as CASL [6,2]. There is no necessity for these specifications to be finite but in practice we would expect them to be so.

---

[8] The restriction to global formulae is merely for convenience. Replace local formulae Δ in a locality l by the set of global formulae $\{\varphi^l : \varphi \in \Delta\}$ and use the rules (Enter) and (Exit).

[9] The proof is as usual except that we have to ensure consistency across localities. This is ensured by the model commonality requirement, see Remark 2 above.

[10] See Remark 1 for the definition of Ξ^l.

CASL stands for "Common Algebraic Specification Language", see [6,2]. It was designed by the Common Framework Initiative (CoFI) for algebraic specification and development. It is a tool for specifying the modular and functional requirements of software, and has first order logic as its base language and as such it may be used for for tier 0. A good overview of CASL from an applied logic standpoint may be found in [18] but we give a very brief review of CASL here. From an ontology point of view, there is a strong reason to use CASL-type languages as ontology languages, primarily because the operations provided by CASL flow over to the operations one might want to do to ontologies, e.g. translate one to another (**with** operation), combine them (**and** operation), hide some parts (**hide** operation), or extend them (**then** operation).

CASL builds other specifications from *basic specifications*. A basic specification is an ordinary first order many-sorted logic of the form $Sp = < \Sigma, Ax >$, where Σ is the *signature* which comprises sorts, functions and predicates, Ax is a set of axiom formulae whose members come from the set of well formed formulae of SP. Models for CASL specifications are ordinary many-sorted models for first order logic. We denote the set of models of SP by $Mod(SP)$.

CASL Algebraic Operations. CASL provides algebraic operations for building specifications. One starts with basic specifications and then uses the operations of translation, union, extension and hiding, which we briefly describe below. We use the *architectural specifications* of CASL so that we preserve the categorical structuring of the set of specifications. In practice this means that we have no problems of clashes of names.

When one views a CASL specification as a description of a theory i.e. a locality or ontology [15], then we readily have ontology operations at our finger tips. The operations that may be performed on CASL specifications are defined by *specification expressions* in CASL literature.

Structured specifications are ways of combining basic specifications. Fuller details of all our constructions may be found in the CASL Manual [6] or [18].

Translation is simply the renaming of constants, predicates and functions in a specification. Formally a translation is the inductive closure of a symbol mapping ρ, which maps the symbols of SP to another specification, preserving sorts, etc..[11] This is written in CASL as SP **with** ρ.

In CASL the union of two specifications (possibly with some amalgamation) is achieved in such a way that the union specification is a conservative extension of the two given specifications[12] and, moreover, the models of the union are always such that they have reducts that are models of the originally given specifications, see e.g. [18] or [7].

Formally we proceed as follows. The *amalgamated union* of two specifications, written SP_1 **and** SP_2 is defined as the pushout in the following diagram.

---

[11] If symbols are in SP but not in the domain of ρ we make the convention that they are left unchanged. However, we also insist that this is done in such a way that there is no clash of names.

[12] I.e. no new sentences in the language of either specification are provable from the theory of the union specification.

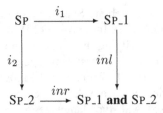

Extensions are defined in a very similar way to unions except that we can extend by a partial specification. The extension of SP by SP_EXT is denoted as SP **then** SP_EXT For examples, see [18].

Hiding may perhaps be regarded as an opposite of taking extensions. Given a SP and a symbol list SL, the operation SP **hide** SL cuts down the signature of SP to SP **hide** SL which is that of SP omitting the symbol list SL. The models of SP **hide** SL are $Mod(\text{SP } \textbf{hide } SL) = \{m|_\sigma : m \in Mod(\text{SP})\}$ where σ is the signature of SP **hide** SL, see [18].

5   The Tiered CASL System

Syntax. We use architectural specifications as localities and we recall that a specification has a language inside it and this we designate as the "local language". We then follow the same model as before (see Section 2). In Tiered CASL, the *strictly local formulae* are simply first order formulae in the syntax of the locality SP. *Basic global formulae* are strictly local *sentences* annotated by superscripts that are specification names. Thus a strictly local sentence, φ, is lifted to the global level as a basic global sentence φ^{SP}. *Local formulae* in a specification (locality) SP are the inductive closure of the strictly local formulae and the global formulae.

Examples: We assume that the language of locality SP_1 has *only* the predicate letter P, that the locality SP_2 has *only* the predicate letters P_1 and P_2, and that locality SP_3 has only the predicate letter Q.

Strictly local formulae: $\forall x : s \bullet P(x)$ in the locality SP_1 and $\forall x : s \bullet P(x)$ in the locality SP_1 **and** SP_2; $\forall x : s \bullet (P_1(x) \to P_2(x))$ in the locality SP_2; and $\exists y : s \bullet P_2(y)$ in the locality SP_1.

Global formulae: $(\forall x : s \bullet P(x))^{\text{SP-1}}$, and
$((\forall x : s \bullet P(x))^{\text{SP-1 and SP-2}} \to (\exists x : s \bullet P_1(x))^{\text{SP-2}})$, $(\exists y : s \bullet P_2(y))^{\text{SP-2}}$.

Local formulae for the locality SP_2:
$(\forall x : s \bullet P(x))^{\text{SP-1}}, \forall x : s \bullet (P_1(x) \to P_2(x)), \exists y : s \bullet ((\forall x : s \bullet P(x))^{\text{SP-1}} \to P_2(y))$,
$(\forall x : s \bullet P(x))^{\text{SP-1}} \to (\forall x : s \bullet Q(x))^{\text{SP-3}}$ and $[(\forall x : s \bullet P(x))^{\text{SP-1}}]^{\text{SP-3}}$.

The first formula, $(\forall x : s \bullet P(x))^{\text{SP-1}}$ is local (even in the locality SP_2) because it is a global formula; the second is local in SP_2 because it is a strictly local formula of SP_2; and the third is local in SP_2, because it is a first order logic combination of a strictly local (and therefore also local) formula, $P_2(y)$, of SP_2, and a global (therefore also local) formula, $(\forall x : s \bullet P(x))^{\text{SP-1}}$. The fourth is a mixture of global formulas from SP_1 and SP_3. The last one is a local formula for it is derived from a global formula.

Examples of bridge axioms: $(\forall x : s \bullet P(x))^{\text{SP-1}} \to (\forall x : s \bullet Q(x))^{\text{SP-3}}, (P(a))^{\text{SP-1}} \to (Q(b))^{\text{SP-3}}$, and $(\exists x : s \bullet P(x))^{\text{SP-1}} \leftrightarrow (\exists x : s \bullet Q(x))^{\text{SP-3}}$.

We define derivations as before using the same schemata, but add rules for structured specifications, viz., the rules of the global system Tiered CASL are given by first order logic at the local level and propositional calculus at the global level with the slightly modified transfer rules in Figure 2, and the structural rules in Figure 3.

Consistency, strictly local, global and local is defined exactly as above in Section 2, and as before we assume that all of the basic specifications, SP, in our system are consistent.[13]

$$\frac{\Gamma \vdash_\gamma A^{\text{SP}}}{\Gamma, \text{SP} \vdash_\lambda A} \ (\text{Enter}) \qquad \qquad \frac{\Gamma, \text{SP} \vdash_\lambda A}{\Gamma \vdash_\gamma A^{\text{SP}}} \ (\text{Exit})$$

provided A is a local SP formula and Γ is a set of global formulae.

Fig. 2. The transfer rules for Tiered CASL: going from global to local and *vice versa*

$$\frac{\Gamma \vdash_\gamma A^{\text{SP}}}{\rho(\Gamma) \vdash_\gamma \rho(A)^{\text{SP with } \rho}} \ (\text{trans})$$

If SL is any symbol list
$$\frac{\Gamma \vdash_\gamma A^{\text{SP}}}{\Gamma \vdash_\gamma A^{\text{SP hide } SL}} \ (\text{hide})$$
provided the signature of $\{A\} \cup \text{SP}$ does not contain SL.

$$\frac{\Gamma \vdash_\gamma A^{\text{SP\_1}}}{\Gamma \vdash_\gamma inl(A)^{\text{SP\_1 \& SP\_2}}} \ (\text{union}_1) \qquad \frac{\Gamma \vdash_\gamma A^{\text{SP\_2}}}{\Gamma \vdash_\gamma inr(A)^{\text{SP\_1 \& SP\_2}}} \ (\text{union}_2)$$

$$\frac{\Gamma \vdash_\gamma A^{\text{SP\_1}}}{\Gamma \vdash_\gamma inl(A)^{\text{SP\_1 then SP\_EXT}}} \ (\text{ext}_1) \qquad \frac{\Gamma \vdash_\gamma A^{\text{SP\_EXT}}}{\Gamma \vdash_\gamma inr(A)^{\text{SP\_1 then SP\_EXT}}} \ (\text{ext}_2)$$

Fig. 3. The structural rules involving specifications

Semantics. Again we define the semantics of our system, strictly local, global and local, exactly as in Section 2, except that the models we are now considering are many-sorted. Global models $\mathfrak{M}$ will now be sets of models $\mathfrak{m}_{\text{SP}}$ such that SP is a specification in our system. However, because of the structural rules of Figure 3, such a global model $\mathfrak{M}$ must also include models for all the specifications constructed from the basic specifications using translation, union, extensions and hiding.

The soundness of Tiered CASL is proved as before, except that we also have to consider the structural rules and the specifications that can be constructed from the basic ones. We take (union$_1$) as an example.

Assume $\mathfrak{M} \models_\gamma A^{\text{SP\_1}}$, then the local model $\mathfrak{m}_{\text{SP\_1}}$ in $\mathfrak{M}$ is such that $\mathfrak{m}_{\text{SP\_1}} \models_\lambda A$. Let $\mathfrak{m}_{\text{SP\_2}}$ be any model of SP_2. Then the amalgamated union of $\mathfrak{m}_{\text{SP\_1}}$ and $\mathfrak{m}_{\text{SP\_2}}$ is a model of $inl(A)$. Since this is true for all such pairs of models we have $\Gamma \models_\gamma inl(A)^{\text{SP\_1 \& SP\_2}}$. The other cases are similar.

---

[13] The categorical nature of the construction of the non-basic specifications guarantees that all of the specifications constructed are consistent (provided the basic ones are!).

The initial idea of the completeness proof was inspired by that in Section 3. However, because changes in basic specifications cause changes in any structural specification constructed from them, we have to modify our strategy.

First recall that localities (i.e. specifications) may be built from other localities, so when we add witnesses to each basic specification, SP to get a new basic specification SP+, this expands the specification at that locality in a trivial way, but it carrries over to constructed specifications, so that for SP_1 and SP_2 we now have extensions SP_1+ and SP_2+, to which we further add new constants to obtain $(\text{SP\_1+ \textbf{and} SP\_2+})+$. Similarly for specifications using the other operations of Section 4: extension, hiding and translation.

In the procedure leading to the construction of the model, the cases for the basic sets of rules go as before. We give just one example for the structural rules.

(union$_1$) Assume that $A^{\text{SP\_1+}} \in \Gamma^{\infty}$. We now test if $inl(A)^{(\text{SP\_1+ \& SP\_2+})+} \in \Gamma_{\infty}$. Suppose not, then we have $\neg(inl(A)^{(\text{SP\_1+ \& SP\_2+})+}) \in \Gamma_{\infty}$ by maximality. Therefore $inl(\neg A)^{(\text{SP\_1+ \& SP\_2+})+} \in \Gamma_{\infty}$ since negation commutes with the locality, by rules (DT) and (K), and also commutes with inl by the definition of inl. But then by (hide) $inl(\neg A)^{inl(\text{SP\_1+})} \in \Gamma_{\infty}$ and $(\neg A)^{\text{SP\_1+}} \in \Gamma_{\infty}$ by (trans) using the (partial) inverse of inl. Finally using (DT) once more $\neg(A^{\text{SP\_1+}}) \in \Gamma_{\infty}$ which is a contradiction. $\qquad \square$

Now, for each specification SP+ we construct a local model $\mathfrak{m}_{\text{SP+}}$ (which will automatically give a model for SP) as in Section 3, and then the global model is $\mathfrak{M} = \{\mathfrak{m}_{\text{SP+}} : \text{SP is a specification}\}$.

Theorem 5 (Completeness of Tiered CASL). *The system of rules and axioms for Tiered CASL is complete (both locally and globally), i.e. if, for every global model $\mathfrak{M}$ and every global sentence Φ we have $\mathfrak{M} \models_{\gamma} \Phi \leftrightarrow \vdash_{\gamma} \Phi$, and similarly for local sentences for each specification.*

6 Future Work

We have described a scheme that provides for global communication between agents in different localities, possibly with different logics, but certainly with different languages. In doing so we have allowed one locality to influence another by *bridge axioms* and *bridge rules*. The new range of bridging allows for much more complex interactions than those in e.g. [11] and [3], since two (or more) localities may affect what happens in another locality.[14] Further, the lifting axioms of [16] and the bridge rules of [11] (and cf. also [4]) are expressible within our systems and do not need to be extraneous as they were in that earlier work.

We have proved completeness and consistency results for a basic system, Tiered First Order Logic, and also for a system, Tiered CASL, which allows the localities to be structured specifications in CASL.

For a practical implementation of our scheme we have built software where the local logic is PROLOG and the global logic is propositional calculus.

---

[14] In the thesis of the first author [8] the bridge rules based on [11] and [3] have been directly simulated, but also strengthened in a description logic context.

There remains one general area that particularly requires further investigation. How do we do quantification at the global level? Quantification over localities was developed in [5], and we see no difficulty in extending our work in that direction. However we would like to imitate Borgida's $C^k \sqsubseteq D^l$ directly , but it does not seem to make sense to write $\forall x (C(x)^k \rightarrow D(x)^l)$ since some elements in locality k may not be in locality l. So we remain like the ancient Chinese mathematician, Liú Huī (see p. 74 of [14]), "... not daring to guess, [we] wait for a capable person to solve it."

References

1. Akman, V., Surav, M.: Steps toward formalizing context. AI Magazine 17(3), 55–72 (1996)
2. Bidoit, M., Mosses, P.D.: CASL User Manual: Introduction to Using the Common Algebraic Specification Language. Springer, Heidelberg (2004)
3. Borgida, A., Serafini, L.: Distributed description logics: Assimilating information from peer sources. In: Spaccapietra, S., March, S., Aberer, K. (eds.) Journal on Data Semantics I. LNCS, vol. 2800, pp. 153–184. Springer, Heidelberg (2003)
4. Bouquet, P., Serafini, L.: On the Difference between Bridge Rules and Lifting Axioms. In: Blackburn, P., Ghidini, C., Turner, R.M., Giunchiglia, F. (eds.) CONTEXT 2003. LNCS, vol. 2680, pp. 80–93. Springer, Heidelberg (2003)
5. Buvač, S., Buvač, V., Mason, I.A.: Metamathematics of contexts. Fundamenta Informaticae 23(2/3/4), 263–301 (1995)
6. COFI: CASL, The Common Algebraic Specification Language, Summary. CoFI Language Design Task Group on Language Design (March 25, 2001),
   http://www.brics.dk/Projects/CoFI/-Documents/-CASL/-Summary/
   (2001)
7. Cengarle, M.V.: Formal Specifications with Higher-Order Parametrization. PhD thesis, Ludwig-Maximilians-Universität, München (1994)
8. Cruz, R.P.: Tiered Logic with Applications to Contextualizing Logics. PhD thesis, Monash University, Melbourne, Australia (in preparation)
9. Cruz, R.P., Crossley, J.N.: Tiered logic method for assisting agents. Technical Report Technical Report 2008/232. Monash University, Clayton School of Information Technology, Clayton, Victoria, Australia,
   http://www.csse.monash.edu.au/publications/2008/
   tr-2008-232-abs.html
10. Gabbay, D., Nossum, R.T.: Structured contexts with fibred semantics. In: Working Papers of the AAAI Fall Symposium on Context in Knowledge Representation and Natural Language, pp. 48–57. American Association for Artificial Intelligence, Menlo Park (1997)
11. Ghidini, C., Serafini, L.: Distributed first order logics. In: Baader, F., Schultz, K.U. (eds.) Frontiers of Combining Systems 2, pp. 121–139. Research Studies Press/Wiley, Berlin (1998)
12. Giunchiglia, F., Ghidini, C.: Local models semantics, or contextual reasoning = locality + compatibility. Artificial Intelligence 127(2), 221–259 (2000)
13. Henkin, L.: The completeness of the first-order functional calculus. Journal of Symbolic Logic 14, 159–166 (1949)
14. Yǎn, L., Shírán, D.: Chinese Mathematics: A Concise History. Oxford University Press, Oxford (1987); Translated from the Chinese by J.N. Crossley and A.W.-C. Lun

15. Lüttich, K., Mossakowski, T.: Specification of ontologies in CASL. In: Varzi, A.C., Vieu, L. (eds.) Proceedings of the International Conference on Formal Ontology in Information Systems (FOIS 2004), pp. 140–150. IOS Press, Amsterdam (2004)
16. McCarthy, J., Buvač, S.: Formalizing Context (Expanded Notes). In: Buvač, S., Iwańska, Ł. (eds.) Working Papers of the AAAI Fall Symposium on Context in Knowledge Representation and Natural Language, pp. 99–135. American Association for Artificial Intelligence, Menlo Park (1997)
17. Nossum, R.T.: A decidable multi-modal logic of context. J. Applied Logic 1, 119–133 (2003)
18. Poernomo, I.H., Crossley, J.N., Wirsing, M.: Adapting proofs-as-programs. Springer, New York (2005)
19. Serafini, L., Bouquet, P.: Comparing formal theories of context in AI. Artificial Intelligence 155, 41–67 (2004/2005))

HomeManager: Testing Agent-Oriented Software Engineering in Home Intelligence

Ambra Molesini[1], Enrico Denti[1], and Andrea Omicini[2]

[1] ALMA MATER STUDIORUM—Università di Bologna
viale Risorgimento 2, 40136 Bologna, Italy
ambra.molesini@unibo.it, enrico.denti@unibo.it
[2] ALMA MATER STUDIORUM—Università di Bologna a Cesena
via Venezia 52, 47023 Cesena, Italy
andrea.omicini@unibo.it

Abstract. Ambient Intelligence is an interesting research application area for Multi-agent Systems, in general, and for Agent-oriented Software Engineering, in particular. In this paper, we focus on the methodological support that agent-oriented methodologies can provide to such kind of systems: we discuss Home-Manager, an application for the control of an intelligent home designed through the SODA agent-oriented methodology. There, the house is seen as an intelligent environment made of independent, distributed devices, each equipped with an agent to support the user's goals and tasks.

1 Introduction

The concept of Ambient Intelligence (AmI) introduces a vision of the Information Society where the emphasis is on greater user-friendliness, more efficient services support, user-empowerment, and support for human interaction [1]. In this scenario, an "intelligent environment" is a space that contains myriad devices that work together to provide users access to information and services [2,3]. So, people are surrounded by intelligent intuitive interfaces that are embedded in all kinds of objects and an environment that is capable of recognising and responding to the presence of different individuals in a seamless, unobtrusive and often invisible way. As noted in [4], AmI implies several challenges concerning the world of physical resources (sensors, actuators), the existence of many forms of task or goal interactions, a natural physical and functional distribution of control, a requirement for actions to be taken in given time intervals, and the integration of potentially conflicting preference specifications.

Other features include *(i)* the adoption of suitable coordination models to manage the physical resources, *(ii)* the support for the security techniques – access control, intrusion detection, etc. –, *(iii)* the support for richer interactions with the users, and *(iv)* a deeper understanding of the structure of the physical space. Due to these requirements, autonomy, distribution, adaptation and proactiveness emerge as key features of the entities populating the intelligent environment [5]. Such features naturally lead to consider agents and Multi-Agent Systems (MASs) as effective metaphors for system design and implementation in AmI scenarios [6].

J. Filipe, A. Fred, and B. Sharp (Eds.): ICAART 2009, CCIS 67, pp. 205–218, 2010.
© Springer-Verlag Berlin Heidelberg 2010

Several works exploiting MASs in the context of home intelligence (or smart home) – an AmI specific scenario – can be found in the literature—among these, IHome [4], C@sa [5] and MavHome [7]. There, the house is seen as a MAS where agents perceive the environment by means of sensors, and act upon it by means of actuators. However, these works mainly present the application from the implementation viewpoint, providing only few guidelines from the methodological viewpoint: so, the underlying process guiding designers from the requirements analysis to the design choices – in particular, to select an architecture among all the possible ones – is somehow hidden and difficult to understand.

In this paper we adopt a different approach, focussing specifically on the system design process. In particular, we discuss how SODA [8], an agent-oriented software engineering (AOSE) methodology, can support the analysis and design of HomeManager, an agent-based application for the control of an intelligent home. Section 2 presents our reference scenario and discusses its main requirements, while Section 3 is devoted to the analysis, the design and an outline of the first prototype of HomeManager. Discussion and comparison with some relevant related work are reported in Section 4. Conclusions and future work follow in Section 5.

2   Scenario

We consider a home automation application, whose goals are limiting the overall energy consumption while supporting people's activities inside the house [9]. The application is supposed to mediate between the desire for higher comfort and the service cost, respecting both technical constraints and users' specifications; it should also be controllable both locally and remotely – e.g. via short messages on PDAs or mobile phones – in a transparent way.

Two basic user categories can be identified: people stably living in the house, and visitors. The latter cannot operate on the system, but should be enabled to exploit some support functionalities. Inhabitants, instead, may have different authorisation levels. So, at least three user types can be identified:

- *administrators* can specify the management policies of any all the systems aspects, and decide (other) users' privileges;
- *standard users* can express their preferences and give commands via SMS, but cannot act on power consumption nor can they change the privilege level of other users;
- *visitors* are occasional, unregistered users, which can only receive basic assistance.

All users can specify their preferences about the available devices—e.g. the desired temperature in a given room, the automatic switch-on of TV when entering the room, etc.; the system should do its best to meet such requirements, adopting suitable criteria to solve possible conflicts. For instance, different priorities could be assigned to different users based on temporal precedence (i.e., who asked first), or following ad-hoc policies imposed by administrators (e.g., a hi-fi stereo and a TV set cannot be switched on in the same room at the same time). The resulting plan will take care both on users' preferences and policies, and power consumption policies: for the sake of simplicity we assume that administrators can choose whether to specify either the maximum power consumption allowed, or the period of the day where to concentrate most of the power consumption. Of course, many other strategies could be devised, without affecting the structure of the envisioned scenario.

2.1   Problem Analysis

The environment to be controlled is a family house: its rooms are supposedly provided with input/output sensors for identifying each (registered) user upon entry/exit into/from that room, via suitable identification techniques; unrecognised users should be offered the chance to identify explicitly via suitable system terminals. If they remain unidentified, they should be treated as visitors, with only a minimal set of actions allowed (such as, for instance, turning on/off the room light or the radio). These assumptions make it possible to know the exact position of each user at any time, which is needed for the system to perform correctly (e.g., to turn on the radio or TV in the right room).

Given that the house is a private building, we choose not to limit the number of people that can stay in a room at any time, and to leave free access to any room even without identification: the idea behind this assumption is that the system should assist its users in a constructive and pro-active way, rather than complicating their life with annoying constraints. This choice does not affect generality, since adding further security requirements has no impact on the other system requirements—specifically, on the user preferences and on the system planning choices.

As outlined in the requirements, the system should support users in three ways:

- satisfying the users' requests whenever possible, and trying to anticipate their needs by suitably pre-programming the available devices according to their preferences (e.g. room temperature);
- allowing users to give commands both directly and remotely, via short messages;
- minimising the house total power consumption.

Of course, several kinds of conflicts could arise: different users might express conflicting desires (for instance, one might want to turn on the heater, another the air conditioner), or their requests might involve more power than it is actually available. The conflict resolution policy is to be determined by administrators, and should be able to define priorities in a clear way. In this case, it should at least:

- decide whether to keep into higher consideration either the power consumption or the user requests;
- give different priorities to the different users—for instance, assigning higher priority to the (more tired) family member who comes back home later from work;
- keep into account the temporal order of users' requests;
- allow the system to postpone some non-critical activities when needed—for instance, in order to keep the overall power below the limit.

In order to carry out its management policy, the system must obviously operate on the available devices—to turn them on/off, adjust their parameters, etc; our hypothesis is that such devices are never accessed directly, but, by asking them to perform some services via some suitable interaction protocol. Each device is supposed to handle its low-level details, so that the higher-level coordination system needs not be aware of them, also enhancing scalability.

2.2 SODA

SODA (Societies in Open and Distributed Agent spaces) [10,11] is an agent-oriented methodology for the analysis and design of agent-based systems, which adopts the Agents & Artifacts (A&A) meta-model [12] and introduces *layering* as the main tool for scaling with the system complexity, applied throughout the analysis and design process [13].

The SODA abstractions (explained below) are logically divided into three categories: *i)* the abstractions for modelling/designing the system active part (task, role, agent, etc.); *ii)* the abstractions for the reactive part (function, resource, artifact, etc.); and *iii)* the abstractions for interaction and organisational rules (relation, dependency, interaction, rule, etc.). Moreover, the SODA *process* is organised in two phases, structured in two sub-phases: the *Analysis phase*, which includes the Requirements Analysis and the Analysis steps, and the *Design phase*, including the Architectural Design and the Detailed Design steps. Each sub-phase models (designs) the system exploiting a subset of SODA abstractions: in particular, each subset always includes at least one abstraction for each of the above categories – that is, at least one abstraction for the system active part, one for the reactive part, and another for interaction and organisational rules.

Figure 1 shows an overview of the methodology structure: as we will show in the case study (Section 3), each step is technically described as a set of relational tables (listed in Figure 1 for completeness).

3 Home Manager

In this section we present the analysis (Subsection 3.1), the design (Subsection 3.2) and the first prototype (Subsection 3.3) of HomeManager. Due to the obvious limitations in space, only a few tables are actually reported in the article.

3.1 Analysis

Requirements Analysis. The first SODA step is the Requirements Analysis, where the system requirements and the world of the legacy systems are identified. Figure 2 shows the system requirements at the core layer – SODA supports a layering principle: a system can be seen at different levels of detail – so that only three coarse-grained requirements are present, one for each "macro" category identified: namely users, plans/policies, and devices. The legacy systems devised out are the devices placed in the rooms: each is in some relation with the three requirements above. This coupled relationship between requirements and legacy systems is tied to the particular kind of system, as in the home automation scenario the devices represents a very important portion of the system to be controlled.

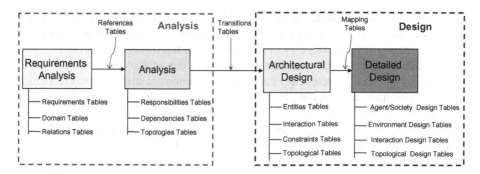

Fig. 1. An overview of the SODA process

Requirement	Description
Users management	management of the user data and preferences
Plans management	management of the plan
Devices management	management of the devices

Fig. 2. Requirement table

Analysis. The transition between Requirements Analysis and Analysis is expressed by the References Tables: as an example, Figure 3 reports the mapping between the requirements and the tasks they generate. Even though such tasks are more fine-grained than requirements, they are still represented at a high level of abstraction, so they could be in-zoomed in a more detailed layer.

Requirement	Task
Users Management	show_data, show_preferences, update_preference, add_user, canc_user, update_user
Plans Management	show_plane, show_policy, load_policy, accomplish_policy, elaborate_plan, execute_plan, detect_conflict, solve_conflict, check_request, check_activity
Devices Management	show_status, add_device, remove_device, check_people

Fig. 3. Reference Requirement-Task table

During the Analysis step also the environmental functions that derive both from requirements and legacy-systems and from the topological structure of the environment are individuated. This latter is already identified by both the physical structure of the controlled house and the position of the devices inside the rooms. In this step the dependencies among the tasks and functions are devised out, too. These relationships are exploited to determine the interaction spaces of the entities, the rules that govern interactions, and the related social aspects like organisation management, coordination and system security.

The reason why dependencies are derived from requirements, legacy systems and topology is that some access control rules are often expressed implicitly as topological "requirements"—as for instance in "The double room has a private bathroom". If we analyse such a requirement, we see that there are two connected rooms – a double room and a bathroom – that represent a portion of the environment structure, and recognise a clear, yet implicit, access control rule: "only the people allowed to enter the double room can access the bathroom". In turn, this can be seen as a prohibition for visitors to enter the bathroom, leading to an access control rule like "the role Visitor is not allowed to enter in the double room bathroom", ready to be translated into an RBAC (Role-Based Access Control) [14] rule.

Coordination plays a key role in this kind of applications, because the right "implementation" of the plan derives from the coordination of the different entities that mediate between the users preferences and the limitation of the energy consumption. The next subsection focuses on the design of such rules.

3.2 Design

Architectural Design. During the Architectural Design, the system is designed in terms of roles, resources, actions, operations, interactions, rules and spaces. These entities derive form the abstractions outlined in the previous step: roles are responsible for the achievement of tasks by executing actions, while resources offer the functions outlined in the previous step by providing the corresponding operations. Typically the task-role and function-resource mappings are not one-to-one, as a role/resource is able to complete/accomplish several different tasks/functions. For example, the role "Preference Manager" is responsible for the "show_preference" and "update_preference" (Figure 3) tasks. The SODA spaces are easily derived from the topology, and map one-to-one the physical rooms that compose the controlled house.

As noted in the Analysis, the core of the system is the coordination among the entities, which is captured by the SODA abstract entities "interactions" and "rules" derived from the dependencies. So, in order to design such aspects, a reference model for coordination [15] has to be chosen. Given the intrinsic features of this application, where a multiplicity of user preferences have to be considered altogether, prioritised and enforced at any time, while users enter/exit the house rooms, an approach based on mediated interaction seems more suitable than, for instance, a bid-based approach. In fact, mediated interaction makes it possible to delegate the coordination medium for policy store and enforcement, thus reducing the time needed to enforce the policy, and freeing roles from taking care of the coordination burden. Also, any policy change affects only a specific place, with no impact on roles—and therefore transparently with respect to them.

This model leads to design interactions as if no policies exist, delegating their enforcement and the access control to the design of the coordination laws expressed by the SODA rules. More concretely, if a user asks HomeManager to switch on the dishwasher, the request is managed by the role designed for achieving this task ("request dispatcher") by sending the request to the "dishwasher manager" role. Such a delivery, however, is not performed directly: rather, it is managed by a coordination medium, that is supposed to react according to the policy and thus dispatching the request to the dishwasher manager or send a notification to the request dispatcher if the action is not allowed by the policy (for instance, if the energy consumption is close to the top limit— see the "MaxEnergy Rule" in Figure 4). The coordination medium is "transparent" for request dispatcher and dishwasher manager, as it represents a sort of "infrastructure layer" not directly perceived by roles.

For space reasons, Figure 4 shows only an excerpt of the Rule table that lists and describes all rules—namely, some representative rules for each "macro category". So, the first block includes the rules devoted to the general house policy – energy consumption and users priority –, the second represents the rules concerning the use of devices, while the third block lists the access control rules.

Rule	Description
MaxEnergy Rule	The electrical power cannot exceed 3 KW
Priority Rule	The preferences of the priority user have to be satisfied
Default Rule	Turn on lights when Visitor goes inside room
Heating Rule	Heating is more priority then other devices
Visitor Rule	Visitor cannot access to the system
Role Rule	User cannot modify its role
User Rule	User can modify only its data and preferences
. . .	. . .

Fig. 4. Rule table

Moving from Architectural Design to Detailed Design, the engineer should now operate a carving operation, so as to choose the most adequate representation level for each architectural entity: this leads to depict one (detailed) design from the several potential alternatives outlined by the layering. In our case, since the core layer is placed at a high level of abstraction, and no in-zoom operation is present, the carving operation is simply not necessary, and the core layer becomes the only layer of the Detailed Design.

Detailed Design. Detailed Design is expressed in terms of agents, agent societies, composition, artifacts, aggregates and workspaces for the abstract entities; interactions, in their turn, are expressed by means of concepts such as uses, manifests, speaks to and links to.

In HomeManager, agents play the roles individuated in the previous step, while resources are mapped onto suitable environmental artifacts—a kind of artifact [16] aimed at wrapping the physical devices that constitute the MAS external environment, or realising functions derived by requirements but not available in the external environment. Since the design level is very abstract, we devise out just one agent society. Moreover, spaces and their connections are then mapped one-to-one onto workspaces and the respective workspace connections. Interactions are mapped onto different detailed abstractions in order to capture the different "nature" of the interaction itself: more precisely, agent-to-agent interaction is mapped onto *speaks to*, agent-to-artifact onto *uses*, artifact-to-agent onto *manifests*, and artifact-to-artifact onto *links to*.

In addition, in SODA each agent is associated to an individual artifact [16] that handles the interaction of a single agent within a MAS, and is used to shape of admissible interactions of MAS agents. Such individual artifacts play an essential role in engineering both organisational and security concerns: in fact, access control rules (third block of rules in Figure 4) are mapped precisely onto agents' individual artifacts. The rules in the first and second block are mapped onto social artifacts [16], as they rule the social interactions—even though indirectly, since they technically mediate interactions between individual, environmental, and possibly other social artifacts.

Social artifacts in SODA play the role of the coordination artifacts – i.e., coordination media – that embody the rules around which agent societies are built. Among the possible different social artifacts "organisations" (discussed in Section 4), we choose the one where each workspace is associated to a social artifact that rules the agents and the devices inside a room so as to manage the user priorities and control the access to the devices. We also easily individuate an aggregate of such social artifacts that can "enforce" the global rules of the house—like, for example, the "MaxEnergy Rule" in Figure 4. Figure 5 shows only an excerpt of the Artifact-UsageInterface table, which actually lists all the operations provided by each artifact in detail.

Artifact	Usage Interface
Room manager	receive_command send_command send_notification set_policy, get_policy get_plan, get_user...
Home manager	set_homepolicy, get_homepolicy send_command get_homeplan, get_userpos...
Device manager	switch_on switch_off...
...	...

Fig. 5. Artifact-UsageInterface table

As mentioned above, this is a simplified version of the system we actually designed and implemented (Subsection 3.3). In fact, in a real scenario, each room should be expected to contain a multitude of agents and devices, which could not be reasonably managed by a single social artifact for obvious performance and reliability reasons—preventing bottlenecks, avoiding delays in the system responses, etc. So, the social artifact should rather be seen as an abstract view of an aggregate of other social artifacts, each enforcing some given kind of rules.

3.3 Prototype

Following the above analysis and design process, we realised a prototype simulator [9] for HomeManager based on the TuCSoN infrastructure [17,18]. TuCSoN is well

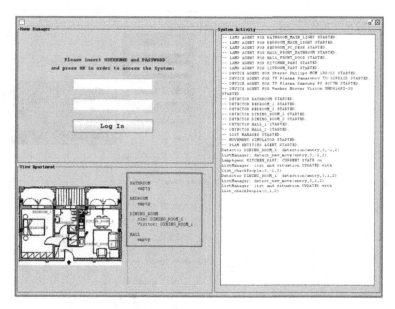

Fig. 6. The HomeManager main window

suited for this kind of application, since it provides "tuple centres" as coordination media, which allow a one-to-one map with the SODA social artifacts. In short, a tuple centre is a tuple space enhanced with behaviour specification: it is still perceived by agents as a standard tuple space [19], but its behaviour can be specified by suitable coordination laws. Since the behaviour specification of TuCSoN tuple centres is expressed in the ReSpecT language [12], the SODA rules enforced by the social artifacts take the form of suitable ReSpecT reactions.

In our prototype we simulated the control of a house as composed of four rooms: the main entrance, the dining room, the bedroom and the bathroom (Figure 6). Each room has several devices, like a washing machine in the bathroom, a hi-fi and a tv set in the dining room, another tv set in the bedroom, as well as various devices for light management. The main interface (Figure 6) is organised in three areas [9]: the login area (top-left) where users authenticate and modify the HomeManager parameters; the view-house area (bottom-left) which reports the map of the house (in the left side) and the current position of the people inside the house (right side): this information, highlighted in the figure by the red rectangle, is currently expressed as a textual description, but will be reported directly on the map in a future version; finally, the system activities area (right side) shows both the activities of the system and the messages sent by each device when working.

So, for instance, when an (authorised) user goes into the dining room, HomeManager reacts by signalling the presence of the user in the room, turning on the light if necessary, and applying the user's preferences – temperature, devices turned on/off – to that room. If there are two or more people inside the same room with different preferences, HomeManager reacts according to the preferences of the most important (priority) user balancing her/his preference with the energy consumption. A simple

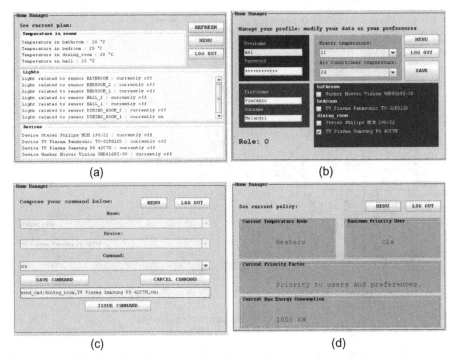

Fig. 7. Some screenshots of HomeManager: a) policy-management interface, b) profile-management interface, c) command interface and d) current-policy interface

default policy is applied to visitors (users with no profile), which just switches the light on/off when he/she enters the room, as no assumptions can be made about the visitor's preferences.

Figure 7 shows four further views of HomeManager. View (a) shows a screenshot of the policy-management interface: for each room, the GUI reports the temperature, the state of the lights and of the other room devices. View (b) shows the profile-management interface: there, each user can change his/her profile and preferences about the temperature and devices. However, he/she cannot change his/her role in the house, as this operation is reserved to administrators. The screenshot in view (c) is about the command interface: the authorised user operates on the system by choosing the room and the device (first two rows), and then issuing the desired command. HomeManager replies by starting a new action plan for the room where the device is placed. The GUI in view (d) shows the current policy for the selected room: in particular, the GUI reports the current temperature mode, the maximum priority user, the current priority factor and the current max energy consumption.

4   Discussion

The intelligent home scenario represents a good application domain for testing the MAS approach in general and the agent-oriented methodologies in particular. In fact, this scenario introduces several different challenges from the methodology viewpoint, such as, for instance, the management of the security aspects and the energy consumptions limitation, that should be balanced with the system fast reaction and user's comfort, wellness and entertainment.

In order to obtain the best balancing among these requirements, the methodology should support the design of suitable coordination models so that the agents can collaborate and fulfil the house control task. As highlighted in Subsection 3.2, mediated coordination seems the best solution for this kind of application. This choice leads to structure the design of interactions and coordination rules in a specific way, yet without imposing any specific "architecture" for the system. In particular, the model considers a "coordination medium", but makes no hypotheses on the entities that should "implement" this role.

In SODA the coordination rules are naturally mapped onto social artifacts, so the coordination artifacts are delegated to enforce the coordination rules; in IHome [4], instead, the centrally-managed resources (IHome also considers decentralised resources) are controlled by an "agentified" coordination medium. Coordination is managed by agents also in Cas@ [5] and Intelligent Room [3].

Even though both approaches present strengths and drawbacks, from the design-for-change perspective delegating coordination to a coordination artifact seems a better choice, as it leads to encapsulate a coordination service in a specific entity, allowing user agents to abstract from how the service is implemented. As such, a coordination artifact is perceived as an individual entity, but can be actually distributed on different nodes of the MAS infrastructure, depending on its specific model and implementation [20]. In addition, coordination artifacts are meant to support coordination in open agent systems, characterised by unpredictable events and dynamism. For this purpose, they have to support a specific form of artifact malleability—namely, they should allow their coordinating behaviour to be adapted and changed both by engineers (humans) willing to sustain the MAS behaviour, and by agents responsible for managing the coordination artifact. So, coordination artifacts can be seen as engineering abstractions for designing, building and supporting at runtime coordination in agent societies, suitably instrumenting their dynamic working environment [20].

Of course, one may also adopt agents working so as to mimic coordination media, but such an approach seems less flexible in this context, for a change in a house policy would imply a modification of the coordination rules and probably also of the coordination protocols between the coordinator agent and the coordinated agents, spreading the impact of the change nearly everywhere. Instead, a coordination artifact helps keeping such changes inherently encapsulated.

With respect to the choice of the social artifact "organisation", it should be noted that the design solution mentioned in Subsection 3.2 is not the only possible. In fact, we exploited a social artifact – or an aggregate of social artifacts – as the room manager, so that all the room managers form an aggregate that coordinates the whole house: however, a different architecture, not-so-tied to the physical environment structure, could also be possible. For instance, an organisation based on the system functionalities could be based on a social artifact devoted to coordinate the entertainment devices – tv set, hi-fi, ... –, another for heating and air conditioning, a third one for lights, etc.—all forming an aggregate to coordinate the whole house.

Currently, we do not have enough experimental results convincing us to prefer an architecture to another. We chose the first – a social artifact manager for each room – for the prototype mainly because it naturally bounds the "scope" of each social artifact to the physical structure of the environment, making it easier to define the coordination rules holding inside each room.

5 Conclusions and Future Work

HomeManager is an agent-based application for controlling an intelligent home. We discussed its analysis and design with special regard to the coordination and access control rules which motivate the architectural design choices, from the coordination model to the social artifact organisation. In particular, we stressed the relevance of a mediated coordination model and of delegating coordination to a coordination artifact, so that agents can exploit coordination as any other service, without being concerned about the possible changes in the house policies.

The experiment was an interesting testbed for AOSE in general, and for the SODA methodology in particular, highlighting some benefits as well as some limitations that we mean to address in the near feature. In particular, the tabular representation is clearly more suitable for an automatic tool than for a human designer, due to the large amount of tables to be filled in at each stage: so, we plan to develop tools for supporting designers in doing so in a consistent and complete way. Another interesting extension could be the definition – or the adoption – of a language for specifying SODA rules in a more precise and formal way, overcoming the implicit limitations of the natural language which is currently adopted in the tabular representation. We also plan to evaluate whether to enrich SODA with methods for the internal design of agents and artifacts—which are not yet covered, since SODA currently does not deal with intra-agent (and more generally with "internal") issues.

Further work will be devoted to improve the current version of the prototype, implementing all the access control checks and comparing the two architectures discussed in Section 4. Moreover, as a form of protection against thiefs, HomeManager could inform administrators by SMS/email if an unidentified visitor is in the house: RBAC policies of the role visitor should then be revised accordingly, splitting the visitor role in sub-roles such as "closely related", "friends", "authorised visitor", and "unknown". Of course, this would imply the installation of some user (possibly biometrical) authentication mechanism. Moreover, we plan to consider some fuzzy-set theory methods to find a better compromise between the profiles of different people, instead of the simple priority policy which is currently adopted, and to test our prototype in some real environment.

Acknowledgements. Authors would like to thank Dr. Claudia Fontan for her contribution to the project and her work in the prototype implementation.

This work has been supported by the *MEnSA* project (*Methodologies for the Engineering of complex software Systems: Agent-based approach*) funded by the Italian Ministry of University and Research (MIUR) in the context of the National Research 'PRIN 2006' call.

References

1. Ducatel, K., Bogdanowicz, M., Scapolo, F., Leijten, J., Burgelman, J.C.: Scenarios for ambient intelligence in 2010. final. IPTS – European Commission's Joint Research Centre (2001)
2. Brumitt, B., Meyers, B., Krumm, J., Kern, A., Shafer, S.A.: Easyliving: Technologies for intelligent environments. In: Thomas, P., Gellersen, H.-W. (eds.) HUC 2000. LNCS, vol. 1927, pp. 97–119. Springer, Heidelberg (2000)
3. Kulkarni, A.: Design Principles of a Reactive Behavioral System for the Intelligent Room. ACM/IEEE Bitstream: The MIT Journal of EECS Student Research, 22–26 (2002)
4. Lesser, V., Atighetchi, M., Benyo, B., Horling, B., Anita, R., Vincent, R., Wagner, T., Xuan, P., Zhang, S.X.: The UMASS intelligent home project. In: Etzioni, O., Müller, J.P., Bradshaw, J.M. (eds.) 3rd International Conference on Autonomous Agents (Agents 1999), pp. 291–298. ACM, New York (1999)
5. De Carolis, B., Cozzolongo, G.: C@sa: Intelligent home control and simulation. In: Okatan, A. (ed.) International Conference on Computational Intelligence (ICCI 2004), Istanbul, Turkey, pp. 462–465. International Computational Intelligence Society (2004)
6. Zambonelli, F., Omicini, A.: Challenges and research directions in agent-oriented software engineering. Autonomous Agents and Multi-Agent Systems 9(3), 253–283 (2004)
7. Cook, D.J., Youngblood, M., Heierman, E.O.I., Gopalratnam, K., Rao, S., Litvin, A., Khawaja, F.: MavHome: An agent-based smart home. In: 1st IEEE International Conference on Pervasive Computing and Communications (PerCom 2003), Washington, DC, USA, pp. 521–524. IEEE CS, Los Alamitos (2003)
8. SODA: Home page (2008), http://soda.apice.unibo.it//
9. Fontan, C.: Tecnologie ad agenti per una casa intelligente. Master's thesis, Università di Bologna, Italy (2006)
10. Omicini, A.: SODA: Societies and infrastructures in the analysis and design of agent-based systems. In: Ciancarini, P., Wooldridge, M.J. (eds.) AOSE 2000. LNCS, vol. 1957, pp. 185–193. Springer, Heidelberg (2001)
11. Molesini, A., Omicini, A., Denti, E., Ricci, A.: SODA: A roadmap to artefacts. In: Dikenelli, O., Gleizes, M.-P., Ricci, A. (eds.) ESAW 2005. LNCS (LNAI), vol. 3963, pp. 49–62. Springer, Heidelberg (2006)
12. Omicini, A.: Formal ReSpecT in the A&A perspective. ENTCS 175(2), 97–117 (2007)
13. Molesini, A., Omicini, A., Ricci, A., Denti, E.: Zooming multi-agent systems. In: Müller, J.P., Zambonelli, F. (eds.) AOSE 2005. LNCS, vol. 3950, pp. 81–93. Springer, Heidelberg (2006)
14. RBAC: American National Standard 359-2004 – Role Base Access Control home page (2004), http://csrc.nist.gov/rbac/
15. Papadopoulos, G.A., Arbab, F.: Coordination models and languages. Advances in Computers 46, 330–401 (1998)
16. Omicini, A., Ricci, A., Viroli, M.: *Agens Faber*. Toward a theory of artefacts for MAS. ENTCS 150(3), 21–36 (2006)
17. TuCSoN: Home page (2008), http://tucson.apice.unibo.it/

18. Omicini, A., Zambonelli, F.: Coordination for Internet application development. Autonomous Agents and Multi-Agent Systems 2(3), 251–269 (1999)
19. Gelernter, D., Carriero, N.: Coordination languages and their significance. Communications of the ACM 35(2), 97–107 (1992)
20. Omicini, A., Ricci, A., Viroli, M.: Coordination artifacts as first-class abstractions for MAS engineering: State of the research. In: Garcia, A., Choren, R., Lucena, C., Giorgini, P., Holvoet, T., Romanovsky, A. (eds.) SELMAS 2005. LNCS (LNAI), vol. 3914, pp. 71–90. Springer, Heidelberg (2006)

Developing Multi-Agent Systems through Integrating Prometheus, INGENIAS and ICARO-T

Antonio Fernández-Caballero and José M. Gascueña

Universidad de Castilla-La Mancha, Departamento de Sistemas Informáticos &
Instituto de Investigación en Informática de Albacete, 02071-Albacete, Spain
{caballer,jmanuel}@dsi.uclm.es

Abstract. A great number of methodologies to develop MAS systems have been proposed in the last few years. But, a perfect methodology that satisfies all the developer necessities cannot be found. This is the reason why different methodologies are studied to create a new one. In this article, a methodology that includes all steps from the capture of requirements to the implementation and deployment of an agent-based application is proposed. In first place, an Analysis Overview Diagram is created to obtain an initial sketch of the application. Afterwards, the model obtained - by following the two first stages proposed by Prometheus methodology – could be integrated into INGENIAS through UML-AT language. Next, the modeling goes on with INGENIAS. Finally, code is generated for the ICARO-T platform.

Keywords: Multi-agent systems, Agent development methodologies.

1 Introduction

Multi-agent systems (MAS) technology is adequate for developing open, complex, and distributed systems, and they offer a natural way of operating with legacy systems [19]. A great number of methodologies to develop MAS systems have been proposed in the last few years. Gaia, Tropos, MaSE, MESSAGE, Prometheus, and INGENIAS are just a few examples. Nonetheless, a unique methodology cannot be general enough to be useful for everyone without some level of customization [4]. Usually, techniques and tools proposed in different methodologies to provide a solution to the specific problem that is being approached are combined. The result is a new methodology fruit of combining several proposals of the analyzed methodologies. In fact, in the literature, methodologies can be found that are influenced by other methodologies that already were proposed previously. For instance, fragments of PASSI, Gaia, and Tropos are considered to define the MAR&A methodology [3].

In this article, INGENIAS is chosen as the basis, due to its recent direction towards model-driven development (MDD) [24] in order to define a new methodology to develop MAS. But, the two first stages proposed in Prometheus [21], namely *system specification* and *architectural design*, are previously integrated in order to solve some current deficiencies in INGENIAS (see section 2). The language used by Prometheus is different from the INGENIAS language. Therefore, in order to use INGENIAS, it is necessary to transform the model obtained with Prometheus into an

J. Filipe, A. Fred, and B. Sharp (Eds.): ICAART 2009, CCIS 67, pp. 219–232, 2010.

equivalent INGENIAS model. This transformation can be performed with language UML-AT [7], [8]. Later we propose to continue modeling with INGENIAS. Finally, code is generated for the ICARO-T platform [11]. The result of using the mentioned technologies, Prometheus, INGENIAS, UML-AT and ICARO-T, turns into a new methodology to develop MAS. The process followed in the methodology assists the MAS developer from the capture of requirements to the implementation and deployment of the application.

The article structure is as follows. Section 2 describes the contributions made by INGENIAS and the deficiencies that it presents. Methodologies Prometheus and IN-GENIAS, as well as the tools that support them, are compared in section 3. In section 4 the phases of the integrative methodology to develop MAS are proposed and described. Finally, some conclusions are offered.

2   Why Start with INGENIAS?

In the initial INGENIAS proposal [23] there are several contributions to develop MAS. First, it offers a meta-model to specify MAS. A MAS is considered from five complementary viewpoints: organization, agent, goals and tasks, interaction, and environment. Second, it adopts the unified software development process (USDP) [17] as a guideline to define the steps necessary to develop the elements and diagrams of MAS during the analysis and design phases. Third, INGENIAS Development Kit (IDK) is a tool that supports the methodology. IDK has integrated a set of utilities that allow model edition, verification, validation, and automatically generate code and documentation.

Now, INGENIAS is being reformulated in terms of the MDD paradigm [24]. Nowadays the use of model-driven engineering (MDE) techniques along the life cycle of software development is gaining more and more interest [28]. The key idea underlying this paradigm is that if the development is guided by models there will be important benefits in fundamental aspects such as productivity, portability, interoperability and maintenance. Therefore, in the MAS field, it seems quite useful to use a methodology such as INGENIAS, which supports this approach. There are some other works using MDE in the area of MAS [27], [18], [16], [10], among others.

Indeed, there are other reasons for studying the methodology INGENIAS and the tools created around. The INGENIAS engineer, connoisseur of the INGENIAS meta-model, can (a) define the meta-model for the domain of a concrete application, (b) personalize the IDK for a specific application domain, and, (c) create transformations to generate source code for the final platform on which the agents will run. There exist some previous experiences to adapt the INGENIAS language to more specific systems. For example, the IDK framework has been used to construct an editor for Holonic Manufacturing Systems. Also the INGENIAS language has been adapted for social simulation environments.

Unfortunately, in our opinion, the process followed in INGENIAS during the analysis and design phases of MAS is very complex and difficult to understand, because it is not clear how the different models are being constructed along the phases, despite the documented general guidelines. Moreover, INGENIAS does not provide any mechanism to discover which will be the agents of the system and their interactions. Thus, it is necessary to raise a process of alternative development that makes system development simpler. In order to make the MAS methodology easy to use for non expert people in the development of such systems, it is necessary that it offers a collection of detailed guidelines, including examples and heuristics, which help better understanding what is required in each step of the development process used in the methodology. These guidelines also serve as a help to the experts in MAS development. They will be able to transmit their experience to other users explaining why and how they have obtained the different elements (agents, interactions, etc.) of the agent-based application.

3   Comparison of Prometheus and INGENIAS

INGENIAS has several advantages as opposed to Prometheus (see Table 1): (a) it follows an MDD approach, (b) it facilitates a general process to transform the models generated during the design phase into executable code. The advantages of Prometheus can be used (following the process to discover which be the agents of the system and its interactions) to enhance INGENIAS. In Table 2 the Prometheus Design Tool (PDT) [22] and INGENIAS Development Kit (IDK) [15] tools are compared. It may be observed that PDT only has one advantage with respect to IDK: it has a mechanism to prioritize parts of a project. In the rest of considered characteristics, IDK equals or surpasses PDT. Thus, the tool used to support the new methodology proposed is IDK as it is independent from the development process and it may be personalized for the application under development.

Table 1. Comparing Prometheus and INGENIAS

	Prometheus	INGENIAS
Proper development process	YES	NO: Based in the USDP (analysis and design phases)
General process to generate code from the models	NO: Only obtains code for JACK language	YES: Based in template definitions
Iterative development	YES	YES
Model-driven development (MDD)	NO: Only proposes a correspondence between design models and JACK code	YES
Requirements capture	YES: A version of KAOS is used to describe the system's goals [30] complemented with the description of scenarios that illustrate the operation of the system. In addition, in [5] guidelines appear to generate the artifacts of the Prometheus system specification from organizational models expressed in i*	YES: Performed by means of use case diagrams. Then, use cases are associated to system goals, and a goals analysis is performed to decompose them into easier ones; and finally tasks are associated to get the easiest goals
Meta-model	YES [6]	YES
Mechanisms to discover agents and interactions	YES: Groups functionalities through cohesion and coupling criteria	NO
Agent model	BDI-like agents	Agents with mental states

Table 2. Comparing PDT and IDK

	PDT	**IDK**
Supported methodology	Prometheus	INGENIAS
Interface references the development process	YES: Diagrams are grouped in three levels according to the three Prometheus phases	NO: Possibility to create packets that correspond to the diverse phases of the process. Models of each phase are added to the corresponding packet
Mechanisms to prioritize parts of the project	YES: Three scope levels (essential, conditional and optional) [26]	NO
Code generation	YES: JACK http://www.agent-software.com/	YES: JADE http://jade.tilab.com/
Report generation of the MAS specification in HTML	YES	YES
Model fragmenting in various pieces	NO: For instance, only one diagram may be created to in order to gather all the objectives of the system	YES
Save a diagram as an image	YES	YES
Deployment diagrams	NO	YES
Agent communication	Defined in basis of messages and interaction protocols. Does not use a specific communication language. For JACK, there is a module compliant with FIPA [32]	Defined in accordance with communication acts of the agent communication language (ACL) proposed by FIPA http://www.fipa.org/specs/fipa00061/
Utility to simulate MAS specifications before generating the final code	NO	YES: Realized on the JADE platform. It is possible to manage interaction and tasks, and to inspect and modify the agents' mental states

4 Phases of the New Methodology

First Phase. In the first stage of the methodology proposed an *analysis overview diagram* is created. This diagram is used to develop a high level view of the system requirements [29]. This diagram will specify, in main lines, which are the actors - entities (human or software/hardware) external to the system – that interact with our system, where the perceptions that enter the system come from, which are the responses of the system (actions), an initial proposal of which might be the system roles, what messages are sent, and some used data. This kind of diagram appeared for the first time in PDT version 2.5.

Second Phase. Prometheus defines a proper detailed process to specify, implement and test/debug agent-oriented software systems. This process incorporates three phases: (1) *system specification* identifies the basic goals and functionalities of the system, develops the use case scenarios that illustrate the functioning of the system, and specifies which are the inputs (percepts) and outputs (actions) – it obtains the scenarios diagram, goal overview diagram, and system roles diagram; (2) *architectural design* uses the outputs produced in the previous phase to determine the agent types that exist in the system and how they interact – it obtains the data coupling diagram, agent-role diagram, agent acquaintance diagram, and system overview diagram; and, (3) *detailed design* centers on developing the internal structure of each agent and how each agent will perform its tasks within the global system – it obtains agent overview and capability overview diagrams. Finally, Prometheus details how to obtain the implementation in the agent-oriented programming language JACK.

The two first phases proposed in Prometheus (*system specification* and *architectural design*) are used to be the next phase of the new integrative methodology. The user identifies the agents and their interactions following the guidelines offered by Prometheus in these phases. In general terms, the mechanism provided by Prometheus to identify agents consists in identifying the goals during the system specification phase, and then in grouping the goals to obtain functionalities. Next, in the architectural design phase, functionalities are grouped to obtain the system agents, using cohesion and coupling criteria to decide which the best groupings are. These two concepts are essential in Software Engineering to obtain a good software development (the one that has maximum cohesion and minimum coupling) and to ease its further maintenance. On the other hand, the process for obtaining interactions between agents consists in changing the column role in the scenarios by the agent associated with each role. After that, if there are scenario steps with different agents then it means that there should be an interaction between the agents. Thus, the MAS developer gets an initial model according to Prometheus, following its two first stages (*system specification* and *architectural design*).

Next, an example of a moving robot application for the detection and following of humans is introduced in order to depict some diagrams obtained in the two Prometheus phases mentioned previously. In this application [14], a robot is moving randomly around the environment while the images collected are shown to the guard (state *wandering*). After some elapsed time (*Timer_P*) the robot stops in order to analyze the images captured in that instant (state *detecting*). After that, if movement has been detected, (1) information about the detected blob is obtained, and, (2) the guard is warned to decide if the robot should follow the blob or not. The process to follow persons is started (state *following*) if he chooses to follow it (*Follow_P*). When the robot is wandering, the guard may perceive that something is moving in the environment, according to the images displayed on his interface. In that case, the guard orders (*Detect_P*) that the images are analyzed to check if there is or not movement. If the image analysis does not detect movement, then the robot goes on moving randomly. In order to achieve tracking an object correctly (state *following*) the images are captured, displayed, and analyzed continuously in order to obtain blob information. The object is followed until the tracking phase finishes. This condition can be satisfied by three different reasons: (1) the guard has decided not to continue to follow the target (*Follow stop_P*), (2) the target is out of the field of vision, or, (3) it is impossible to follow it because some physical inaccessibility is encountered in the environment (for example, the target takes a staircase). After that, the robot wanders again.

In *system specification* phase, after obtaining the goals and scenarios diagrams, the roles are identified by clustering goals and linking perceptions and actions (see Fig. 1). *Start System_R* role handles the guard's request to start the robot devices. *Control Collision_R* role is responsible for achieving *Control Collision* goal, for which it needs inputs detected by the physical bumper device. *Observe environment sonar_R* uses the sonar to perceive distances to obstacles in order to avoid them. *Management guard order_R* aims to meet the guard's orders to control the system operation, which has already been started. These orders correspond to perceptions that allow to start/stop the tracking phase (*Follow_P, Stop follow_P*), and to analyze the images (*Detect_P*). *Wander_R* objective is to control the robot "wandering" process. It consists in randomly moving the robot around the environment, avoiding obstacles and controlling situations when a collision has been detected. *Follow_R* is responsible for controlling the robot's movement when the system is following an object. *Follow_R/Wander_R* roles do not include perceptions from the environment or actions on the environment, but it uses information obtained from physical sensors different from the camera, and therefore they need to "communicate" with the roles responsible for achieving *Follow object/Wander* sub-goals (*Avoid obstacle, Move robot, Control collision*). *Detect_R* is responsible for the goals of analyzing images captured by the camera, getting information from the detected moving blob, and performing an action to display results to the guard. *Capture Image_R* perceives images from the environment (*Image_P* percept), and moves the camera to set the camera focus (*Set_focus_a* action) to capture images in an optimum way (*Capture image* goal). *Show Image_R* is responsible for displaying the camera field of view to the guard. To satisfy this goal, *Show Image_a* action is executed when no movement is detected. *Motion_R* uses wheels to move the robot around the area (*Move robot* goal). This is controlled by actions that allow to stop, move and set the motion direction of the robot (*Stop_a, Move_a; Set motion direction_a*).

One task carried out in *architectural design* phase is to decide the agent types (as collections of roles). This is drawn in the agent-role grouping diagram. In our case we have grouped (1) *Start System_R* and *Management guard order_R* roles into *Central* agent, (2) *Wander_R* and *Follow_R* roles into *Motion Manager* agent, and, (3) *Show image_R* and *Detect_R* roles into *Image Manager* agent. Finally, *Control Collision_R, Observe environment sonar_R, Motion_R, Capture Image_R* roles are related with *Bumper, Sonar, Wheels,* and *Camera* agents, respectively. An agent is responsible for the functionalities – roles – related. Once roles have been grouped into agents, information about percepts and actions related to roles, depicted in system roles diagram, it is automatically propagated and linked with the agents in the system overview diagram (see Fig 2). After that agent conversations (interaction protocols - IP) are defined in order to describe what should happen to realize the specified goals and scenarios. For instance, *Sonar_IP* includes messages exchanged between *Sonar, Motion Manager* and *Wheels* agents as a result of using information provided by the physical device sonar (it measures the distance from an obstacle to the robot). *Central_IP* protocol contains messages sent from the *Central* agent to manager agents (*Motion Manager* and *Image Manager*) to monitor the robot's state (*wandering, following, detecting*) according to the orders provided by the guard (*Detect_P, Follow_P, Follow stop_P* percepts) or end of a time slice (*Timer_P* percept).

It has been shown in previous figures that there are entities, such as percepts and actions, which appear in several diagrams. This means that updating some diagram may lead to the need of updating another diagram when taking an iterative approach.

Mapping Prometheus into INGENIAS. Afterwards, mappings are used to obtain an equivalent model in INGENIAS. From this point on the advantages offered by

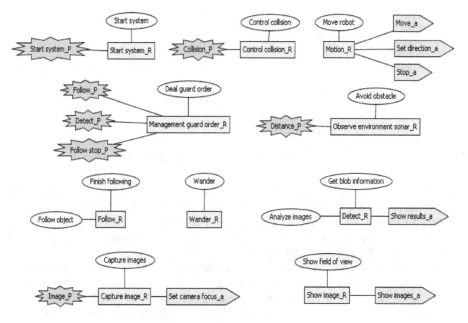

Fig. 1. System Roles Diagram

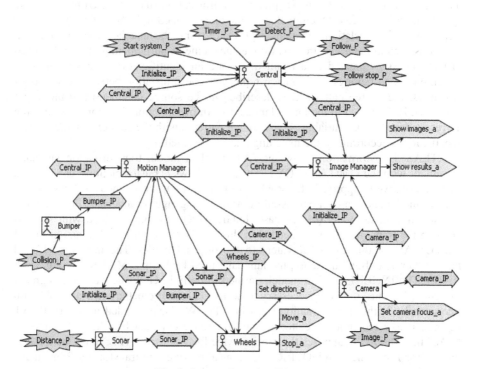

Fig. 2. System Overview Diagram

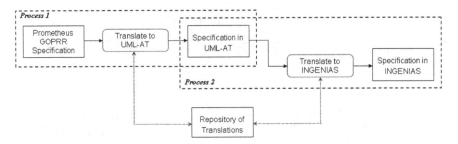

Fig. 3. Mapping Prometheus into INGENIAS

model-driven software development are used. The mappings are defined leaning in an intermediate language denominated Unified Modeling Language for Activity Theory (UML-AT) [8]. UML-AT allows establishing bidirectional transformations between models of different languages. There exists a previous experience in integrating two methodologies, Tropos and INGENIAS, using UML-AT [7].

The process of transforming Prometheus into INGENIAS methodology elements, with the help of intermediate language UML-AT, is shown in Fig. 1. In first place (process 1), we start from the Prometheus meta-model specified with language GOPRR (Graph Object Property Relationship Role) [20]. Translation rules to obtain elements, expressed in UML-AT language, equivalent to the ones selected in Prometheus, are created and used. Next (process 2), translation rules are used to obtain the specification in INGENIAS equivalent to the one obtained in UML-AT language. The Repository of Translations contains tuples indicating the matches and instantiation functions used in the translation, as well as the elements participating in it (either used in the process or created as a result of it) and an identifier of the specification to which each one belongs to. A match represents the translation between two sets of structures, the source pattern (it is described in the source language) and the target pattern (it is described in the target language). An instantiation function describes the correspondence of the variables in the source patterns with the elements in the current specifications according to the matching that is presented.

ATLAS Transformation Language (ATL) [2], a model transformation language compliant with the OMG MOF/QVT (Queries / Views /Transformation), can be used as an alternative to approach the problem of transforming Prometheus to INGENIAS. A meta-model and a model expressed in original language (in this case, Prometheus), a destination meta-model (in this case, INGENIAS) and rules defined with ATL to perform the transformation are needed in this case. The result is a destination model (INGENIAS, in this case) equivalent to the original model. The meta-models and models can be defined in Ecore, the language used by Eclipse Modeling Framework (EMF). The INGENIAS meta-model, defined initially in GOPRR, has been migrated to Ecore [9]. Thus, it is only necessary to define: (1) the Prometheus meta-model with Ecore, and, (2) the transformation rules in ATL. We have decided to use UML-AT because it is a technology related to the research group that has developed INGE-NIAS. In addition, it supposes the same service load as using ATL: to define a meta-model (for Prometheus in GOPRR) and translation rules (to transform a Prometheus specification into a UML-AT specification). The corresponding part to transform UML-AT into INGENIAS is solved in [8]. A tool called Activity Theory Assistant

(ATA) has been developed to help using the techniques based on the theory of the activity and to support the translation process. ATA is embedded in a plug-in of the IDK. Notice, however, that the current IDK version available in SourceForge, http://sourceforge.net/projects/ingenias, does not include it.

In Prometheus, in order to describe the interactions among agents, interaction protocols using a reviewed version of Agent UML (AUML) denominated AUML-2 are developed. UML-AT has already been applied to establish correspondences between FIPA protocols designed with AUML models [1] and INGENIAS models [8]. This work could be taken as departure point to transform interaction protocols obtained with Prometheus into the equivalent notation used in INGENIAS. The IDK tool, which provides support to INGENIAS, allows representing protocols according to the AUML annotation [15]. This means that the interaction protocols created with Prometheus could be used directly in INGENIAS, with no need to use any transformation. Nevertheless, its development has not evolved enough. In fact, in version 2.6 of the IDK this utility no longer appears.

At present, an informal approach is followed to transform Prometheus models into the equivalent INGENIAS models [12], [13]. The transformations are carried out manually. The next paragraphs explain how in the robot application some information obtained in Prometheus models are transformed into INGENIAS models.

A percept is a piece of information from the environment received by means of a sensor. In Prometheus, percepts must at least belong to one functionality, and, thus, to the agent associated to that functionality, too. The relations among percepts and roles (Percept → Role) and the relations among percepts and agents (Percept λ → Agent) appear in the system roles diagram and the system overview diagram, respectively. The percepts of a Prometheus agent can be modeled in INGENIAS by specifying a collection of operations in an application. In the INGENIAS environment model, an *EPerceivesNotification* relation between the agent and the corresponding application will be established. In a Prometheus percept descriptor, there is a field, Information carried, where it is specified the information transported as part of the percept. In INGENIAS, this information is included with an *ApplicationEventSlots* type of event associated to *EPerceivesNotification* relation. In the same way, in Prometheus, also every action must at least belong to one functionality and the agent associated to the functionality must execute it. The relations among actions and roles (Role → Action) and the relations among actions and agents (Agent → Action) appear in the system roles diagram and the system overview diagram, respectively. An action represents something that the agent does to interact with the environment. In INGENIAS, actions on the environment are assumed to be calls to operations defined in the applications. Therefore, an action present in Prometheus will be transformed into an application operation present in the environment model in INGENIAS. *EPerceives* will be used to establish the relation between the agent and the application.

In the environment model, an Agent - ApplicationBelongs To → Application relation will be also established to express that an agent uses an application. Fig. 4 shows the environment model obtained in INGENIAS to the *Camera* agent after applying the described equivalences.

A goal that appears in the goal overview diagram used in Prometheus will correspond to a goal in the goals and tasks model in INGENIAS. *AND* and *OR* dependencies between goals can be established in both models; therefore, it is possible

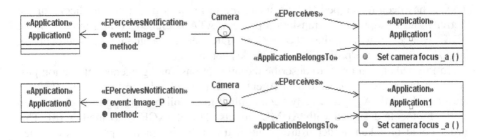

Fig. 4. Environment model for the *Camera* agent

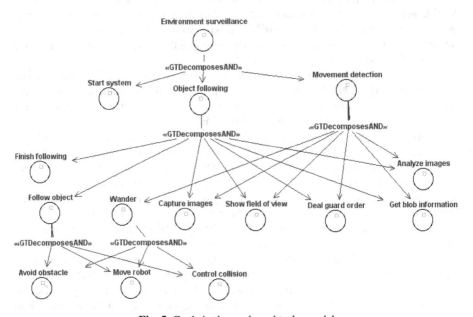

Fig. 5. Goals in the goals and tasks model

to directly transfer these relations from one model to another. In the goals and tasks model, a *GTDecomposeAND* relation and a *GTDecomposeOR* relation will be established to reflect an *AND* and *OR* relation between goals, respectively. When a goal has only one sub-goal, *GTDecomposes* is used. Fig. 5 summarizes the goals that appear in INGENIAS goals and tasks model and that have been obtained from the Prometheus goal overview diagram.

In the system roles diagram of Prometheus methodology, relations between goals and functionalities are established. The latter will be grouped to determine the types of agents in the system - the relation among agents and roles, Agent → Role, appear in the agent-role grouping diagram, whilst the relation among roles and goals, Role → Goal, appear in the system roles diagram. Therefore, implicitly, there is a relation between goals and agents. So, there is a Agent - *GTPursues* → Goal relation in the agent model. IDK only supports roles to generate the code that corresponds to an interaction. Therefore, for every agent identified in Prometheus, an associated role is

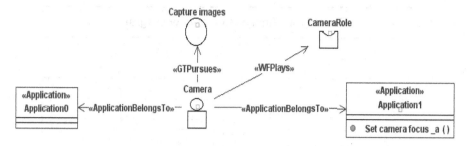

Fig. 6. Fragment of *Camera* agent model

necessary to state its participation in an interaction. So, we will establish an Agent -
WFPlays → Role relation. Fig. 6 depicts a fragment of the INGENIAS agent model
corresponding to the *Camera* agent.

Third Phase. The new methodology does not reuse the last phase of Prometheus
(*detailed design*) because it is too much centered in BDI-like agents. Moreover, Prometheus also describes how obtained entities are transformed in the design phase into
the concepts used for a specific implementation language (JACK). These two aspects,
centering in a single type of agent and defining a mechanism to generate code for a
particular implementation language, suppose, in principle, a loss of generality. In the
new methodology, once the equivalent model in INGENIAS has been obtained, the
architecture of each type of identified agent is provided. The possible types of agents
are the ones available in ICARO-T: cognitive agents and reactive agents. In this phase
the necessary guidelines for completing all the INGENIAS models already exist.

Fourth Phase. With respect to code generation, the INGENIAS proposal is followed.
INGENIAS generalizes a process to transform the models, generated in the phase of
design, in running code for any destination platform [7]. It is based in the definition of
templates for each destination platform and procedures for extracting information
present in the models. Once the code has been obtained, the developer refines the
resulting code completing any information that was not contained in the specifications
(models) or in the templates. Finally, the application is deployed.

   ICARO-T is the platform selected for running the agents [11], [25]. It offers four
categories of reusable component models: *agent organization models* to describe the
overall structure of the system, *agent models*, *resource models* to encapsulate computing entities providing services to agents, and *basic computing entities*. There are
several reasons for selecting the multi-agent platform ICARO-T. The use of its components has allowed to significantly reducing time and effort in the design and
implementation phases by an average of a 65 percent. In the phases of testing and
correction cycles the errors are also reduced. Consequently, the applications require
less resources and lower implementation time [25].

   The ICARO-T components have been used in Spanish telecommunications company Telefonica for developing several voice recognition services. At the moment, it
is also being used by other research teams. This is the case, for example, in an e-
learning project denominated ENLACE [31]. Fig. 7 shows the technology and tools
used in the integrative methodology proposed.

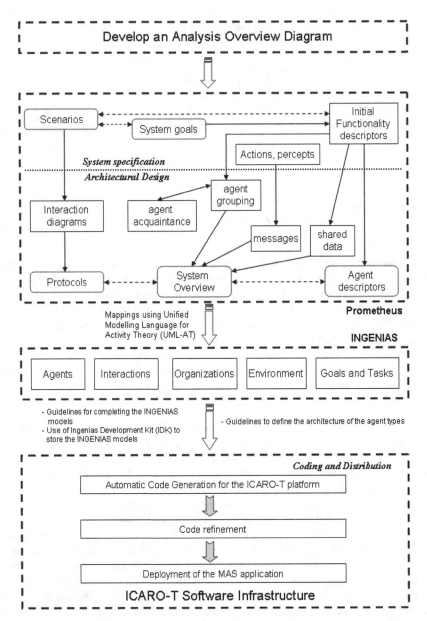

Fig. 7. Multi-agent system development methodology

5   Conclusions and Future Work

The combination of current technologies (Prometheus, INGENIAS, UML-AT and ICARO-T) has given rise to a new integrative methodology for the development of agent-based systems. It uses the guidelines offered by Prometheus to identify agents and their interactions. Later, the obtained model, following Prometheus methodology, is transformed into INGENIAS to continue the development. This transformation is performed under UML-AT. Once modeling has ended up, code is generated for the ICARO-T platform.

In order to use this methodology definitively it is necessary: (1) to specify the Prometheus meta-model in GOPRR, (2) to create the rules to translate Prometheus concepts in UML-AT language, and, (3) to define templates for the IDK to generate code for the ICARO-T platform. A first proposal in order to transform Prometheus models into equivalent INGENIAS models using an informal language has recently been considered [12].

Acknowledgements. This work is partially supported by the Spanish Ministerio de Ciencia e Innovación under project TIN2007-67586-C02-02, and by the Spanish Junta de Comunidades de Castilla-La Mancha under projects PII2I09-0069-0994 and PEII09-0054-9581.

References

1. Bauer, B., Odell, J.: UML 2.0 and agents: How to build with the new UML standard. Journal of Engineering Applications in Artificial Intelligence 18(2), 141–157 (2005)
2. Bézivin, J., Jouault, F., Touzet, D.: An introduction to the ATLAS model management architecture. Research Report LINA (05-01) (2005)
3. Cabri, G., Puviani, M., Leonardi, L.: The MAR&A methodology to develop agent system. In: First International Conference on Agents and Artificial Intelligence, pp. 501–506 (2009)
4. Cossentino, M., Gaglio, S., Garro, A., Seidita, V.: Method fragments for agent design methodologies: From standardisation to research. International Journal of Agent-Oriented Software Engineering 1(1), 91–121 (2007)
5. Cysneiros, G., Zisman, A.: Refining Prometheus methodology with i*. In: Third International Workshop on Agent-Oriented Methodologies (2004)
6. Dam, K.H., Winikoff, M., Padgham, L.: An agent-oriented approach to change propagation in software evolution. In: Australian Software Engineering Conference, pp. 309–318 (2006)
7. Fuentes, R., Gómez-Sanz, J.J., Pavón, J.: Integrating agent-oriented methodologies with UML-AT. In: Fifth International Joint Conference on Autonomous Agents and Multiagent Systems, pp. 1303–1310 (2006)
8. Fuentes, R., Gómez-Sanz, J.J., Pavón, J.: Model integration in agent-oriented development. International Journal of Agent-Oriented Software Engineering 1(1), 2–27 (2007)
9. García-Magariño, I., Gómez-Sanz, J.J., Pavón, J.: Representación de las relaciones en los metamodelos con el lenguaje Ecore. In: Desarrollo de Software Dirigido por Modelos, DSDM 2007 (2007)
10. Garcia-Magariño, I., Gómez-Sanz, J., Fuentes, R.: INGENIAS development assisted with model transformation By-Example: A practical case. In: Seventh International Conference on Practical Applications of Agents and Multi-Agent Systems (2009)
11. Garijo, F., Polo, F., Spina, D., Rodríguez, C.: ICARO-T User Manual. Internal Report, Telefonica I+D (May 2008)
12. Gascueña, J.M., Fernández-Caballero, A.: Prometheus and INGENIAS agent methodologies: A complementary approach. In: Luck, M., Gomez-Sanz, J.J. (eds.) Agent-Oriented Software Engineering IX. LNCS, vol. 5386, pp. 131–144. Springer, Heidelberg (2009)
13. Gascueña, J.M., Fernández-Caballero, A.: Towards an integrative methodology for developing multi-agent systems. In: International Conference on Agents and Artificial Intelligence (2009)

14. Gascueña, J.M., Fernández-Caballero, A.: Agent-based modeling of a mobile robot to detect and follow humans. In: Håkansson, A., Nguyen, N.T., Hartung, R.L., Howlett, R.J., Jain, L.C. (eds.) Agent and Multi-Agent Systems: Technologies and Applications. LNCS, vol. 5559, pp. 80–89. Springer, Heidelberg (2009)
15. Gómez Sanz, J.J., Pavón, J.: INGENIAS Development Kit (IDK) Manual. Version 2.5.2 (2008), http://heanet.dl.sourceforge.net/sourceforge/ingenias/ingeniasmanual.v2.5.pdf
16. Hahn, C., Madrigal-Mora, C., Fischer, K.: A platform-independent metamodel for multi-agent systems. Autonomous Agents and Multi-Agent Systems 18(2), 239–266 (2009)
17. Jacobson, I., Booch, G., Rumbaugh, J.: The Unified Software Development Process. Addison-Wesley, Reading (1999)
18. Jarraya, T., Guessoum, Z.: Towards a model driven process for multi-agent system. In: Burkhard, H.-D., Lindemann, G., Verbrugge, R., Varga, L.Z. (eds.) CEEMAS 2007. LNCS (LNAI), vol. 4696, pp. 256–265. Springer, Heidelberg (2007)
19. Jennings, N.R., Wooldridge, M.: Applying agent technology. International Journal of Applied Artificial Intelligence 9(4), 351–359 (1995)
20. Kelly, S., Lyytinen, K.S., Rossi, M.: METAEDIT+ - A fully configurable multi-user and multi-tool CASE and CAME environment. In: Constantopoulos, P., Vassiliou, Y., Mylopoulos, J. (eds.) CAiSE 1996. LNCS, vol. 1080, pp. 1–21. Springer, Heidelberg (1996)
21. Padgham, L., Winikoff, M.: Developing Intelligent Agents Systems: A Practical Guide. John Wiley and Sons, Chichester (2004)
22. Padgham, L., Thangarajah, J., Paul, P.: Prometheus Design Tool. Version 2.5. User Manual (2008), http://www.cs.rmit.edu.au/agents/pdt/docs/PDT-Manual.pdf
23. Pavón, J., Gómez-Sanz, J.J., Fuentes, R.: The INGENIAS methodology and tools. Agent-Oriented Methodologies. Idea Group Publishing, USA (2005)
24. Pavón, J., Gómez-Sanz, J.J., Fuentes, R.: Model driven development of multi-agent systems. In: Rensink, A., Warmer, J. (eds.) ECMDA-FA 2006. LNCS, vol. 4066, pp. 284–298. Springer, Heidelberg (2006)
25. Pavón, J., Garijo, F., Gómez-Sanz, J.: Complex systems and agent-oriented software engineering. In: Weyns, D., Brueckner, S.A., Demazeau, Y. (eds.) EEMMAS 2007. LNCS (LNAI), vol. 5049, pp. 3–16. Springer, Heidelberg (2008)
26. Perepletchikov, M., Padgham, L.: Systematic incremental development of agent systems, using Prometheus. In: Fifth International Conference on Quality Software, pp. 413–418 (2005)
27. Perini, A., Susi, A.: Automating model transformations in agent-oriented modelling. In: Müller, J.P., Zambonelli, F. (eds.) AOSE 2005. LNCS, vol. 3950, pp. 167–178. Springer, Heidelberg (2006)
28. Schmidt, D.C.: Guest Editor's Introduction: Model-Driven Engineering. Computer 39(2), 25–31 (2006)
29. Sokolova, M.V., Fernández-Caballero, A.: Facilitating MAS complete life cycle through the Protégé-Prometheus approach. In: Nguyen, N.T., Jo, G.-S., Howlett, R.J., Jain, L.C. (eds.) KES-AMSTA 2008. LNCS (LNAI), vol. 4953, pp. 63–72. Springer, Heidelberg (2008)
30. van Lamsweerde, A.: Goal-oriented requirements engineering: A guided tour. In: Fifth IEEE International Symposium on Requirements Engineering, pp. 249–263 (2001)
31. Verdejo, M.F., Celorrio, C.: A multi-agent based system for activity configuration and personalization in a pervasive learning framework. In: Third IEEE International Workshop on Pervasive Learning, pp. 177–181 (2007)
32. Yoshimura, K.: FIPA JACK: A plugin for JACK Intelligent AgentsTM. Technical Report, RMIT University (2003)

An Efficient Winner Approximation for a Series of Combinatorial Auctions

Naoki Fukuta[1] and Takayuki Ito[2,3]

[1] Shizuoka University, Hamamatsu Shizuoka 4328011, Japan
[2] Nagoya Institute of Technology, Gokiso-cho Nagoya 4668555, Japan
[3] Massachusetts Institute of Technology, Cambridge, MA 02142 U.S.A.
fukuta@cs.inf.shizuoka.ac.jp
http://whitebear.cs.inf.shizuoka.ac.jp/

Abstract. In this paper, we show an analysis about approximated winner determination algorithms for iteratively conducted combinatorial auctions. Our algorithms are designed to effectively reuse last-cycle solutions to speed up the initial approximation performance on the next cycle. Experimental results show that our proposed algorithms outperform existing algorithms when a large number of similar bids are contained through iterations. Also, we show an enhanced algorithm effectively avoids undesirable reuses of the last solutions in the algorithm without serious computational overheads.

1 Introduction

Combinatorial auctions [1], one of the most popular market mechanisms, have a huge effect on electronic markets and political strategies. For example, Sandholm et al. [2][3] proposed real markets using their innovative combinatorial auction algorithms. The FCC tried to employ combinatorial auction mechanisms to assign spectrums to companies [4]. Also [1] shows other realistic examples that utilize combinatorial auction mechanisms.

Demand exists to utilize combinatorial auction mechanisms that cannot be covered by existing approaches due to hard time constraints and the limitations of usable computational resources. Resource allocation for agents in ubiquitous computing environments is a good example for understanding the needs of the short-time approximation of combinatorial auctions [5]. In such an environment, agents must provide specific services to their users using various available resources. However, in ubiquitous computing environments, since such resources as sensors and devices are typically limited, they do not satisfy all the needs of all agents. For various reasons including physical limitations and privacy concerns, most of the resources cannot be shared with other agents. Furthermore, agents will simultaneously use two or more resources to realize desirable services for users. Since agents provide services to their own users, agents might be self-interested. Therefore, a combinatorial auction mechanism is a good option for such situations since it provides effective resource allocation to self-interested agents.

In order to utilize combinatorial auctions on the above situation, we need to complete winner determination within a very short time. Consider 256 resources and 100 agents, where each agent places, for example, from 200 to 1,000 of combinations for the items

J. Filipe, A. Fred, and B. Sharp (Eds.): ICAART 2009, CCIS 67, pp. 233–246, 2010.
© Springer-Verlag Berlin Heidelberg 2010

as complex bids to an auction, they will be expanded to, from 20,000 to 100,000 of atomic bids. Here, to avoid occupying a set of resources for a long time by a certain agent, we consider a resource allocation scenario based on fixed time slice assignment model. In the scenario, those resources are auctioned and allocated from a system to agents for a fixed time period. After a certain time has passed, those are once all returned to the system and then they are auctioned again for the next period. When an agent prefers to continue using the same resources at the next period, the agent will place a higher price for the resources to increase the possibility to win them.

In ubiquitous computing scenarios, since physical locations of users are always changing, resources should be reallocated in a certain period to catch up with those changes. For better usability, the resource reallocation time period will be 0.1 to several seconds depending on services provided. Furthermore, since we should complete whole resource allocation procedure that includes pricing and communication to devices for actual resource assignment, we must determine auction winners within an extremely short time period that is far less than the actual resource allocation period.

In general, the optimal winner determination problem of combinatorial auctions is NP-hard[1]. Thus, much work focuses on tackling the computational costs for winner determination [6][1][2]. Although some works try to achieve *approximate* solutions in winner determination[7][8][9], there is a demand to enhance the performance to be able to handle much larger problems more quickly.

In this paper, we show an analysis about enhanced approximation algorithms of winner determination on combinatorial auctions that are suitable for the purpose of iterative reallocation of items mentioned above. Since the above-mentioned existing algorithms are *offline* algorithms, we need to re-calculate the winners when bids are added to or deleted from the auction even when the modification of bids is only slight. Intuitively, it could be helpful to reuse results of past similar auctions for faster approximation of the current auction. However, those algorithms did not consider reusing past approximated results for performance improvement since such reuse may cause serious performance down in certain cases. Our enhanced algorithms have mechanisms to reuse past approximation results but avoid such performance down with very small overhead.

The rest of the paper is organized as follows. In section 2, we show some preliminaries about combinatorial auctions and winner determination problem. In section 3, we describe our algorithms that reuse past approximation results and eliminate undesirable reuses. Section 4 shows evaluation and analysis about our approach by experiments. Here, we show our experiment settings, experimental results of comparisons to existing algorithms including a sophisticated linear programming(LP) solver, and an analysis about the shown results. In section 5, we discuss about some limitations in our approach. Finally, section 6 concludes the paper.

2 Preliminaries

2.1 Winner Determination Problem

In this paper, to keep simplicity of discussion, we only focus on utility-based resource allocation problems[10], rather than generic resource allocation problems with numerous complex constraints. Utility-based resource allocation problem is a problem that aims to maximize the sum of utilities of users for each allocation period, but does not consider other factors and constraints (i.e., fair allocation [11] , security and privacy concerns[12], uncertainty[13], etc). Also we only consider a scenario that is based on fixed time slice assignment model.

Combinatorial auction is an auction mechanism that allows bidders to locate bids for a bundle of items rather than single item[1]. When we apply combinatorial auction mechanism for utility-based resource allocation problems, the problem can be transformed to solve a winner determination problem on combinatorial auctions.

The winner determination problem on combinatorial auctions is defined as follows[1] : The set of bidders is denoted by $N = 1, \ldots, n$, and the set of items by $M = \{1, \ldots, m\}$. $|M| = m$. Bundle S is a set of items : $S \subseteq M$. We denote by $v_i(S)$, bidder i's valuation of combinatorial bid for bundle S. An allocation of the items is described by variables $x_i(S) \in \{0, 1\}$, where $x_i(S) = 1$ if and only if bidder i wins the bundle S. An allocation, $x_i(S)$, is feasible if it allocates no item more than once, i.e.,for all $j \in M$,

$$\sum_{i \in N} \sum_{S \ni j} x_i(S) \leq 1$$

The winner determination problem is the problem to maximize total revenue. For feasible allocations $X \ni x_i(S)$,

$$\max_X \sum_{i \in N, S \subseteq M} v_i(S) x_i(S)$$

Here, we used simple *OR-bid* representation as our bidding language. Substitutability can be represented by a set of atomic *OR-bids* with dummy items[1].

Even when we only focus on utility-based resource allocation problems, they enforce us to solve winner determination problem with really hard-time constraint for realizing fine-grained resource allocation. Here, we have to consider that, in such resource allocation procedures, we need to spend much time for pricing and communications for actual resource allocation protocols. Therefore, we need a fast winner determination algorithm for auctions with a large number of bids. In this paper, primarily we focus on solving this problem.

2.2   Lehmann's Greedy Winner Determination

Lehmann's greedy algorithm [7] is a very simple but powerful linear algorithm for winner determination on combinatorial auctions. Here, a bidder declaring $< s, a >$, with $s \subseteq M$ and $a \in \mathcal{R}_+$ will be said to put out a bid $b = < s, a >$. Two bids $b = < s, a >$ and $b' = < s', a' >$ conflict iff $s \cap s' \neq \emptyset$. The greedy algorithm can be described as follows: (1) The list of bids L is sorted by some criterion. In [7], a method to sort the list L by descending average amount per item is proposed. More generally, they proposed sorting L by a criterion of the form $a/|s|^c$ for some number $c \geq 0$, which possibly depends on the number of items, m. (2) A greedy algorithm generates an allocation. L is the sorted list in the first phase. The algorithm walk down the list L, accepting bids if the items demanded are still unallocated and unconflicted.

In [7], Lehmann et, al . argued that $c = 1/2$ is the best parameter for approximation when the norm of the worst case performance is considered[1]. Also they showed that the mechanism is truthful when single-minded bidders are assumed and their proposed pricing scheme is used.

2.3 Hill-Climbing Search

In [14] we proposed a preliminary idea of our hill-climbing approach, and in [5] and [15] we showed our hill-climbing approach performs well when an auction has an enormous number of bids. In this section, we summarize our proposed algorithms.

Lehmann's greedy winner determination typically performs well and the lower bound of the optimality has been analyzed[7]. A straightforward extension of the greedy algorithm is to construct a local search algorithm that continuously updates the allocation to increase optimality. Intuitively, one allocation corresponds to one state of a local search.

The inputs are $Alloc$ and L. L is the bid list of an auction. $Alloc$ is the initial greedy allocation of items for the bid list.

```
 1: function LocalSearch(Alloc, L)
 2:    RemainBids:= L ∩ Alloc̄;
 3:    for each b ∈ RemainBids as sorted order
 4:      if b conflicts Alloc then
 5:        Conflicted:=Alloc ∩ consistentBids({b}, Alloc);
 6:        NewAlloc:= (Alloc ∩ Conflicted̄) ∪ {b};
 7:        ConsBids:=
 8:          consistentBids(NewAlloc, RemainBids);
 9:        NewAlloc:=NewAlloc ∪ ConsBids;
10:      if price(Alloc) < price(NewAlloc) then
11:        return LocalSearch(NewAlloc,L);
12:    end for each
13:    return Alloc
```

Function $consistentBids$ finds consistent bids for the set $NewAlloc$ by walking down the list $RemainBids$. Here, since a new inserted bid will wipe out some bids that are conflicting with the inserted bid, free items will appear to be allocated to other bidders after the insertion. Function $consistentBids$ tries to find out potential winner bids that do not conflict to the specified allocation.

2.4 Parallel Search for Multiple Weighting

The optimality of allocations obtained by Lehmann's algorithm (and the subsequent hill-climbing) deeply depends on which value was set to c in the bid weighting function. Lehmann et al. reported that $c = 1/2$ guarantees lower bound of approximation.

---

[1] Note that, in [2], Sandholm et,al. determined experimentally that $c \in [0.8, 1]$ yields best performance in their approach.

However, the optimal values for each auction are varied from 0 to 1 depending on the auction problem.

In [14], an enhancement has been presented for local search algorithm to parallel search for different bid weighting strategies (e.g., doing the same algorithm for both $c = 0$ and $c = 1$) In the algorithm, the value of c for Lehmann's algorithm is selected from a pre-defined list. Selecting c from neighbors of $1/2$ is reasonable, namely, $C = \{0.0, 0.1, \ldots, 1.0\}$. The results are aggregated and the best one (with the highest revenue) is selected as the final result. More detailed analyses about parallel greedy approach and bid weighting strategies are shown in [15] and [16].

2.5 Other Approximation Approaches

Zurel and Nisan[8] proposed a very high performance approximate winner determination algorithm for combinatorial auctions. The main idea is a combination of approximated positive linear program algorithm for determining initial allocation and stepwise random updates of allocations.

Hoos[9] proposed Casanova algorithm, and showed that a generic random walk SAT solver may perform well for approximation of combinatorial auctions. In Casanova, each search state is scored by revenue-per-item of the corresponding allocations, and random walk search is applied for seeking better state.

3 Enhanced Approximation

3.1 Fast Partial Reallocation by Last Result

In the setting of the periodical resource re-allocation scenario, winner determination occurs when some bids are revised. Theoretically, we need to recalculate winners even if only one bid is changed in the auction. However, in some cases, reusing the winners of previous auctions is useful when the change is small so that it has small effects to the next winner determination process. The following simple algorithm reuses the approximation result of the last cycle when recalculation is needed due to changes of the bids in the auction[2]. Here, we assume that the bids won at the last cycle ($LastWinners$) and the all bids at the last cycle ($LastBids$) are known.

```
1: Function PartialReallocationA(
2:         LastBids,LastWinners,CurrentBids)
3:    AddedBids :=
4:      CurrentBids ∩ (LastBids ∩ CurrentBids);
5:    DeletedBids :=
6:      LastBids ∩ (LastBids ∩ CurrentBids);
7:    Winners := LastWinners;
8:    foreach d ∈ DeletedBids
9:      if d ∈ Winners
```

---

[2] A preliminary idea has been proposed in [17].

```
10:    then Winners := Winners ∩ {d̄};
11:  foreach a ∈ AddedBids
12:    foreach w ∈ Winners
13:      if w and a are bids placed for the exactly same items
14:        and price({w}) < price({a})
15:        then Winners := (Winners ∩ {w̄}) ∪ {a};
16:  Winners := LocalSearch(Winners,CurrentBids);
17:  return Winners
```

First, the algorithm deletes winners that no longer valid due to deletion of bids. Then, some winner bids are replaced by newly added bids. Note that we only replace a bid when the bids are placed for exactly the same items, i.e., for two bids $b_i(X)$ and $b_j(Y)$, $X = Y$, to avoid the ordering problem of newly added bids. Modification of a bid through cycles is treated as a combined operation of the deletion of previous bid and the addition of the renewed bid.

3.2 Eliminating Undesirable Reallocations

Generally speaking, the performance of reusing the partial results of similar problems depends on the problem. Therefore, in some cases, reusing the last result may cause performance decreases. To avoid such a situation, we slightly modified our algorithm to switch the initial allocation by evaluating its performance.

Here, the modified algorithm simply compares the reused result with greedy allocation. Then, the better one is used as the seed of hill-climbing improvement. Note that both our reallocation and greedy allocation algorithms complete their executions in very short time. Therefore, computational overhead for them is expected to be negligible.

```
1:  Function PartialReallocationX(
2:            LastBids,LastWinners,CurrentBids)
3:    AddedBids :=
4:    CurrentBids ∩ (LastBids ∩ CurrentBids);
5:    DeletedBids :=
6:    LastBids ∩ (LastBids ∩ CurrentBids);
7:    Winners := LastWinners;
8:  foreach d ∈ DeletedBids
9:    if d ∈ Winners
10:     then Winners := Winners ∩ {d̄};
11: foreach a ∈ AddedBids
12:   foreach w ∈ Winners
13:     if w and a are bids placed for the exactly same items
14:       and price({w}) < price({a})
15:       then Winners := (Winners ∩ {w̄}) ∪ {a};
16:   GreedyWinners := GreedySearch(CurrentBids);
17:   if price(Winners) ≤ price(GreedyWinners)
```

18:    **then** $Winners := GreedyWinners$;

19:    $Winners :=$ LocalSearch($Winners, CurrentBids$);

20:    **return** $Winners$

4 Evaluation

4.1 Experiment Settings

We implemented our algorithms in a C program for the following experiments. We also implemented the Casanova algorithm in a C program. For Zurel's algorithm, we used Zurel's C++ based implementation that is shown in [8]. Also we used CPLEX Interactive Optimizer 11.0.0 (32bit) in our experiments[3]. The experiments were done with above implementations to examine the performance differences among algorithms. The programs were employed on a Mac with Mac OS X 10.4, a CoreDuo 2.0GHz CPU, and 2GBytes of memory.

We conducted several experiments. In each experiment, we compared the following search algorithms: greedy(C=0.5) uses Lehmann's greedy allocation algorithm with parameter ($c = 0.5$). greedy-3 uses the best results of Lehmann's greedy allocation algorithm with parameter ($0 \le c \le 1$ in 0.5 steps). HC(c=0.5) uses a local search in which the initial allocation is Lehmann's allocation with $c = 0.5$ and conducts the hill-climbing search shown in section 2.3. HC-3 uses the best results of the hill-climbing search with parameter ($0 \le c \le 1$ in 0.5 steps). We denote the Casanova algorithm as casanova and Zurel's algorithm as Zurel Also we denote results of 1st stage of Zurel's algorithm as Zurel-1st. Note that Zurel's algorithm does not produce any approximation result until completing its 1st stage. cplex is the result of CPLEX with the specified time limit.

In the following experiments, we used 0.2 for the epsilon value of Zurel's algorithm. This value appears in [8]. Also we used 0.5 for np and 0.15 for wp on Casanova that appear in [9]. Note that we set $maxTrial$ to 1 but $maxSteps$ to ten times the number of bids in the auction.

We conducted detailed comparisons among our past presented algorithms and the other existing algorithms mentioned above. The details of the comparisons are shown in [18] and [5]. In [18] and [5], we prepared datasets with 20,000 bids in an auction. The datasets were produced by CATS[19] with default parameters in 5 different distributions. They contain 100 trials for each distribution. Each trial is an auction problem with 256 items and 20,000 bids.[4]

However, since CATS common datasets only provide static bids for an auction, we prepared extended usage for those datasets to include the dynamic changes of bids in an auction.

Procedure. In each auction, the bid set is divided into k blocks by the order of bid generation (i.e., bid id). The bid set is modified totally k times and the modification is

---

[3] Although CPLEX is an optimizer that can obtain optimal results, it is reported in [2] that its anytime approximation performance is also good.

[4] Due to difficulty of dataset preparation, we only prepared five distributions. Producing a dataset with other distributions is difficult in feasible time.

done in each second. In each 1-second period, a block is marked as hidden so that bids within these marked blocks are treated as *deleted bids*. For example, at the first period, the first block is marked as hidden so the remaining bids (second to kth blocks) are used for winner determination. After 1 second, the mark is moved to the second block (i.e., the first, and the third to kth blocks are used) and the winner determination process is restarted due to this change. Here, we can see it as the bids in the first block are newly added to the auction and the bids in the second block are deleted from the auction. This process is repeated until the mark has been moved to the kth block. Finally, all marks are cleared and the winner determination process is restarted with full bids in the auction. Ordinary algorithms should be completely restarted in each cycle. However, when we use our proposed reallocation algorithms, some intermediate results can be reused in the next cycle in the same auction.

Since the bid set in the $k + 1$th cycle completely equals the bids of the auction, the results of the $k + 1$th cycle can be compared to our previous experimental results.

4.2   Time Performance

Table 1 shows the experimental result on the datasets with 20,000 bids in an auction focused on execution time of approximation. Due to the difficulty of attaining optimal values, we normalized all values as Zurel's results equal 1 as follows.

Let A be a set of algorithms, $z \in A$ be the zurel's approximation algorithm, D be a dataset generated for this experiment, and $revenue_a(p)$ such that $a \in A$ be the revenue obtained by algorithm a for a problem p in a dataset, the average revenue ratio $ratin A_a(D)$ for algorithm $a \in A$ for dataset D is defined as follows:

$$ratioA_a(D) = \frac{\sum_{p\in D} revenue_a(p)}{\sum_{p\in D} revenue_z(p)}$$

Here, we use $ratioA_a(D)$ for our comparison of algorithms.

The name of each distribution is taken from [19]. We prepared the cut-off results of Casanova and HC. For example, casanova-10ms denotes the results of Casanova within 10 milliseconds. Also we prepared a variant of our algorithm that has a suffix of -seq or -para. The suffix -seq denotes that the algorithm is completely executed sequentially that is equal to be executed on a single CPU computer. For example, greedy-3-seq denotes that the execution time is the sum of execution times spent by three threads. The suffix -para denotes that the algorithm is completely executed in a parallel manner, the three independent threads are completely executed in parallel. Here, we used ideal value for -para since our computer has only two cores in the CPU. The actual execution performance will be between -seq and -para.

Additionally, we added results with names AHC or XHC in the same table. They are the average approximated results of the $k+1$th cycle of auctions with our proposed algorithms PartialReallocationA and PartialReallocationX, respectively.

In most distributions, Zurel-1st takes more than one second but the obtained optimality is lower than greedy-3-seq. However, our proposed HC-3 performs better or slightly lower although their computation times are shorter than Zurel-1st and Zurel, excluding L3. Surprisingly, in most cases, the results of XHC-3-seq-100ms are better than HC-3-seq-1000ms while their spent computation time is only 1/10. This fact

Table 1. Time Performance of (k+1)th cycle on 20,000bids-256items (k=10)

	L2	L3	L4	L6	L7	average
greedy(c=0.5)	1.0002 (23.0)	0.9639 (19.0)	0.9417 (23.0)	0.9389 (23.4)	0.7403 (22.1)	0.9170 (22.1)
greedy-3-seq	1.0003 (69.1)	0.9639 (59.2)	0.9999 (72.9)	0.9965 (67.8)	0.7541 (66.8)	0.9429 (67.2)
greedy-3-para	1.0003 (26.4)	0.9639 (20.9)	0.9999 (28.4)	0.9965 (26.0)	0.7541 (25.5)	0.9429 (25.4)
HC(c=0.5)-100ms	1.0004 (100)	0.9741 (100)	0.9576 (100)	0.9533 (100)	0.8260 (100)	0.9423 (100)
HC-3-seq-100ms	1.0004 (100)	0.9692 (100)	1.0000 (100)	0.9966 (100)	0.8287 (100)	0.9590 (100)
AHC-3-seq-100ms	1.0004 (100)	0.9690 (100)	1.0006 (100)	0.9974 (100)	1.0225 (100)	0.9980 (100)
XHC-3-seq-100ms	1.0004 (100)	0.9813 (100)	1.0005 (100)	0.9987 (100)	1.0217 (100)	1.0005 (100)
HC-3-para-100ms	1.0004 (100)	0.9743 (100)	1.0001 (100)	0.9969 (100)	0.9423 (100)	0.9828 (100)
AHC-3-para-100ms	1.0004 (100)	0.9741 (100)	1.0006 (100)	0.9977 (100)	1.0249 (100)	0.9995 (100)
XHC-3-para-100ms	1.0004 (100)	0.9820 (100)	1.0006 (100)	0.9988 (100)	1.0249 (100)	1.0013 (100)
HC(c=0.5)-1000ms	1.0004 (1000)	0.9856 (1000)	0.9771 (1000)	0.9646 (1000)	1.0157 (1000)	0.9887 (1000)
HC-3-seq-1000ms	1.0004 (1000)	0.9804 (1000)	1.0003 (1000)	0.9976 (1000)	1.0086 (1000)	0.9975 (1000)
AHC-3-seq-1000ms	1.0004 (1000)	0.9795 (1000)	1.0007 (1000)	0.9982 (1000)	1.0266 (1000)	1.0011 (1000)
XHC-3-seq-1000ms	1.0004 (1000)	0.9830 (1000)	1.0006 (1000)	0.9991 (1000)	1.0266 (1000)	1.0019 (1000)
HC-3-para-1000ms	1.0004 (1000)	0.9856 (1000)	1.0006 (1000)	0.9987 (1000)	1.0240 (1000)	1.0019 (1000)
AHC-3-para-1000ms	1.0004 (1000)	0.9847 (1000)	1.0008 (1000)	0.9990 (1000)	1.0272 (1000)	1.0024 (1000)
XHC-3-para-1000ms	1.0004 (1000)	0.9853 (1000)	1.0008 (1000)	0.9996 (1000)	1.0272 (1000)	1.0027 (1000)
Zurel-1st	0.5710 (11040)	0.9690 (537)	0.9983 (2075)	0.9928 (1715)	0.6015 (1796)	0.8265 (3433)
Zurel	1.0000 (13837)	1.0000 (890)	1.0000 (4581)	1.0000 (4324)	1.0000 (3720)	1.0000 (5470)
casanova-10ms	0.2583 (10)	0.0069 (10)	0.0105 (10)	0.0202 (10)	0.2577 (10)	0.0632 (10)
casanova-100ms	0.2583 (100)	0.0069 (100)	0.0105 (100)	0.0202 (100)	0.2577 (100)	0.1107 (100)
casanova-1000ms	0.5357 (1000)	0.1208 (1000)	0.0861 (1000)	0.1486 (1000)	0.7614 (1000)	0.3305 (1000)
cplex-100ms	0.0000 (288)	0.0000 (121)	0.0299 (111)	0.0000 (150)	0.0000 (119)	0.0060 (158)
cplex-333ms	0.0000 (489)	0.0000 (393)	0.9960 (497)	0.9716 (354)	0.0000 (487)	0.3935 (444)
cplex-1000ms	0.0000 (1052)	0.0000 (1039)	0.9960 (1143)	0.9716 (1140)	0.0000 (2887)	0.3935 (1452)
cplex-3000ms	0.0000 (9171)	0.9338 (3563)	0.9964 (3030)	0.9716 (3077)	0.0000 (3090)	0.5804 (4386)

(each value in () is time in milliseconds).

Table 2. Time Performance of (k+1)th cycle on 100,000bids-256items (k=10)

	L2	L3	L4	L6	L7	average
HC-3-para-100ms	1.1098 (100)	0.9836 (100)	1.0003 (100)	1.0009 (100)	0.8688 (100)	0.9927 (100)
AHC-3-para-100ms	1.1098 (100)	0.9836 (100)	1.0003 (100)	1.0009 (100)	0.9941 (100)	1.0177 (100)
XHC-3-para-100ms	1.1098 (100)	0.9880 (100)	1.0003 (100)	1.0010 (100)	0.9939 (100)	1.0186 (100)
HC-3-para-1000ms	1.1098 (1000)	0.9880 (1000)	1.0003 (1000)	1.0010 (1000)	0.9814 (1000)	1.0161 (1000)
AHC-3-para-1000ms	1.1098 (1000)	0.9880 (1000)	1.0003 (1000)	1.0010 (1000)	0.9991 (1000)	1.0197 (1000)
XHC-3-para-1000ms	1.1098 (1000)	0.9889 (1000)	1.0003 (1000)	1.0011 (1000)	0.9990 (1000)	1.0198 (1000)
zurel-1st	0.8971 (74943)	0.9827 (2257)	0.9998 (5345)	0.9987 (4707)	0.7086 (8688)	0.9174 (19188)
Zurel	1.0000 (91100)	1.0000 (6036)	1.0000 (30568)	1.0000 (44255)	1.0000 (17691)	1.0000 (37930)
cplex-100ms	0.0000 (2022)	0.0000 (232)	0.0000 (143)	0.0000 (133)	0.0000 (852)	0.0000 (676)
cplex-333ms	0.0000 (2021)	0.0000 (559)	0.9998 (1084)	0.0000 (412)	0.0000 (852)	0.2000 (986)
cplex-1000ms	0.0000 (2021)	0.0000 (1045)	0.9998 (1085)	0.0000 (1328)	0.0000 (1285)	0.2000 (1353)
cplex-3000ms	0.0000 (3496)	0.0000 (3286)	0.9998 (5207)	0.9965 (3092)	0.0000 (15667)	0.3993 (6149)

(each value in () is time in milliseconds).

shows that our **XHC-3** could effectively reuse the approximated results of previous cycles.

In many settings of CPLEX, the values are 0. This is because CPLEX could not generate initial approximation result within the provided time limit. Only L4 and L6 have results for CPLEX. For them, CPLEX spends around 400 msec for the computation but the results are still lower than **greedy-3**. For L3, CPLEX could prepare results in 3.8 sec of computation, however, the result is still lower than **greedy-3**. This is because the condition we set up gave extremely short time limit so therefore CPLEX could not generate sufficient approximation results in such hard time constraint.

Table 3. Time Performance of (k+1)th cycle on 20,000bids-256items (k=2,5,10,20,40)

	k=2	k=5	k=10	k=20	k=40
HC-3-para-100ms	0.9828 (100)	0.9828 (100)	0.9828 (100)	0.9828 (100)	0.9828 (100)
AHC-3-para-100ms	0.9952 (100)	0.9979 (100)	0.9995 (100)	1.0003 (100)	1.0009 (100)
XHC-3-para-100ms	0.9952 (100)	0.9998 (100)	1.0013 (100)	1.0021 (100)	1.0028 (100)
HC-3-para-1000ms	1.0019 (1000)	1.0019 (1000)	1.0019 (1000)	1.0019 (1000)	1.0019 (1000)
AHC-3-para-1000ms	1.0019 (1000)	1.0021 (1000)	1.0024 (1000)	1.0026 (1000)	1.0027 (1000)
XHC-3-para-1000ms	1.0019 (1000)	1.0025 (1000)	1.0027 (1000)	1.0031 (1000)	1.0035 (1000)

(each value in () is time in milliseconds)

Table 4. Time Performance of intermediate cycles on 20,000bids-256items (k=2,5,10,20,40)

	k=2	k=5	k=10	K=20	K=40
HC-3-para-100ms	0.9889	0.9847	0.9829	0.9826	0.9818
AHC-3-para-100ms	0.9889	0.9805	0.9838	0.9874	0.9897
XHC-3-para-100ms	0.9892	0.9917	0.9943	0.9951	0.9966

(values are normalized as HC-3-para-1000msec equals 1)

Table 2 shows the experimental result on the datasets with 100,000 bids in an auction focused on execution time of the approximation. The settings are identical as Table 1 excluding the difference of number of bids in an auction. Due to hard time constraint, results of -seq-100ms (sequential execution with a cutoff time of 100ms) are excluded from the table since they could not complete their execution within the cutoff time. Here, our proposed methods (AHC-3,XHC-3) clearly have a certain advantage of their performance time ratio. HC-3,AHC-3,and XHC-3 produced acceptable approximated results within 100 to 1000 msec that are 2 to 443 times faster than Zurel's approximation. Especially, in most cases, our AHC-3-para-100ms outperforms HC-3-seq-1000ms and HC-3-para-1000ms.

On above experiments, we used $k = 10$. Table 3 shows average time performance of our algorithms on $k = 2, 5, 10, 20, 40$, respectively. At same cutoff time, XHC-3 obtains higher or at least same performance compared to HC-3. Furthermore, XHC-3-para-100ms outperforms HC-3-para-1000ms when $k \geq 20$, while its computation time is 10 times shorter.

On above experiments, for direct comparison to other existing algorithms, we have shown results on final(e.g., $(k + 1)$th) cycle in our procedure. We also confirmed performance improvement on intermediate cycle in our procedure. Table 4 shows average results on intermediate cycles for three algorithms (HC-3,AHC-3,and XHC-3). Here, since we do not have approximation results for those intermediate cycles on Zurel's algorithm, instead of using $ratioA$, we normalized all values as HC-3-para-1000msec equals 1. Since our algorithms improve results much more for latter cycles by cumulative reuse of the last cycle, we used first four cycles in this comparison. For results on $k = 5, 10, 20, 40$, we used an average value for first four intermediate cycles (e.g., from 2nd to 5th). Note that, only for results on $k = 2$, we used the results on 2nd cycle since we do not have other intermediate cycles when $k = 2$. Here, results of XHC-3 constantly better than AHC-3 and HC-3 and the differences are bigger when k is increased.

5 Limitations

The communication overhead problem is a problem of the communication overheads taken via an auctioneer and bidders to exchange bid information(e.g., [8]). In [20], Sandholm pointed out that it is relatively easy to solve a winner determination problem when it has a huge number of bids but a small number of items. It can be interpreted that the problem is rather communication overhead for gathering too much amount of bids[21]. Especially, it takes a certain overhead when we use a kind of agent communication protocol via the Internet to gather bid information for an auction.

In our experiments, we used CATS format file (a simple text file) to store information about bids in an auction. Typically, the program spent 200msec of CPU time to load 20,000bids, and 1000msec for 100,000bids. This is relatively large and hard since our cutoff time was set around 100msec to 1000msec for each approximation. When we gather such a number of bids via network rather than from a local storage, it will take much time. However, considering a differential bid updating approach for iteratively conducted combinatorial auctions, the overhead will be much smaller than loading all bid information for each time of iteration when there are small number of updates of bids. Therefore, it is meaningful to realize such fast approximation algorithms for a large number of bids in an auction. Furthermore, it is possible to make a concurrent mechanism that has a thread for gathering bid information for next auction and another thread for approximating winners in the current auction. In this case, there will be negligible overhead for loading bid data since it will be processed simultaneously. For above reason, we excluded communication overhead (e.g., overhead of loading data from a file) from the recorded computation time in our experiments. However, all differential bid updating overheads are included in the results since they are negligibly small.

Using sequential auctions[22] is another approach to overcome this communication cost problem. Koenig et,al. proposed a multiple-round auction mechanism that guarantees the upper bound of communication cost as fix size k, that is independent from the number of agents or items in the auction[23]. In our approach, we only assume that there is a small number of updated bids from the last round of auction. Although our algorithm itself can approximate winners within a very short time with huge number of updated bids, the communication cost problem still remains there. This is a limitation of our approach.

In this paper, we focused on winner determination problem so we eliminated other important issues. Pricing mechanism is an important part in auction mechanism. A number of pricing mechanisms are proposed for different goals and situations(e.g., [7],[24],etc.). Although it is possible to treat our mechanism as a simple 'ascending' combinatorial auctions, many issues will be left unsolved. In this paper, we left pricing problem as a future work.

6 Related Work

There have been a lot of works on the optimal algorithms for winner determination in combinatorial auctions[25]. Recently, Dobzinski et, al. proposed improved approximation algorithms for auctions with submodular bidders[26]. Lavi et, al, reported an LP based algorithm that can be extended to support the classic VCG[27]. Those researches are mainly focused on theoretical aspects. In contrast to those papers, we rather focus on experimental and implementation aspects. Those papers did not present experimental analysis about the settings with large number of bids we presented in this paper. Also, Guo[28] proposed local-search based algorithms for large number of bids in combinatorial auction problems. However, they did not present experiments with such a huge number of bids we used in our experiments.

CPLEX is a well-known, very fast linear programming solver system. In [8], Zurel et al. evaluated the performance of their presented algorithm with many data sets, compared with CPLEX and other existing implementations. While the version of CPLEX used in [8] is not up-to-date, the shown performance of Zurel's algorithm is approximately 10 to 100 times faster than CPLEX. In this paper, we showed direct comparisons to the latest version of CPLEX we could prepare. Our approach is far better than latest version of CPLEX for large-scale winner determination problems. Therefore, the performance of our approach is better than CPLEX in our settings. This is natural since Zurel's and our approaches are specialized for combinatorial auctions, and also focus only on faster approximation but do not seek optimal solutions. In case we need optimal solutions, it is good choice to solve the same problem by both our approach and CPLEX in parallel.

The above approaches are based on offline algorithms and therefore there are no considerations about addition and deletion of bids in their approximation processes. Although our algorithms are not strict online algorithms, it is possible to reuse the last results when bids are modified and recalculation is necessary.

7   Conclusions

In this paper, we showed an analysis about enhanced approximation algorithms for combinatorial auctions that are suitable for the purpose of iterative reallocation of items. We showed that our algorithms effectively reuse the last solutions to speed up initial approximation performance. The experimental results showed that our proposed algorithms outperform existing algorithms in some aspects. Furthermore, we showed that they outperform CPLEX, a sophisticated LP solver product. We found that in some cases reusing the last solutions may worsen performance compared to ordinary approximation from scratch. We showed an enhanced algorithm that effectively avoids the undesirable reuse of the last solutions in the algorithm. We showed this is especially effective when a non-negligible number of existing bids are deleted from the last cycle.

References

1. Cramton, P., Shoham, Y., Steinberg, R.: Combinatorial Auctions. The MIT Press, Cambridge (2006)
2. Sandholm, T., Suri, S., Gilpin, A., Levine, D.: Cabob: A fast optimal algorithm for winner determination in combinatorial auctions. Management Science 51(3), 374–390 (2005)
3. Sandholm, T.: Expressive commerce and its application to sourcing: How we conducted $35 billion of generalized combinatorial auctions. AI Magazine 28(3), 45–58 (2007)

4. McMillan, J.: Selling spectrum rights. The Journal of Economic Perspectives (1994)
5. Fukuta, N., Ito, T.: Periodical resource allocation using approximated combinatorial auctions. In: Proc. of The 2007 WIC/IEEE/ACM International Conference on Intelligent Agent Technology (IAT 2007), pp. 434–441 (2007)
6. Fujishima, Y., Leyton-Brown, K., Shoham, Y.: Taming the computational complexity of combinatorial auctions: Optimal and approximate approarches. In: Proc. of the 16th International Joint Conference on Artificial Intelligence (IJCAI 1999), pp. 548–553 (1999)
7. Lehmann, D., O'Callaghan, L.I., Shoham, Y.: Truth revelation in rapid, approximately efficient combinatorial auctions. Journal of the ACM 49, 577–602 (2002)
8. Zurel, E., Nisan, N.: An efficient approximate allocation algorithm for combinatorial auctions. In: Proc. of the Third ACM Conference on Electronic Commerce (EC 2001), pp. 125–136 (2001)
9. Hoos, H.H., Boutilier, C.: Solving combinatorial auctions using stochastic local search. In: Proc. of the Proc. of 17th National Conference on Artificial Intelligence (AAAI 2000), pp. 22–29 (2000)
10. Thomadakis, M.E., Liu, J.C.: On the efficient scheduling of non-periodic tasks in hard real-time systems. In: Proc. of IEEE Real-Time Systems Symp., pp. 148–151 (1999)
11. Andrew, L.L., Hanly, S.V., Mukhtar, R.G.: Active queue management for fair resource allocation in wireless networks. IEEE Transactions on Mobile Computing, 231–246 (2008)
12. Xie, T., Qin, X.: Security-aware resource allocation for real-time parallel jobs on homogeneous and heterogeneous clusters. IEEE Transactions on Parallel and Distributed Systems 19(5), 682–697 (2008)
13. Xiao, L., Chen, S., Zhang, X.: Adaptive memory allocations in clusters to handle unexpectedly large data-intensive jobs. IEEE Transactions on Parallel and Distributed Systems 15(7), 577–592 (2004)
14. Fukuta, N., Ito, T.: Towards better approximation of winner determination for combinatorial auctions with large number of bids. In: Proc. of The 2006 WIC/IEEE/ACM International Conference on Intelligent Agent Technology(IAT 2006), pp. 618–621 (2006)
15. Fukuta, N., Ito, T.: Fine-grained efficient resource allocation using approximated combinatorial auctions–a parallel greedy winner approximation for large-scale problems. Web Intelligence and Agent Systems: An International Journal 7(1), 43–63 (2009)
16. Fukuta, N., Ito, T.: Performance analysis about parallel greedy approximation on combinatorial auctions. In: Bui, T.D., Ho, T.V., Ha, Q.T. (eds.) PRIMA 2008. LNCS (LNAI), vol. 5357, pp. 173–184. Springer, Heidelberg (2008)
17. Fukuta, N., Ito, T.: Fast partial reallocation in combinatorial auctions for iterative resource allocation. In: Proc. of 10th Pacific Rim International Workshop on Multi-Agents (PRIMA2007), pp. 196–207 (2007)
18. Fukuta, N., Ito, T.: Short-time approximation on combinatorial auctions – a comparison on approximated winner determination algorithms. In: Proc. of The 3rd International Workshop on Data Engineering Issues in E-Commerce and Services (DEECS 2007) pp. 42–55 (2007)
19. Leyton-Brown, K., Pearson, M., Shoham, Y.: Towards a universal test suite for combinatorial auction algorithms. In: Proc. of ACM Conference on Electronic Commerce (EC 2000), pp. 66–76 (2000)
20. Sandholm, T.: Algorithm for optimal winner determination in combinatorial auctions. Artificial Intelligence 135, 1–54 (2002)
21. Lehmann, D., Müller, R., Sandholm, T.: The winner determination problem. In: Cramton, P., Shoham, Y., Steinberg, R. (eds.) Combinatorial Auctions, pp. 507–538. MIT Press, Cambridge (2006)
22. Boutilier, C., Goldszmidt, M., Sabata, B.: Sequential auctions for the allocation of resources with complementarities. In: Proc. of International Joint Conference on Artificial Intelligence (IJCAI 1999), pp. 527–534 (1999)

23. Koenig, S., Tovey, C., Zheng, X., Sungur, I.: Sequential bundle-bid single-sale auction algorithms for decentralized control. In: Proc. of International Joint Conference on Artificial Intelligence (IJCAI 2007), pp. 1359–1365 (2007)
24. Parkes, D.C., Cavallo, R., Elprin, N., Juda, A., Lahaie, S., Lubin, B., Michael, L., Shneidman, J., Sultan, H.: Ice: An iterative combinatorial exchange. In: The Proc. 6th ACM Conf. on Electronic Commerce, EC 2005 (2005)
25. de Vries, S., Vohra, R.V.: Combinatorial auctions: A survey. International Transactions in Operational Research 15(3), 284–309 (2003)
26. Dobzinski, S., Schapira, M.: An improved approximation algorithm for combinatorial auctions with submodular bidders. In: Proc. of the seventeenth annual ACM-SIAM symposium on Discrete algorithm (SODA 2006), pp. 1064–1073. ACM Press, New York (2006)
27. Lavi, R., Swamy, C.: Truthful and near-optimal mechanism design via linear programming. In: 46th Annual IEEE Symposium on Foundations of Computer Science (FOCS 2005), pp. 595–604 (2005)
28. Guo, Y., Lim, A., Rodrigues, B., Zhu, Y.: A non-exact approach and experiment studies on the combinatorial auction problem. In: Proc. of the 38th Hawaii International Conference on System Sciences (HICSS 2005), p. 82.1 (2005)

How to Integrate Personalization and Trust in an Agent Network

Laurent Lacomme[1], Yves Demazeau[1], and Valérie Camps[2]

[1] Laboratoire d'Informatique de Grenoble, Grenoble, France
[2] Institut de Recherche en Informatique de Toulouse, Toulouse, France
{Laurent.Lacomme,Yves.Demazeau}@imag.fr,Valerie.Camps@irit.fr

Abstract. Trust and personalization are two important notions in social network that have been intensively developed in multi-agent systems during the last years. But there is few works about integrating these notions in the same network of agents. In this paper, we present a way to integrate trust and personalization in an agent network by adding a new dimension to the calculus of trust in the model of Falcone and Castelfranchi, which we will call a similarity degree. We first present the fundamental notions and models we use, then the model of integration we developed and finally the experiments we made to validate our model.

Keywords: Trust, Agent network, Personalization, Social network.

1 Introduction

From Web Services to experimental negotiating agendas, many multi-agent systems have been developed to implement links between people or organizations in order to enable them to interact indirectly through agents that represent them. In such *social networks*, each agent stands for a person or a group of people.

These networks have often some particular properties. The first we are interested in is *openness*. In open networks, agents can be added or removed from the network at any time. This implies that the network evolves, while each agent needs to adapt its own behavior to the appearance or disappearance of partners. The second property is *partial representation*. In social networks, agents often only have little knowledge about others and about the network itself. For instance, when an agent is added to the network, it usually only knows a few other agents that we call its neighborhood. A third interesting property is *heterogeneity*. That is, in such networks, agents are not always homogeneous. Every agent can have individual skills that others do not have, and each agent is free to cooperate or not with known agents. So each agent has to choose cleverly its partners in this kind of networks, because these partners must fulfill some requirements for the partnership to be useful.

Hence such networks need some protocols for the agents to be able to act correctly while knowing only a few facts about a constantly evolving environment. One way to fulfill this requirement is to add a trust model to their reasoning abilities. This trust model enables them to take decisions, such as which agent to ask for doing a task, from the little knowledge they have. This is done by first computing probable

J. Filipe, A. Fred, and B. Sharp (Eds.): ICAART 2009, CCIS 67, pp. 247–259, 2010.

behaviors of others and the results of such behaviors, and then selecting the best ones for the agent.

On another hand, as the agents in these networks are used to represent human beings, users often want to have some control over them. Indeed, when agents are faced to choices, their reactions should be the closest possible to the users' own preference. One way to realize it is to add to the agents' reasoning methods a personalization model, which checks for alternatives and selects the one the user would prefer.

As a result of the two previous remarks, there is a need to include both trust and personalization models. But as both are reasoning methods that can lead to contradictory conclusions, we need a way to integrate them in the agent's global reasoning protocol. From our best knowledge, such an integration does not seem to exist yet.

In the remaining of this paper, we will first present the notions of trust and personalization in a network of agents, the theoretical criteria we will develop on these notions for our integration work and the trust model we chose as a foundation to our integration model. Then, we will describe the integration criteria and the solution we are proposing. And finally, we will present an experimental validation of our solution.

2 Positioning

There are many ways for agents to represent and compute trust they have in other ones, and there are also many ways for them to represent and exploit user's preferences. So, to understand how trust and personalization should interact in an agent's reasoning schema, we firstly need to describe what they are, how they work, and what their different possible models are.

2.1 Trust in Agent Networks

A trust model describes how an agent can use its past experience and others' experience to take decisions about future plans. It involves a facts storing and a reasoning method over this memory. As we are interested in user-representing agents, the most common way to describe agents' reasoning methods is through the *Beliefs, Desires, and Intentions* – BDI – paradigm (Rao & Georgeff, 1995).

Trust is based on trust evidences (Melaye & Demazeau, 2006), which are facts that are relevant to the question of trusting an agent or not, which can come from different sources. Common trust sources are: *direct experience*, which can be positive or negative, *reputation*, which is an evaluation that a third-party agent provides about another one, and *systemic trust*, which is the trust an agent has in a group of other agents, without necessarily knowing specifically each member of the group. These evidences are stored in a way so that they can later be used by a reasoning process, *i.e.* beliefs in the case of BDI agents. Moreover, all trust knowledge is contextual, *i.e.* it is related to an action or a goal Ω the agent wants to perform or achieve.

Most commonly, these beliefs are split into a small number of categories that are considered as trust dimensions. The most used dimensions can be described as *ability*, *willingness* and *dependence* beliefs (Castelfranchi & Falcone, 1998). The belief of *ability* means that an agent A believes that an agent B is able to do what A wants it to do in the context Ω. The belief of *willingness* means that A believes that B will do what A wants it to do if A asks it to do that. The belief of *dependence* means that A believes that he relies on B to achieve its goals in context Ω. There exist other dimensions, but trust can often be easily simplified to retain only these three ones without losing any accuracy.

Trust is then learnt through experience, interaction and reputation transmission, stored as agents' beliefs and then used in decision-making processes when agents have to make choices involving other agents.

Amongst the large amount of existing trust models, we need to rely on some criteria to make our choice and ground our work on an adequate model.

2.1.1 Some Theoretical Criteria for Trust Models

A trust model has to fulfill some criteria to be able to be used in an agent network to create what is usually called a *trust network*.

The first obvious criterion is *optimization*. This comes from the fact that a trust model is made to help agents to adapt their behavior to the network configuration. So the trust model must be able to improve the network's global performance – the ability for each agent to achieve its goals. *Optimization* can only be tested experimentally, because we are not able to foresee the network improvement given a particular theoretical trust model.

The second one is a practical requisite: the trust model should be easily calculable, in order for the agent to compute it in real-time without any significant lack of reactivity; we call it *calculability*.

The third one is related to the fact that the agents we describe are related to users. In many of these systems, users often want to be able to understand how the entity that represents them reacts. So, the *intelligibility* criterion describes the ability of the reasoning process and the semantic of stored beliefs to be explained to the user and understood by him.

The other criteria we take care about are four properties of the trust values (Melaye & Demazeau, 2006): *observability, understandability, handlability* and *social exploitability*. They describe the ability of the agent to apprehend other agents' mutual trust, to compute multi-dimensional trust values, to combine these values into a global trust level and to use this trust level to make decisions.

We will use these criteria altogether for both the choice of the trust model to ground our integration work and, later, for the integration model itself.

2.1.2 Falcone and Castelfranchi Trust Model

The trust model we used is the one introduced by Falcone et al. (Castelfranchi & Falcone, 2004). It is a BDI-based model – which corresponds perfectly to the social network requisites – and uses numerical representation for trust beliefs. Numerical representations are better for a trust network than logical representations, because these latest rely quite always on complicated and high-complexity modal logics, and so cannot be implemented.

This model uses three contextual values to represent trust dimensions (cf. figure 1): the *Degree of Ability* ($DoA_{Y,\Omega}$), the *Degree of Willingness* ($DoW_{Y,\Omega}$) and the *Environment Reliability* ($e(\Omega)$). The latest is the only one that does not depend on the agent B that is considered by A for the trust evaluation. It measures the intrinsic risk of failure due to the environment. The *Degree of Ability* measures the competence that B has to

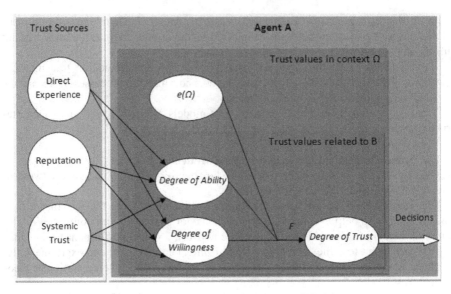

Fig. 1. Falcone and Castelfranchi Trust Model

accomplish the task or to help A to accomplish the task Ω. The *Degree of Willingness* measures the will of B to help A to achieve its goals. All these three values are defined in [0;1] and are combined into a global *Degree of Trust* ($DoT_{Y,\Omega}$) that describes the trust that A has in B in context Ω through a function F. This function is not specified in the model, but has to preserve monotonicity and to range in [0;1].

$$DoT_{Y,\Omega} = F(DoA_{Y,\Omega}, DoW_{Y,\Omega}, e(\Omega)) \tag{1}$$

Both of the ability and willingness beliefs – *DoA* and *DoW* values – can be learnt from any trust source. To learn these values, the agent uses a reinforcement learning process that uses new trust evidences to review its knowledge about others.

We chose this model because it satisfies very well the criteria we have for a trust model. The fact that trust is computed from values representing its dimensions with a simple formula guarantees the respect of *observability, understandability, handlability* and *intelligibility*. *Social exploitability* is also respected because trust values provide a ranking for potential partners that can be used to make a decision. And then, this model is *calculable*, because it is based on simple mathematical formula and numerical values.

2.2 Agents Personalization

The other fundamental notion in this work is personalization. Personalization is the ability for an agent acting on behalf of a user to acquire and to learn his preferences, his centre of interests and to use them during its decision-making process.

While preferences nature is quite domain-related, preferences representation has some universal methods and properties.

2.2.1 Preferences Representation Models

The notion of personalization handles a couple of distinct concepts. It is both a way to represent users' preferences, a way to learn them from the user and a method to use them in various contexts to improve the agent's behavior to the user's point of view.

There are two distinct ways to represent users' preferences (Endriss, 2006). They can be represented by a valuation function giving a note to alternatives the agent is faced to – and called *cardinal preferences*. They can also be described as a binary relation between each two of the alternatives – and called *ordinal preferences*.

There are many ways to represent these two kinds of preferences. But the most known and useful are probably the weighted conjunction of literals for *cardinal preferences* and the prioritized goals for *ordinal preferences*. These models describe a way to store user preferences but also a way to use them in the reasoning process by evaluating and choosing one between several alternatives.

All these preferences representation models can be combined with several well-known reinforcement learning techniques (Gauch et al, 2007), which will enable them to improve the precision of the user profile (*i.e.* the set of all represented user's preferences) and adjust it to the user's real preferences. The learning process can use an *explicit feedback*, which can be, for instance, a form that the agent presents to the user. It can use an *implicit feedback*, which is the analysis of the user activity, for instance, the user web history for web navigation assistants. Or it can use *hybrid feedback*, which is a combination of both (Montaner et al, 2003).

2.2.2 Some Criteria for Personalization

As for trust models, we proposed a set of important criteria that a personalization model has to respect in order to be useful in an agent network.

In our work, we have kept four usual criteria about personalization models (Endriss, 2006) and added a new one. Firstly, the *expressive power* is defined as the amount of preferences structures that the model is able to represent. Secondly, *succinctness* is the amount of information about these preferences which can be stored in a given place. Then, we have to take care about *elicitation*, which represent the ease with which a user can formulate his preferences in the model's representation language. This is an important criterion, especially when a user is able to see directly his profile and to modify or correct it on his own. And finally, as it is also the case for trust model, the *complexity* of the model is important in order to be able to be computed in real-time by agents.

Since the preferences learning mechanism is a dynamic process, we have to describe the ability of the system to react to any change in the user's preferences. This is why we add another criterion, *reactivity,* which measures how much time the model takes to adapt the profile to a change in the user's behavior it represents.

3 Integration Work

Concepts of trust and personalization having been studied in a network of agents, we will now see why there is a necessity to find a way to integrate these two reasoning processes into a single one.

3.1 Personalization Integration in a Trust Network

In a lot of networks – agent-based social networks, B2B applications and negotiating calendars for instance – the agents have to make decisions about which other agents to interact with, which ones to ask information to, and with which to form teams or to take contracts and partnerships. Agents often take these decisions via a trust mechanism, but in such networks, the user is often able to express preferences that should influence these choices. So, as trust and personalization needs to coexist in those networks, we obviously need a way to make them function altogether.

3.1.1 The Necessity of Integration
We can first believe that simply putting both models on the same agents will be enough to make the network work well. But this cannot be true, because both being reasoning processes that cost much time and resources and that leads to conclusions, there will be two main problems happening. The first one is that the cost of both inferences will be high for an agent. The second one, and most important, is that the two inferences can lead to different and perhaps incompatible conclusions. And the problem will be: how to handle these two conclusions and act while taking both into account.

So the way to solve this problem is to create a single reasoning process for agents, that takes into account all the knowledge they possess, about both the network (other agents) and user's preferences.

3.1.2 Related Work
The only work we have found in the literature that tries to integrate an approach of personalization and trust (Maximilien & Singh, 2005) proposes a model of multi-criteria trust in which the user has some control over the importance of each evaluation following a particular criterion in the final trust calculus. This is a very limited and particular sort of personalization, and this approach is not applicable for the kind of networks we are interested in, as the preferences are related to the trust model itself and not to the domain the agents are concerned with.

So, to the best of our knowledge, no work exists that tackles the interaction of these two notions in an agent in the way of providing a single reasoning process that handles both notions.

3.2 Considered Agents and Network

To be able to explain clearly how the solution we propose works, we first need to describe the agents and the types of networks in which it will be applied.

3.2.1 Agents' Architecture and Capacities
The agents are based on the BDI architecture. This means that (*i*) they possess some *beliefs* about their environment – including the users of the system – and other agents and (*ii*) that they all have some goals, called *desires*, which are states – personal or of the environment – they wish to be true. In order to make these goals true, (*iii*) they use plans to make decisions that become *intentions* – things they plan to do.

As every agent does not have all the ability needed to achieve every one of its goals, it has to cooperate with some other agents in the network. To minimize the cost of this required cooperation and to avoid losing time and resources asking wrong agents for help – wrong agents are those which cannot help or will not help – it uses some trust process to determine which agents are the best partners for a specific task by the mean of the previously described Falcone and Castelfranchi trust model.

Every agent should also be related to a single user, and should be able to stock and use a preference profile related to this user. We will see later how the agents are able to do that.

3.2.2 The Network Structure

The network is merely an evolving set of agents which are able to communicate one with another through a message protocol that enables them to exchange data, requests, answers and perhaps beliefs and plans.

When a new agent is added to the network, it knows a few other agents – its neighborhood – and can learn the knowledge of other agents by interacting with them. It order to make agents able to learn the existence of unknown other agents, we must include in the network a mechanism that makes agents who are not able to process an information or answer a request forward this request to another agent. This mechanism involves only agents that are not concerned by a request or information but are cooperative – they are ready to help the request sender or to distribute information in the network. When such an agent receives an irrelevant message from its viewpoint it does not ignore it but forwards it to one or several agents of its own neighborhood which it considers as the most able to answer to this message; this mechanism is called *restricted relaxation* (Camps & Gleizes, 1995). The number of times a message can be forwarded in the entire network is obviously limited to avoid cycles and thus network overload.

3.3 Integration Constraints

As for trust and personalization models, we propose some theoretical criteria needed to be fulfilled by an integration model.

The first criterion derives directly from the fact that we want to integrate a personalization model in a trust model: this operation needs to result in an improvement of the correlation between the user's expectations and the agent's behavior. Hence the criterion of *accuracy* will be the measure of proximity between user's desires and agent's observable behavior and results.

Two other criteria will simply describe facts that are linked with the operation we are trying to do. The *target correlation* describes the fact that the alternatives considered by the preferences profiles are defined by the integration model we choose. In other words, the personalization model must be able to evaluate the kind of alternatives that the trust model will require it to evaluate. On the other hand, the *type correlation* criterion describes the fact that the personalization model should give as a result of an alternative evaluation a value of a type that is useful to the trust model.

Finally, because both of the two models we are trying to integrate are contextual models, we have to be sure that the contexts defined in both of the models are compatible – which means that they are the same or at least one is a subdivision of the other. We will call it the *context compatibility*.

3.4 Towards an Integration Model

Taking into account all the listed criteria, and basing our work on the chosen trust model, we developed a solution for an integration of a personalization model into this trust model. This integration has been done in order to create a reasoning process that handles information about both user's preferences and other agents' behavior.

3.4.1 Description of the Proposed Integration Model

The solution we propose simply consists in defining a new dimension of trust in an already multi-dimensional trust model (cf. figure 2). Indeed, in order to take into account the personalization evaluations, we considered that it was quite the best solution to keep the preferences representations and learning methods as separate agent's ability. We made this choice because it seemed very difficult and confusing – both for programmers and users – to incorporate it to the trust model reasoning process.

So this process is only going to use the evaluator from the personalization model to rank alternatives between other agents' behaviors or results. It is also going to learn a new context-dependant, agent-dependant belief that represents the proximity this

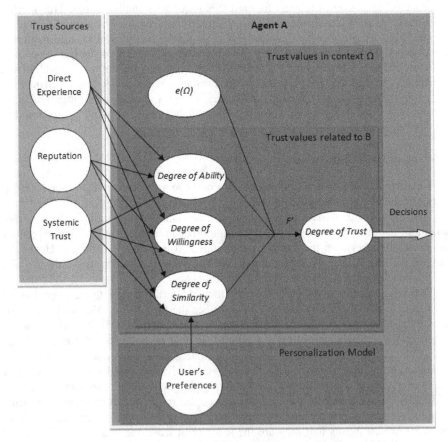

Fig. 2. Modified trust model with integrated personalization model

agent's behaviors or results have with the user's expectations concerning a particular context.

This belief, defined as a numerical value ranging in [0;1], as other trust values, is also going to be learnt by reinforcement learning methods using personalization model's evaluators. This will be called the *Degree of Similarity $DoS_{Y,\Omega}$*. Then, the global trust computation will have to be redefined as another monotonic function F' that ranges in [0;1], which takes as parameters not only the parameters of the function F from the trust model but also the newly defined *Degree of Similarity* :

$$DoT_{Y,\Omega} = F'(DoA_{Y,\Omega}, DoW_{Y,\Omega}, DoS_{Y,\Omega}, e(\Omega)) \qquad (2)$$

Moreover, this new dimension of trust can be learnt from every trust source that is available in the trust model we use. However, to learn it from reputation an agent will have to take into account the similarity level between itself and the evaluator, as they do not have the same user profiles.

The constraints that this integration model makes on the choice of a preferences representation model can be explained through the *type correlation* and *target correlation* criteria: the Falcone and Castelfranchi trust model uses numerical values in [0;1] as trust dimensions values. So, as our new belief will also have to be represented as such a numerical value, and because, for learning, the evaluatedelements will be the results of one interaction with another agent, the personalization model's evaluation functions will have to evaluate a single result and give a numerical value as an evaluation. So the personalization model should be a *cardinal preferences* model.

3.4.2 Criteria Applied to the Proposed Model

As previously shown, the original trust model we used fulfills all required criteria. So, as we have just added a single trust dimension, with its own meaning and its own learning methods, these criteria will not be broken. The *intelligibility* will also be respected, because the meaning of this dimension is easily explainable to the user; it represents the proximity between the real behavior of the target agent and the theoretical behavior it should have, taking into account the user's preferences. In fact, only the *optimization* criterion needs to be experimentally tested.

The criteria related to the personalization model could easily be satisfied, because the choice of the model is quite free between all the cardinal preferences models. They can be satisfied, for instance, by choosing the weighed conjunctions of literals model, which is a light, low-complexity and powerful model which can handle every domain of application and is perfectly compatible – if correctly implemented – with the contexts of the trust model.

Amongst the integration criteria, the two *correlation* criteria are easily respected, as seen before. The *context compatibility* can also be satisfied by correctly implementing the personalization evaluators. So, only the *accuracy* criterion should be experimentally tested.

4 Tests and Experiments

In order to first experiment the integrated trust and personalization model we proposed and then to check the two criteria that we can only validate experimentally – optimization and accuracy – we implemented a simplified version of the model, using the agent programming language and IDE Jack (http://www.agent-software.com/shared/products/).

4.1 A Simplified Model

We firstly simplified the model to test only the two experimental criteria we exposed – the goal of the test was not to determine the efficiency of the Falcone and Castelfranchi trust model nor of any personalization model, but to experiment if the integration model itself is viable.

The first simplification we decided was the implementation of a mono-source trust model. The only source that we considered was direct experiment. So the only trust evidences that were taken into account by the trust learning process were answers given by other agents to the sent requests.

The second simplification we decided concerns the environment; it was supposed to be sure – every message reaches its addressee – and resourceful – if an agent has both the ability and the willingness to perform an action, then the action is performed. So we considerer $e(\Omega)=1$.

We also selected simple functions for trust computation and for learning. F' (for complete model) and F (for trust only model) are defined as simple multiplications between each dimension.

$$DoT_{B,\Omega}= F'(DoA_{B,\Omega}, DoW_{B,\Omega}, DoS_{B,\Omega}) = DoA_{B,\Omega} \times DoW_{B,\Omega} \times DoS_{B,\Omega} \qquad (3)$$

$$DoT_{B,\Omega}= F(DoA_{B,\Omega}, DoW_{B,\Omega}) = DoA_{B,\Omega} \times DoW_{B,\Omega} \qquad (4)$$

Learning method is defined as a weighted mean of current and new values with fixed ratios – let DoX be DoA or DoW, and $a,d \in [0,1]$.

For positive trust evidence:

$$DoX(t+1) = DoX(t) + a * (1-DoX(t)) \qquad (5)$$

For negative trust evidence:

$$DoX(t+1) = DoX(t) - d * DoX(t) \qquad (6)$$

And for the similarity value – in case of positive or negative trust evidences:

$$DoS(t+1) = (DoS(t) + mod_{DoS}(B,\Omega))/2 \qquad (7)$$

Finally, we decided to use a static representation of preferences described by a very simple user profile that enables an agent to rate every result it receives in [0;1]. We emulated a preferences model in that way, because preferences dynamics was not very important for these tests and, given the high number of possible preferences representations, this would not be significant anyway. So we faked a preferences model that would have reached a stable state by attributing a simple static profile to every agent.

4.2 The Experimental Protocol

We experimented in a network of 100 homogeneous agents able to possess 3 basic capacities A, B and C. Each newly created agent randomly receives the ability to use each one of the 3 capacities with a certain probability – $p = 0.6$ for most of the tests.

All the agents are able to communicate through a message protocol defined in the Jack interface, and all use *limited relaxation* paradigm – with a maximum of 5 successive relaxations for a message.

We used randomly generated initial neighborhood for each agent with a probability of knowing each other agents equal to 0.1.

Then, the process continues step by step. At each step, a goal is generated for each agent, which consists in using a random capability: A, B or C. Obviously, when the agent does not possess this particular capability, it has to cooperate with other agents to achieve its goal.

To emulate preferences evaluation, each result of a capability A, B or C, is a document, which is assessable by the personalization model of any agent, according to a simple user profile it possesses. So, when an agent uses one of its capabilities or gets a result from another agent, it is able to rate this result according to its own preferences profile. We thus defined $mod_{DoS}(B,\Omega)$ as the average pertinence of documents given to A by B as an answer to a request from A.

4.3 Experimental Results

We evaluated the number of messages exchanged between agents and the number of goals that were not achieved by agents to validate the *optimization* criterion. We also evaluated the average pertinence of results for the *accuracy* criterion.

Each value is measured at step 1 and step 100 for 4 different networks: (*i*) a simple network without any model, (*ii*) a network with the simplified trust model – without personalization –, (*iii*) a network with the trust and personalization integrated model, and (*iv*) a network with a model that only takes into account the personalization value. Then the results between step 100 and 1 are compared to measure improvement. Expected results (cf. table 1) are an increase for pertinence and a decrease for the two other values.

The results fit with our expectations: the number of messages and the number of failures decrease for all networks where there exist trust models, and the average pertinence of results increases significantly in networks where personalization is taken into account.

Table 1. Experimental results summary: evolution between steps 1 and 100

	Number of messages	Number of goal failures	Average pertinence
No trust nor personalization	-3.6%	+1.6%	-3.3%
Trust model only	-13.7%	-77%	+3.5%
Trust and personalization model	**-21.5%**	**-76%**	+11.9%
Personalization only	+34.3%	+27%	**+42.3%**

Complementary observations can be made; for instance, while in much cases network *optimization* is the same for trust only and for trust and personalization networks, we can observe that when the pertinence results are too often very low, the network obtains worse results, because of the unbalanced importance of the different trust dimensions; we can notice that this effect should probably be corrected by adjusting correctly the global trust computation function.

But, globally, we can conclude from these experiments that the two experimental criteria that we had expressed are satisfied, and so, that our model seems to be an adequate solution to the problem we wanted to address.

5   Conclusions and Perspectives

In this work, we have explored the possibility of associating trust and personalization paradigms in an agent network. We have done this in order to give to agents the ability to handle both the intrinsic uncertainty of a partial-knowledge, evolving network, and the also evolving requirements of a user's set of preferences. Indeed, agents would have to face them both in a social network in which each user has one or more agent to represent him.

Knowing that just putting together the two reasoning methods leads to heavy problems of optimization but also to problems for mixing the results given by each one, we have looked for a solution to integrate both notions in a single reasoning process. We first gave some theoretical criteria to choose every component of a global agent's reasoning method that could handle both trust and personalization: the trust model, the personalization model and the integration model. We then proposed a complete solution that is acceptable according to those criteria.

Our solution involves the Falcone and Castelfranchi trust model, to which we added a new trust dimension, that we called *degree of similarity*. It also involves a cardinal preferences model such as weighted conjunction of literals, which is used by agents to evaluate results and alternatives and learn the *degree of similarity* they have with other ones.

The obtained experimental results for *optimization* and *accuracy* criteria seemed to validate these criteria. That is why even if these experimentations were done on a simplified version of the trust and personalization integration model, we can say that the solution we proposed seems to be viable and to be applicable to the kind of networks we described. This model was developed in order to improve the behavior of agents in these open, partial-knowledge and user centered networks, and it seems to achieve this goal.

Future work on this solution is to test it with a full and multi-source implementation with dynamic personalization from real users. As the Falcone and Castelfranchi model is very powerful and because of the large scale of different cardinal preference implementations that can fit in the theoretical criteria of this solution, we can foresee very different solutions for various domains and the need to find the adequate personalization evaluation and trust evaluation functions to each model.

References

1. Camps, V., Gleizes, M.-P.: Principes et évaluation d'une méthode d'auto-organisation. In: 3èmes Journées Francophones IAD & SMA, St Baldoph, pp. 337–348 (1995)
2. Castelfranchi, C., Falcone, R.: Principles of trust for mas: cognitive anatomy, social importance, and quantification. In: 3rd Int. Conf. on Multi-Agent Systems, ICMAS 1998, Paris, pp. 72–79 (1998)
3. Castelfranchi, C., Falcone, R.: Trust dynamics: How trust is influenced by direct experiences and by trust itself. In: 3rd Int. J. Conf. on Autonomous Agents and Multiagent Systems, AAMAS 2004, New-York, vol. 2 (2004)
4. Endriss, U.: Preference Representation in Combinatorial Domains. Institute for Logic, Language and Computation, Univ. of Amsterdam (2006)
5. Gauch, S., Speretta, M., Chandramouli, A., Micarelli, A.: User Profiles for Personalized Information Access. The Adaptive Web, 54–89 (2007)
6. Maximilien, E.M., Singh, M.P.: Agent-Based Trust Model Involving Multiple Qualities. In: 4th Int. J. Conf. on Autonomous Agents and Multiagent Systems, AAMAS 2005, Utrecht (2005)
7. Melaye, D., Demazeau, Y., Bouron, T.: Which Adequate Trust Model for Trust Networks? In: 3rd IFIP Conference on Artificial Intelligence Applications and Innovations, AIAI 2006, Athens. IFIP (2006)
8. Montaner, M., López, B., De La Rosa, J.L.: A Taxonomy of Recommender Agents on the Internet. Artificial Intelligence Review 19, 285–330 (2003)
9. Rao, A., Georgeff, M.: BDI Agents: From theory to practice, Tech. Rep. 56, Australian AI Institute, Melbourne (1995)

Modeling Two Stage Preventive Medical Checkup Systems with Social Science Approaches

Andreas Martischnig[1], Siegfried Voessner[1], and Gerhard Stark[2]

[1] Department of Engineering- and Business Informatics, Graz University of Technology
Kopernikusgasse 24, 8010 Graz, Austria
[2] Department of Internal Medicine, LKH Deutschlandsberg
8530 Deutschlandsberg, Austria
{andreas.martischnig,voessner}@tugraz.at,
gerhard.stark@lkh-deutschlandsberg.at

Abstract. Modeling preventive medical checkup systems (PMCS) is an important part to predict the future demand for healthcare coverage. In this paper we show how to model a two stage interdependent System as it applies to basic cancer prevention. Starting with a short introduction of the two used social science modeling techniques we show the basic principle of the preventive cancer checkup process (PCCP) and how it was modeled with these opposing approaches. We then extract the key benefits from each technique and also their shortcomings when applying it onto the PCCP. Furthermore we show at what level of detail which method should be used to gain the most valuable insight into those complex checkup systems.

Keywords: Agent based modeling, System dynamics, Healthcare, Preventive medical checkup, Preventive cancer checkup.

1 Introduction

In medical science, especially health care, computer simulation is still a relatively young field. In contrast to that social sciences use computer simulation as a well-established domain of research, to gain insight to a system and make predictions for the future. Troitzsch [19] divided prediction into two parts: (1) qualitative prediction, which is prediction of behavior modes, and (2) quantitative prediction, which is to predict a certain system state in timeline. Currently there are two major schools, System Dynamics and Agent Based Modeling, which use computer simulation to gain insight into non-linear social and socio-economic systems [13]. Both approaches have a broad overlap in research topics, but have been quite unnoticed by each other. [15]

There are only a few publications about health care systems concerning prevention frameworks. The health care system itself is complex and large and it is quite hard to understand all the dependencies and influences in this system. Because of the constantly growing demand for preventive cancer checkups, the main purpose of this paper is to show how to model those systems with both approaches.

There are two major facts concerned with futures healthcare coverage. People are constantly getting older and therefore need more and longer treatment. Western

J. Filipe, A. Fred, and B. Sharp (Eds.): ICAART 2009, CCIS 67, pp. 260–269, 2010.

industrial countries are facing an over aging of their population. These facts make it necessary to model future health care scenarios to get valid answers to problems arising from these systems because media seems to continuously bombard us with one horror scenario of health care issues after the other. For example the amount of people in Austria above the age of 60 will grow till 2030 by 54% although the whole population will just grow by 8% [16]. Is this significant increase in older people an indication requiring 50% more medical specialists to cope the demand of preventive medical checkup in this age group? This is just one pressing question concerning preventive medical checkups for the future. In this paper we will discuss the main modeling differences of the two approaches based on the preventive cancer checkup process (PCCP) and give a first short answer to the question above.

1.1 The System Dynamics Approach

System Dynamics is an approach that has been developed by Jay W. Forrester, an electrical engineer, in the mid 1950s and was originally called Industrial Dynamics since the initial applications, which he described in the book of the same title, were all in private industry [7]. Later works focused on urban dynamics [8] and on social systems, with the probably most popular publication "Limits to growth" [6]. In 1983 the International System Dynamics Society (SDS) has been established, and within it a special interest group on health issues was organized in 2003 [9]. Although many papers dealing with health care systems have been published, in a variety of journals worldwide, since then very few of them focused on prevention frameworks. [11]

The basic concept behind System Dynamics is that the complex behaviors of organizational and social systems are the result of both reinforcing and balancing feedback mechanisms. The central observation point when modeling a system in SD is to describe its feedback loops, which consist of the real-world processes, called stocks, and the flows between these stocks. These generated computerized models can then be used to test alternative scenarios and policies in a systematic way to answer both "what if" and "why" questions [3], [17].

1.2 The Agent Based Modeling Approach

Agent Based Modeling (ABM) is a relatively new computational modeling paradigm. Although it had been developed in the late 1940s, it did not become widespread until the 1990s, because compared to SD significantly more computational power is required. The increase of available and powerful computational resources in the last years and the inherent parallel nature of ABM approaches contributed to their popularity. The roots of ABM can be traced back to Cellular Automata (CA) and the field of Complex Adaptive Systems (CAS) with the underlying notation to build a system from the ground-up in contrast to the top-down view of SD. There are three different fields of research for ABM: (1) artificial intelligence, (2) object oriented programming and concurrent object-based systems, and (3) human-computer interface design [10]. The concept of agents can be tracked through many different disciplines, but using agents on designing simulation models is mainly applied in complexity science and game theory [13]. In contrast to SD there is no universally accepted definition of ABM and this makes it much more difficult to identify the basic concept and assumptions underlying this paradigm. An Agent is basically an independent component that has individual rules and is able to interact with its environment or not. The behavior can range from primitive reactive decision rules to complex adaptive intelligence [12]. The global System behavior emerges as a result of the agents following their rules and doesn't need to be known at the beginning of the modeling session.

That's why ABM is often called bottom-up modeling [3]. Agent Based Modeling is used in a wide range in medical health care but mostly to simulate patient scheduling and workflow management [14]. Estimating the medical demand of equipment and specialists for the future is quite a new area for ABM.

1.3 Short Comparison of the Approaches

To characterize both approaches, the major differences are summarized in Table 1 and described below [13], [18].

Table 1. System Dynamics versus Agent Based Modeling

	System Dynamics	Agent Based Modeling
Basic building Block	Feedback loop	Agents
Level of modeling	Macro	Micro
Mathematical formulation equations	Differential equations	Logic, Differential equations
Perspective	Top-down	Bottom-up
Unit of analysis	Structure	Rules

The core building blocks:
The main behavior of a System Dynamics model is generated by its interacting feedback loops that consist of Stocks and Flows. In Agent Based Models the behavior emerges from the interaction rules of the Agents. These elements can therefore be considered as the basic building blocks of their approaches.

Level of modeling:
In macro simulations, individuals are viewed as a structure that can be characterized by a number of variables, whereas in micro simulations the structure is viewed as emergent from the rules and the interacting individuals. [5]

Mathematical formulation:
The basic principle behind SD is to couple non-linear first-order differential equations. This is done by Levels that accumulate the difference between the Flows (in- and outflows). In ABM there are many diverse methodologies from logic-based to emergent equations and that's why no universally accepted formalism for the mathematical description of a model exists. [13]

Perspective:
In SD the structure of the basic system phenomenon is modeled and in ABM this evolves in the simulation.

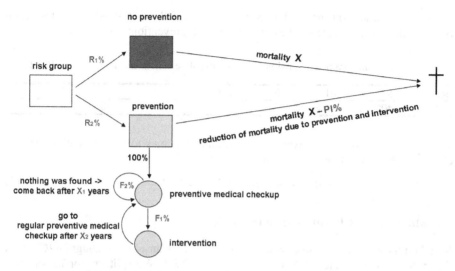

Fig. 1. Basic principle of a preventive cancer checkup process (PCCP)

Unit of analysis:
SD models behavior is determined by the structure that is fix and has to be defined before simulation. In ABM the focus lies on the rules an agent obeys to, to interact with other ones.

2 The Basic Preventive Cancer Checkup Process (PCCP)

Modern preventive cancer checkups can diagnose cancer risks at a very early stage making necessary treatment easier, more effective, and more efficient. Most of the common malignant diseases, if detected in an early stage, can successfully be cured, due to tremendous progress in treatment possibilities. That's why regular checkups can prolong a healthy life.

The basic preventive cancer checkup process that is shown in Figure 1 can be applied to all of the malignant diseases for example (colon cancer, prostate cancer, gynecological tumors, skin tumors, etc.). There is always a risk group in a population, normally being addressed by age and gender. This group can then be divided into two parts (percentage R1 and R2): the ones that will never go to a preventive medical checkup and the other ones that go to a preventive medical checkup at least once in their lifetime after entering the specific risk group. Once entering the prevention path there will be a medical checkup. If an indication for the specific cancer is found during the checkup an intervention will be performed and the patient will be send back to regular preventive medical checkup after some years (indicated by X2). If no indication is detected the patient will also be sent back to regular preventive medical checkup after some years (indicated by X1). Once being in the prevention cycle the normal mortality for the specific cancer will decreases with a given percentage (indicated by PI). The basic PCCP will now be applied onto the colon carcinoma one of the most common cancer type of men and women.

To demonstrate both principles we assumed the following standard values, taken from literature [1], [2], [4] for the colon carcinoma prevention:

Table 2. System parameters for simulating a preventive cancer checkup process (PCCP)

R1	R2	F1	F2	X1	X2	PI	X
60%	40%	10%	90%	7	3	80%	$0,45 * 10^{-3}$

With this given values the average year a patient comes to the preventive medical checkup is 6.6 according to equation (1).

$$\text{Average year} = X1 * F2 + X2 * F1. \tag{1}$$

2.1 Modeling PCCP with System Dynamics

Based on the basic PCCP process we designed a first Causal Loop Diagram (CLD) of the system and simulated it in Powersim Studio 2005. We split up populations age groups into those within the risk group and those outside. Because of the intuitive user Interface of Powersim the model was quickly built but the output did not quite match real systems data because SD averaged all the Stocks representing the age groups. Population distributions in Western industrial countries are more like bulbs or apples than rectangles and because of the two world wars and the baby boom generation Austria's population distribution has two abnormal spikes. And these two spikes are completely filtered in the standard SD model.

So we split up the age groups into one year groups and added both prevention cycles to the simulation to get a more detailed output. A simplified version of the extended basic Causal Loop Diagram (CLD) is shown in Figure 2.

The implemented model now was an "Array Model" with all the different probabilities for each group and the output was qualitatively quite near to real data.

To look at the consequences of another cancer prevention model, for example prostate cancer, we added a second cycle for this disease. This was really a challenging problem because of the arrays and global death rates and at the end we weren't able to complete it because of cyclic references. Both prevention models affect the death rate of the population and are also affected by this rate. When you think in stock and flows you get cyclic references between these rates. Our basic SD model can only capture the qualitative behavior well but lacks realistic quantitative output. The extended model is able to produce a realistic quantitative output but is due to the specialization not able to handle more than one prevention model.

2.2 Modeling PCCP with Agent Based Modeling

In the ABM solution we first had to decide what defines an agent to produce an output like the real data. So we decided to model an agent with the basic attributes like number, age and gender and some medical attributes we needed for the preventive checkup process as shown in Figure 3. In this first solution we modeled non interacting agents, because it was not necessary for the concerning question.

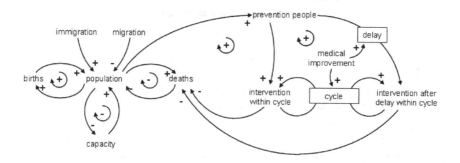

Fig. 2. Basic principle of a preventive cancer checkup process (PCCP)

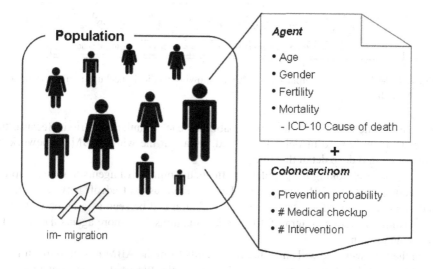

Fig. 3. Population of agents with migration effects

Before implementing the PCCP into our agent framework we had to calibrate our agents to build up a population that was quite similar to the real one in each age group. That's why we had to add immigration, migration and fertility data to each agent. In this case we took statistical data rates from the past decades and added them to the framework. This data can now be loaded from several input files into the framework. Furthermore this attributes can be changed in time to get a similar characteristic as data from the past. Mortality is divided into the main parts of the ICD-10 (International Classification of Diseases endorsed by the WHO in 1990) code and can also be changed in time. Due to this classification the framework is able to handle all different types of classified diseases. To add a specific prevention model one first has to define the ICD-10 category it belongs to and then add the needed attributes to an agent. In our case this new "disease data sheet" that is connected to an agent contains the number of performed interventions, the waiting period till next check is performed, the new death probability, and so forth. Depending on the input data that

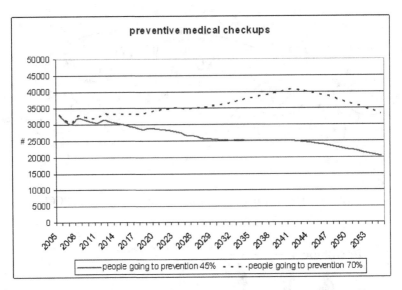

Fig. 4. ABM-Framework output for the PCCP, showing medical checkups as a consequence of different policies

is linked to the agents they act on probabilities each simulation period. Because this paper is about how to model a PCCP and not about the whole ABM framework we will not go into deep detail this time.

Since we are looking for population effects the number of agents that make up this population has to be sufficiently high. There is obviously a tradeoff between accuracy and computational effort. Agent Based simulation can be seen as a numerical solver to Dynamic System's system of differential equations. The more agents the smoother is the integration.

In the following we will show the first results from the ABM model to illustrate the great level of detail our framework is able to handle. We used 1.5 million agents and 50 simulation runs to get a robust estimate of mean and standard deviation.

The output for the PCCP with the given values for the colon carcinoma was really astonishing for us and is shown in the Figures 3 and 4. Although more people are entering than leaving the risk group the demand for preventive checkup will not grow when we assume that the same percentage of people as today will go to checkup in the future. This is because most of the demand is already generated by the people in this two stage cycle. The demand will not grow until the prevention percentage is set up to more than 60% and this is in fact a relatively unrealistic scenario for the future. In Figure 4 we see the absolute difference of people dieing from colon cancer per year. The absolute amount of people that could be saved due to more preventive checkups will not dramatically fall just by doing 55% more of these checkups. This output is really crucial when we think about investing more money in these preventive checkups or advertisement to increase the amount of people going to cancer prevention.

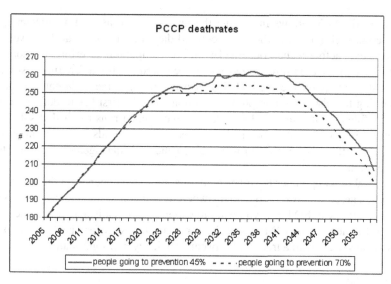

Fig. 5. ABM-Framework output for the PCCP, showing PCCP death rates as a consequence of different policies

3 Discussion

During our modeling sessions we were able to produce the needed output data with both modeling techniques. Building a SD model with realistic real life behavior was really a hard challenge, because of the averaging effect within stocks. Despite all difficulties we found a solution by transferring the initial model into an "Array Model". Due to the specialization of this model it is not possible to simulate more than one PCCP as mentioned above. That's why we had to switch the modeling approach to implement the given PCCP with ABM. After defining the attributes and rules of an agent we implemented our own arbitrary extendable framework. Because of the astonishing answers for the future demand in specialists for colon carcinoma the framework will now be object of further research. Integrating more cancer prevention models, interactions between the agents like transmissibility of diseases, word of mouth advertising for preventive medical checkup, are just a few work packages for the future.

In general both techniques can not be differentiated just by modeling size because both are capable to model small and large-scale systems. They can rather be classified by the problem or perspective and the required output information. One fact that should be considered when deciding for one technique is that with today's modeling tools it is much more complicated to implement a solution in an ABM Framework, when you are not experienced in programming, than implementing a model in one of the intuitive graphic oriented SD tools. Quantifying the parameters of a model is the main difficulty both approaches have in common. In ABM it is tough defining the rules for the agent's behavior and their attributes and in SD it is sometimes quite hard to quantify or find the correlation function between the connections of variables. In contrast to SD ABM allows increasing the level of detail as long as relevant data is available but will not work when this required data does not exist at that level of detail. The next factors to be concerned with are computational effort, memory management, and simulation time. SD provides the output within a few runs lasting only seconds depending on the method that is used to solve the differential equations. When trying to solve the same problem with agents one first has to define the width of the confidence interval and then calculate the needed runs to hit that spread. The simulation time with our model in SD is just a few seconds on an ordinary office computer and there is no need to worry about memory management contrary to our ABM solution.

In general picking one or the other modeling approach depends on the system to be simulated. There are lots of applications where it is much easier and efficient to solve given problems with SD but if you want to capture more realistic real-life phenomena you have to choose the ABM approach. A general decision for one of the two techniques always deals with a trade-off between efficiency and significance.

4   Conclusions

As we could see from our simulation System Dynamics is useful to model the basic system's behavior. With the causal loop diagram SD provides a powerful tool for modeling, to describe a model and its interactions. Combined with Vesters sensitivity analysis [20] one can easily extract the different kinds of elements in the system (active, reactive, buffering, critical, and neutral) to make steering actions more efficient. A substantial advantage of SD is the big number of available Simulation Software and their intuitive and easy use, when needing quick answers about a systems behavior. Generating realistic quantitative output data was quite a challenging problem with SD and we could just manage it by transferring the original model into an "Array Model" but due to the specialization of this model it is not able to cope with more details or other preventive checkups and therefore we had to switch the modeling approach to ABM.

The ABM approach took much more time to implement, but now agents, the primary building block, can easily be extended with more and more details. That is why the ABM approach and our framework can get beyond the limits of SD, especially when the system contains active objects. However it is difficult to decide on attributes and rules of agents in order to get a behavior that is sufficiently similar to the real system and it is much more difficult to get all the data at the needed level of detail for the simulation than just modeling the structure of the system which is where SD ends. Memory management restrictions still become a big issue for the future of our framework when simulating with millions of agents as we experienced it in our simulation.

With the existing framework we are now able to answer questions for the future demand of several preventive checkup systems and we will extend the model as mentioned above to address more crucial questions concerning futures healthcare management.

References

1. Barclay, R.L., Vicari, J.J., Doughty, A.S., Johanson, J.F., Greenlaw, R.L.: Colonoscopic Withdrawal Times and Adenoma Detection during Screening Colonoscopy. The New England Journal of Medicine 355(24) (2006)
2. Barclay, R.L., Vicari, J.J., Doughty, A.S., Johanson, J.F., Greenlaw, R.L.: Prevention Of Colorectal Cancer By Colonoscopic Polypectomy. The New England Journal of Medicine 329(27), 3 (1993)
3. Andrei, B., Alexei, F.: From System Dynamics and Discrete Event to Practical Agent Based Modeling: Reasons, Techniques, Tools (2004)
4. Citarda, F., Tomaselli, G., Capocaccia, R., Barcherini, S., Crespi, M.: Efficacy in standard clinical practice of colonoscopic polypectomy in reducing colorectal cancer incidence (2000)
5. Paul, D.: Agent Based Social Simulation: A Computer Science View. Journal of Artificial Societies and Social Simulation 5(1) (2002)
6. Meadows, D.H., Meadows, D., Randers, J.: Limits to Growth: The 30-Year-Update (2004)
7. Forrester, J.W.: Industrial Dynamics. MIT Press, Cambridge (1961)
8. Forrester, J.W.: Urban Dynamics. MIT Press, Cambridge (1969)
9. Homer Jack, B., Hirsch Gary, B.: System Dynamics Modeling for Public Health: Background and Opportunities. American Journal of Public Health (2006)
10. Jennings, N.R., Wooldridge, M.: Applications of Intelligent Agents (1998)
11. Patrick, K., Schwandt Michael, J.: Health Systems: A Dynamic System – Benefits from System Dynamics (2005)
12. Macal Charles, M., North Michael, J.: Tutorial on Agent-Based Modeling and Simulation (2005)
13. Milling Peter, M., Nadine, S.: Modeling the Forest or Modeling the Trees (2003)
14. John, N., Antonio, M.: Agent-Based Applications in Health Care (2004)
15. Phelan, Steven, E.: A Note on the Correspondence between Complexity and Systems Theory. Systemic Practice & Action Research 12(3), 237–246 (1999)
16. Statistik Austria, Population and Projection for Austria (2005)
17. Sterman, J.: System dynamics modeling: tools for learning in a complex world (2001)
18. Myrjam, S., Andreas, G.: Agentenbasierte Simulation und System Dynamics - Ein Vergleich der Simulationsmethoden anhand eines Beispiels (2004)
19. Troitzsch, K.G.: Social Science Simulation - Origins, Prospects, Purposes. In: Conte, R., Hegselmann, R., Terno, P. (eds.) Simulating Social Phenomena, pp. 41–54. Springer, Berlin (1997)
20. Frederic, V.: Die Kunst vernetzt zu denken, Taschenbuchverlag Mai (2005)

Translating Discrete Multi-Agents Systems into Cellular Automata: Application to Diffusion-Limited Aggregation

Antoine Spicher[1,2], Nazim Fatés[1], and Olivier Simonin[1]

[1] LORIA - INRIA Nancy Grand Est, Campus Scientifique, 54506 Vandœuvre-lés-Nancy,
France
[2] LACL, Université Paris 12, 61 avenue du Général de Gaulle, 94010 Créteil, France
{nazim.fates,olivier.simonin}@loria.fr,
antoine.spicher@univ-paris12.fr

Abstract. This paper deals with the synchronous implementation of situated Multi-Agent Systems (MAS) in order to have no execution bias and to ease their programming on massively parallel computing devices. For this purpose we investigate the translation of discrete MAS into Cellular Automata (CA). Contrarily to the sequential scheduling generally used in MAS simulations, CA are a model for massively parallel computing where the updating of the components is synchronous.

However, CA expressiveness is limited and not always adapted to all types of modeling situations, especially when independent entities move in space. After illustrating these issues on a simple example, we propose a generic method to translate a discrete MAS into a CA, called a *transactional CA*. Our approach consists in using the *influence-reaction model* to perform this translation.

1 Introduction

Multi-agents systems (MAS) are widely used for modeling systems where autonomous entities, the *agents*, move in a virtual space, the *environment*, and act on it. Numerous simulators and platforms have been developed to simulate such systems. However, in most of these tools, the updating of the agents is often left as a hidden procedure, on which the user has no control. The most common updating procedure is the sequential procedure: agents are updated one after the other with a fixed or random order. It is a well-known problem that such scheduling is a potential source of biases, *i.e.*, it may introduce causalities that were not designed by the user but come only from the simulating tool. By contrast, Cellular automata (CA) are a well-known model of massively parallel computing devices where the updating of the components is synchronous: all the cells are updated at once without any priority between them. The advantage of using the CA formalism is simplicity: it involves static homogeneous computing units that are regularly arranged in space. The drawback of expressing a model with cellular automata appears when one needs to build models with independent entities that may move and act on neighbor cells. Indeed, in CA, a cell cannot directly change the state of its neighbor cells, whereas such an ability is usually required to express a MAS model. This paper investigates the translation of discrete MAS models into CA, which is illustrated

J. Filipe, A. Fred, and B. Sharp (Eds.): ICAART 2009, CCIS 67, pp. 270–282, 2010.

on a simple example. The advantage is twofold: (1) to have a synchronous execution of agents and thus to reduce bias due to the update, and (2) to ease the programming of MAS on massively parallel computing devices such as FPGAs or GPUs.

The purpose of our research is thus to find a method to translate "the language of multi-agents" into "the language of cellular automata". In this article, we propose to take advantage of both contexts, the high expressiveness of a MAS specification and the simplicity of a CA implementation. We aim at developing a framework where MAS, that are simply described through the separate specifications of the local agent behaviors and the environment dynamics, are automatically translated as a uniform transition function of a CA.

This article is organized as follows: In Section 2, we compare CA and MAS approaches. Section 3 introduces the concept of *transactional CA*, starting from the study of a paradigmatic example of a MAS model, namely the Diffusion-Limited Aggregation model. Section 4 proposes the first step of a formal description allowing the generic coding of a reactive MAS model into a transactional CA. We finally conclude with discussions on related approaches and future works.

2 MAS *versus* CA

At first sight, the two formalisms look very similar and are often confused. One may find several works where the names "cellular automata" and "multi-agent systems" are used without distinction. This is easily understandable since CA are often used to model the environment of a MAS, and, reciprocally, one may see a CA as a particular kind of MAS where agents do not move.

We now clarify the differences between CA and MAS in the context of our research. We compare them at two levels: (1) the modeling level: *What in a model makes CA or MAS more suitable to express it?* (2) the simulation level: *How intuitive is the implementation of CA and MAS?*

MAS and CA, as modeling tools. In their definition, CA are uniform objects: there is a *unique* neighborhood shape for each cell and a *unique* transition function. As a consequence, CA are fitted to model phenomena that involve *homogeneous* spaces; CA have been used for example to model physical systems [1], biological systems [2], spatially embedded computations [3], etc. Note that it is always possible to take into account inhomogeneities, for example by encoding the heterogeneity in the cell states, but this is generally not straightforward to do so.

MAS are preferred for expressing an heterogeneous population of entities. They necessitate to make a distinction between the agents' behaviors and the *environment* where they are embedded [4]. This distinction allows to focus on the specification of particular and localized events, namely the agents *actions*. They offer a methodology for designing systems, at the level of algorithms, programming languages, hardware, etc. Examples of MAS applications range from the simulation of natural systems, from ants [5] to human behaviors [6], to the design of massively distributed software and algorithms like web-services, peer-to-peer technologies, autonomous robotics [7], etc. Nevertheless, we must note that contrarily to CA, no universal definition of MAS has been accepted so far. From the modeling point of view, *translating MAS in the cellular automata formalism has (at least) the advantage of fixing the mathematical expression of the model and removing ambiguities of formulation.*

MAS and CA, as simulation tools. The key characteristic of complex systems is the difficulty, if not the impossibility, of inferring their global behavior from the local specification of the interactions. Few mathematical tools are available to predict the evolution of complex systems, more especially those which involve self-organization. This gives to simulation a central role to find the mechanisms that explain how complexity emerges from simple local interactions. We thus have to pay attention to the quality of simulations and to detect ambiguities that may be hidden in the way they are implemented.

The agent-based programming style is somehow intuitive and natural as the programmer takes the point of view of the agent. There is a form of anthropomorphism that makes MAS programming particularly attractive. Nevertheless, we emphasize that once all the agents behaviors are individually specified, there are still many ways to make the agents interact and play together in the environment. The implementation of such systems raises many questions, like assessing the importance of the synchronicity in simulations: are the agents updated all together or one after the other? The design of spatially-extended computing devices will require to imagine a new type of computer science, where the computations do not *necessarily* rely on the existence of a synchronization between the components.

By contrast with MAS, CA lead to shift the programmer's point of view from the "eyes" of the agents to their environment. The benefits of this shifting effort are twofold: (1) the CA formalism forces the programmer to solve conflicts between concurrent agents actions at the elementary level of the cell and forbids the use of any global procedure. (2) As a consequence, the implementation on massively distributed devices is eased. Indeed, CA provide the programmer a cell-centered programming style where the set of cells represents computing units that are regularly organized. Recent works have shown that it is possible to have a good efficiency by using parallel architecture to run CA simulations for GPU and for FPGA, *e.g.* [8]. In other words, *CA provide an easy-to-implement framework, but expressing the local rule necessitates a method to "blend" the different components of a complex system.*

3   The DLA Example as a Starting Point

In this section, we introduce our approach through the translation of a simple MAS model into an original kind of CA, called *transactional CA*. For this purpose, we focus on the CA encoding of a *diffusion-limited aggregation* (DLA) system. This example presents a good trade-off between the simplicity of description and the richness of problems risen by this coding.

The DLA model was introduced to study physical processes where diffusing particles, following a Brownian motion, aggregate [9]: for instance, zinc ions aggregate onto electrodes in an electrolytic solution. This process leads to interesting self-organized dendritic fractal structures. Different models of DLA have been proposed; we consider in this article that particles stick together forever and that there is no aggregate formation between two mobile particles.

3.1   MAS Specification of the DLA

The MAS specification of the DLA model describes separately the *agents* and the *environment* where they evolve:

The environment is a 2D finite and toric square grid composed of elements called *patches*. The *exclusion principle* holds: *i.e.*, there cannot be more than one agent on each patch of the grid.

The population of agents, denoted by $\mathcal{A}$, is composed of the particles. Each particle a of $\mathcal{A}$ is localized on a cell ρ_a of the environment and is characterized by a state σ_a: a particle is either Fixed or Mobile.

The initial configuration of the system is composed of a population of Mobile particles and some Fixed particles called the *seeds*. The expected behavior is the aggregation of the Mobile particles to build dendrites from the seeds.

We propose to formulate the agent dynamics using the usual *perception-decision-action* cycle [10]. We first describe the perception and action abilities of an agent. The *perception* consists of two functions:

- Γ_1 returns true if the agent perceives a Fixed *neighbor* particle, and false otherwise;
- Γ_2 computes the set of *directions* that lead to empty *neighbor* patches.

The neighborhood referred in these perceptions corresponds to the four closest positions of ρ_a following North, South, East and West directions. The *set of actions* is:

- Diffuse(d): move following direction d;
- Aggregate: change to the Fixed state;
- Stay: do nothing

Let $\mathcal{U}(S)$ denote the operation of selecting one element in a finite set S with uniform probability, the *decision process* returns an action as a function of the agent perceptions:

$$
\begin{array}{lll}
\text{if} & \Gamma_1 & \text{then Aggregate} \\
\text{else if } \Gamma_2 \neq \emptyset \text{ then Diffuse}(\mathcal{U}(\Gamma_2)) & & \\
\text{else} & & \text{Stay}
\end{array}
\tag{1}
$$

3.2   CA Expression of the DLA Model

We now reach the core of the problem. We first discuss about implementing the agent motion within a synchronous computational model. We then propose our solution, called *transactional CA*, and we finally illustrate it on the DLA example.

The Synchrony Paradox. In the MAS style of programming, emphasis is put on the agents local behaviors. Classically, to avoid collisions between mobile particles, they are introduced one after the other, or in some cases, are introduced simultaneously but updated one after the other using a scheduler. However, two objections can be raised:

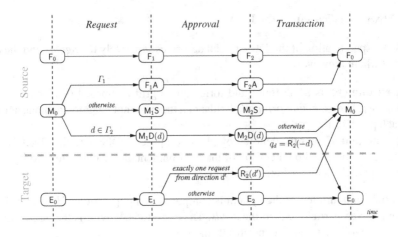

Fig. 1. DLA local evolution rule within a transactional CA. This graph shows the local evolution of a cell from a state to another depending on its neighborhood state. Explanations are given in the text.

1. The implementation of this sequential updating on a massively distributed comput- ing device is not impossible, but it requires the introduction of complex procedures to synchronize the different schedulers.
2. The use of a scheduler introduces an external form of causality that was not spec- ified in the original DLA formulation. This may induce a bias in the formation of dendritic patterns, especially when the density of mobile particles is high.

By contrast, the framework of CA demands an early resolution of the conflicts created by simultaneous moves to a given patch. To achieve that, we propose to establish a dialog between cells.

Transactional CA. A particle move requires a *source* cell (that contains a particle at time t) and a *target* cell (that will contain the particle at time $t + 1$). We propose to elaborate a three-step *transactional* process where cells negotiate their requirements:

1. *Request*: source cells express their needs to their neighbors.
2. *Approval-rejection*: target cells accept or not their neighbors requirements; this de- cision is done with respect to an exclusion principle policy (for example, an empty cell is an available target if and only if there is exactly one particle requesting to move to this cell).
3. *Transaction*: sources and targets separately evolve.

DLA Transaction Model. Figure 1 shows with a graph the local transition function of a transactional CA capturing the agent-based specification of the DLA given in sec- tion 3.1. On this graph, nodes represent the different states of the CA (states E_0, F_0 and M_0 are given twice to clarify the figure) and the arrows specify transitions between states. States are distributed.

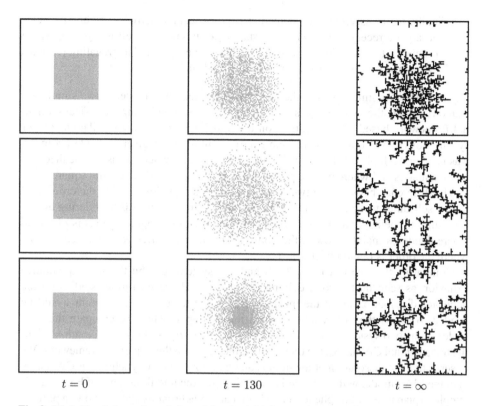

$t = 0$ $t = 130$ $t = \infty$

Fig. 2. DLA Simulations: from left to right, the initial state, the state after 130 simulation time steps and the fixed point, with in black the fixed particles and in gray the mobile particles. Three simulation modes are given: (top) Sequential multi-agent updating where the order of updating is fixed and goes from bottom-left to top-right; (middle) same, but the order of updating is a fixed random permutation on the agents; (bottom) Synchronous updating on a transactional CA. The MAS and CA simulations were obtained with Turtlekit [11] and FiatLux [12] simulators, respectively.

- *vertically*, to segregate the behaviors of sources and targets, and
- *horizontally*, to distinguish the three steps of a transactional CA.

At the beginning, the cells are either empty or contain a particle which is either fixed or mobile: three states are used E_0, F_0 and M_0.

- The request transition consists in deciding an action for each M_0 cell: depending on the perceptions, a mobile particle either aggregates (state $F_1 A$), or requests diffusion following a direction d (state $M_1 D(d)$), or stays at the same position (state $M_1 S$).
- During the approval step, empty cells E_1 decide, by reading their neighbors requirements, if they remain empty (state E_2) or become receptors of particles moving from a direction d' (state $R_2(d')$).

– Finally, the transaction is computed: receptors become particles, mobile particles that target a receptor (*i.e.*, when the state q_d of the pointed cell is $R_2(-d)$, where $-d$ denotes the direction opposite to d), become empty, and aggregating particles become fixed. Other cells remain in their initial state.

Figure 2 presents simulations of the previous described DLA model in two simulation frameworks. On the two first lines, simulations were obtained using a classical sequential framework based on a scheduler, on the third line, we display simulations of our synchronous transactional CA. The same initial configuration, given on the left column, was used on both platforms. It consists of a 100x100 grid where seeds are localized on the boundaries and where mobile particles are gathered in a 40x40 central square.

The difference between the two first lines (MAS simulations) lies in the order used by the scheduler to update agents. On the first line, it corresponds to ordering agents according to their initial positions (from bottom-left to top-right). A bias is observed where dendrites mainly grow at the bottom. This bias is corrected in the second line where the order is randomly choosed.

Comparing the two last lines, MAS and CA systems exhibit the same qualitative behavior, as seen on the right column of Figure 2. However, further studies on the dendrites distribution or on the mean time required to reach a fixed point would be needed to assess the differences between the two approaches. To compare the time scales of the two systems, we define a *simulation time step* as: (a) the three sub-steps of the transactional CA and (b) the update of all the agents in the sequential framework. We observe that the dissolution of the initial square is slower in the synchronous CA than it is in the MAS model with a sequential updating (see the middle column of Figure 2). A simple explanation of this phenomenon is that an asynchronous update allows a particle to move to a just evacuated patch *during a simulation time step*, while the synchronous update forbids this behavior.

4   Towards a Generalization

In this section, we investigate how a generic method could be developed to automatically translate the specification of a MAS into the transition function of a transactional CA. Of course, reducing a MAS to a CA enforces some restrictions on what can be described. More especially, in order to respect the finiteness of CA, we assume that MAS are discrete and finite systems: *i.e.*, the environment is a *discrete* and *regular* grid where a *finite* number of agents are localized on specific parts of this grid (they do not have continuous coordinates).

Our approach is based on the use of the formal *influence-reaction* model [13] to describe a MAS. In fact, the three steps of the transactional CA are similar to the three steps of influence-reaction: (1) agents produce influences that are attempts of actions, (2) influences are combined to avoid conflicts between the corresponding actions, and (3) the environment is updated with respect to the combined influences. In the following, we introduce the influence-reaction model and we finally give the first step of a formal description of an *automatic* translation of an influence-reaction based specification into a transactional CA.

4.1 Influence-Reaction Model

Contrarily to the CA, there is no unique formal description of MAS models. Nevertheless, there exist some generic models that focus on specific kinds of MAS (*e.g.* logic MAS, communicating MAS, etc.). For the sake of clarity, we only consider *situated* MAS that deal with *discrete* environment and *reactive* agents. The term "situated MAS" relates to systems where agents are embedded in a "physical" environment.

In this context, we focus on the *influence-reaction model* that is dedicated to the formal description of situated MAS, allowing, in particular, the simulation of simultaneous actions [13]. In this model, agents release *influences* that will induce *reactions* of the environment. An influence corresponds to attempting an action. The reaction consists in combining the different influences in order to realize the corresponding actions. This modeling principle is inspired by physics where entities react because some forces act on them. Like forces, influences can be combined. For instance, an influence may be an attempt to pull a door. If two agents simultaneously perform this influence with the same intensity from the opposite sides of a door, the combination of both influences vanishes and the resulting action is null.

We assume here that the combination of influences can be computed *locally*. In other words, the evaluation of the combination function can be distributed on each patch of the environment. The dynamics of the three steps of the influence-reaction model may be described as follows:

1. Each agent a of $\mathcal{A}$ separately computes, as a function f_a of its current state σ_a^t and of its perceptions Γ_a, its new state σ_a^{t+1} and the associated set of influences $\mathcal{I}_a$. We denote by $\mathcal{I}$ the set of all the possible influences, $\Sigma_{\mathcal{A}}$ the set of the agent states.
2. Let $\mathcal{I}_\rho$ denote all the influences produced by the agents of $\mathcal{A}$ that could affect the patch ρ of the environment (we denote by $\mathcal{P}$ the set of all the environment patches). Each patch ρ separately computes the set $\prod \mathcal{I}_\rho$ of combined influences affecting this patch.
3. Finally, for each patch ρ of the environment, the new state σ_ρ^{t+1} of ρ is computed as a function $f_{\mathcal{E}}$ of the current position state σ_ρ^t with respect to the set of influences $\prod \mathcal{I}_\rho$. We denote by $\Sigma_{\mathcal{E}}$ the set of the patch states.

Using these notations, the influence-reaction dynamics may be formally summarized by the following two equations:

$$\langle \sigma_a^{t+1}, \mathcal{I}_a \rangle = f_a(\sigma_a^t, \Gamma_a) \quad a \in \mathcal{A} \tag{2}$$

$$\sigma_\rho^{t+1} = f_{\mathcal{E}}(\sigma_\rho^t, \prod \mathcal{I}_\rho) \; \rho \in \mathcal{P} \tag{3}$$

4.2 Generation of a Transactional CA

Formally speaking, a CA is a quadruple $(\mathcal{L}, Q, \mathcal{N}, \delta)$ where:

- $\mathcal{L}$ is the set of *cells* generally taken as a subset of $\mathbb{Z}^{dim}$, dim is the dimension of the space.
- Q is a finite set of *states*. Each cell $c \in \mathcal{L}$ is associated with a value $q_c \in Q$.

- Each cell c is associated with a set of cells $\mathcal{N}(c) \subset \mathcal{L}$ called the *neighborhood* of c. The relationship $\mathcal{N}$ expresses the locality of interactions, *i.e.*, $\mathcal{N}(c)$ is constituted of cells "close" to c.
- The *local transition function* δ returns a value in Q that depends on the current state q_c and on the states of the cells in the neighborhood $\mathcal{N}(c)$.

The generation of a transactional CA consists in defining these four elements using the different components of the MAS specification. For the sake of simplicity, we restrict ourselves to MAS where:

- the set $\mathcal{P}$ corresponds to a 2D regular square grid ($dim = 2$);
- an exclusion principle holds, that limits to one the maximum number of agents on a given patch;
- the set $\mathcal{I}_a$ is always a singleton, that means that an agent decides to attempt only one action at each time step. Note that this case is not restrictive, because any set of influences could be rewritten as a single influence: if the agent does nothing (*i.e.*, $\mathcal{I}_a = \emptyset$), we consider that it releases the special Skip influence, and if there are more than two influences (*e.g.* $\mathcal{I}_a = \{\text{Depose}, \text{Diffuse}(d)\}$), new symbols are considered in $\mathcal{I}$ to capture the corresponding action (*e.g.* $\mathcal{I}_a = \text{DeposeDiffuse}(d)$);
- the combination of influences $\prod \mathcal{I}_\rho$ is always a singleton. In other words, only one action could affect a position ρ.

As shown on Figure 1, the definition of the transition function can be divided into three sub-functions δ_0, δ_1 and δ_2 corresponding to the three steps of a transactional CA. As a consequence, the set of states Q can also be partitioned into three subsets Q_0, Q_1, Q_2, depending on the next step of the transactional CA to be computed.

In the following paragraphs, we detail the key points of the transactional CA generation. We illustrate this translation on the example of the DLA.

Cells and Neighborhood. The set of cells exactly corresponds to the 2D grid defined by the set of patches, so $\mathcal{L} = \mathcal{P}$. The neighborhood relationship $\mathcal{N}$ is defined in such a way that $\mathcal{N}(c)$ gathers the cells that an agent a, localized on c, may read to compute its perceptions Γ_a and its actions $\mathcal{I}_a$.

As an example, the particles of the DLA model may remain on the same position, and perceive or move to the positions following the North, South, West and East directions. As a consequence, the corresponding CA neighborhood relationship $\mathcal{N}$ corresponds to the classical von Neumann neighborhood; for each cell $c \in \mathcal{L}$:

$$\mathcal{N}(c) = \{c' \in \mathcal{L}, \ ||c - c'|| \leq 1\}$$

where $||c - c'||$ denotes the graph distance between two cells.

The Initial States Set Q_0. The set Q_0 corresponds to the initial states of a cell before applying the three steps of the transactional CA. Let c be a cell and ρ its corresponding patch. The state of c is characterized at a given time t by

- the environment state σ_ρ^t at ρ, and
- whether there is an agent a with state σ_a^t on ρ.

Let $\widetilde{\Sigma}_{\mathcal{A}}$ denote the set $\Sigma_{\mathcal{A}} \cup \{\bot\}$, where the symbol $\bot$ represents the absence of agent. Then, the state of Q_0 are couples $(\sigma_a^t, \sigma_\rho^t)$:

$$(\sigma_a^t, \sigma_\rho^t) \in Q_0 = \widetilde{\Sigma}_{\mathcal{A}} \times \Sigma_{\mathcal{E}}$$

For the DLA model, we have $\Sigma_{\mathcal{A}} = \{\text{Mobile}, \text{Fixed}\}$ and $\Sigma_{\mathcal{E}} = \emptyset$: the cells of the environment are here *passive* and holds no information. So we have:

$$Q_0 = \{\text{Mobile}, \text{Fixed}, \bot\}$$

that corresponds to the states M_0, F_0 and E_0 of Figure 1.

The Request Step and the Set Q_1. Compared to the elements of Q_0, the states composing Q_1 are characterized by an additional information: the influence chosen with respect to the specification of f_a (see Eq. 2). In the case of an $\bot$ cell, the particular Skip influence is used. Considering Equation 2 notations, the transition function δ_0 is defined by:

$$\delta_0 : Q_0 \longrightarrow Q_1 = Q_0 \times \mathcal{I}$$
$$(\sigma_a^t, \sigma_\rho^t) \longmapsto \begin{cases} (\sigma_a^t, \sigma_\rho^t, \text{Skip}) \text{ if } \sigma_a^t = \bot \\ (\sigma_a^{t+1}, \sigma_\rho^t, \mathcal{I}_a) \text{ otherwise} \end{cases}$$

Note that the evaluation of the transition function δ_0 depends on the neighborhood state as it requires the perceptions Γ_a to compute $\langle \sigma_a^{t+1}, \mathcal{I}_a \rangle$ using f_a.

In the DLA model, the set of particle actions is $\mathcal{I} = \{\text{Diffuse}(d), \text{Aggregate}, \text{Stay}\}$. If we identify the neutral Skip action to Stay, the definition of δ_0, based on the use of Equation 1, gives:

$$\begin{aligned} \delta_0(\bot) &= (\bot, \text{Skip}) \\ \delta_0(\text{Fixed}) &= (\text{Fixed}, \text{Skip}) \\ \delta_0(\text{Mobile}) &= \begin{cases} (\text{Fixed}, \text{Aggregate}) \\ (\text{Mobile}, \text{Stay}) \quad \text{w.r.t. Eq. 1} \\ (\text{Mobile}, \text{Diffuse}(d)) \end{cases} \end{aligned}$$

These transitions correspond to the five arrows of the "Request" column of the Figure 1. Note that some states of Q_1 are meaningless (*e.g.* the state $(\text{Fixed}, \text{Diffuse}(d))$ would correspond to a diffusing fixed particle).

The Approval Step and the Set Q_2. During this step, each cell c, associated with a patch ρ, computes the set of influences $\mathcal{I}_\rho$ that may affect its state. This computation is done by reading the third element of the states of the neighbor cells. Then, the operator $\prod$ combines these influences into a single influence that will be taken into account during the transaction step. As a consequence, states from Q_2 refer to an additional information: the combined influence $\prod \mathcal{I}_\rho$. The transition δ_1 is then defined by:

$$\delta_1 : Q_1 \longrightarrow Q_2 = Q_1 \times \mathcal{I}$$
$$(\sigma_a^{t+1}, \sigma_\rho^t, \mathcal{I}_a) \longmapsto (\sigma_a^{t+1}, \sigma_\rho^t, \mathcal{I}_a, \prod \mathcal{I}_\rho)$$

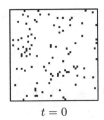  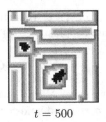 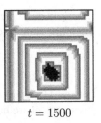

| $t = 0$ | $t = 50$ | $t = 500$ | $t = 1500$ |

Fig. 3. Transactional CA simulation of a virtual amoebae gathering model. Under some environmental conditions, amoebae (in black) release a morphogen (whose concentration is displayed as a gray scale). Then, they follow the gradient generated by reaction-diffusion of that morphogen, until they all gather. Here, each environment patch contains up to two amoebae.

The exclusion principle of the DLA model is specified by the definition of the operator $\prod$. Formally, let ρ be a patch, $\mathcal{I}_\rho$ the set of actions that affect the state of ρ, and $|S|$ the cardinal of a finite set S, we have:

$$\delta_1(\bot, \mathsf{Skip}) = \begin{cases} (\bot, \mathsf{Skip}, \mathcal{I}_\rho) & \text{if } |\mathcal{I}_\rho| = 1 \\ (\bot, \mathsf{Skip}, \mathsf{Skip}) & \text{otherwise} \end{cases}$$
$$\delta_1(\sigma_a, \mathcal{I}_a) = (\sigma_a, \mathcal{I}_a, \mathsf{Skip})$$

The state $(\bot, \mathsf{Skip}, \mathcal{I}_\rho)$ corresponds to the state $\mathsf{R}(d')$ of Figure 1: this state is only reachable from an empty patch with exactly one request of move.

The Transaction Step. The final step consists in computing a new state of Q_0 as a function of a state of Q_2:

$$\delta_2 : Q_2 \longrightarrow Q_0$$

The transition δ_2 computes the new state of the patch ρ and the eventual move of agents from or to ρ. This computation relies on two pieces of information:

- the influence $\prod \mathcal{I}_\rho$ that *has to be realized*, and
- the influence $\mathcal{I}_a$ that is attempted.

As we assume an exclusion principle, we describe separately the case where the cell is empty and the case where an agent a is localized on the cell. If the cell is empty, δ_2 is defined by:

$$\delta_2(\bot, \sigma_\rho^t, \mathcal{I}_a, \prod \mathcal{I}_\rho) = \begin{cases} (\sigma_a^{t+1}, \sigma_\rho^{t+1}) & (1) \\ (\bot, \sigma_\rho^{t+1}) & (2) \end{cases}$$

where (1) and (2) correspond to two possible cases:

Case (1): $\prod \mathcal{I}_\rho$ expresses that ρ is a target cell for an agent a and that this move has to be realized. The state σ_a^{t+1} of this agent comes from the state of the source cell. On Figure 1, the transition from state $\mathsf{R}_2(d')$ to state M_0 illustrates this case.

Case (2): $\prod \mathcal{I}_\rho$ does not allow any agent move to the patch ρ. On Figure 1, the transition from state E_2 to state E_0 illustrates this case.

In both cases, the state of the environment may be affected by the influence $\prod \mathcal{I}_\rho$. The new σ_ρ^{t+1} is computed using Equation 3.

If an agent a is localized on the cell, δ_2 is defined by:

$$\delta_2(\sigma_a^{t+1}, \sigma_\rho^t, \mathcal{I}_a, \prod \mathcal{I}_\rho) = \begin{cases} (\bot, \sigma_\rho^{t+1}) & (3) \\ (\sigma_a^{t+1}, \sigma_\rho^{t+1}) & (4) \end{cases}$$

where (3) and (4) correspond to two possible cases:

Case (3): $\mathcal{I}_a$ expresses a move of the agent a from the patch ρ to another patch ρ' where $\mathcal{I}_a = \prod \mathcal{I}_{\rho'}$ (*i.e.*, the action is allowed by $\prod \mathcal{I}_{\rho'}$). The patch ρ becomes empty. On Figure 1, the transition from state $M_2D(d)$ to state E_0 illustrates this case.

Case (4): $\mathcal{I}_a$ is not allowed by the neighborhood. On Figure 1, the transition from state $M_2D(d)$ to state M_0 illustrates this case.

In both cases, the new σ_ρ^{t+1} is computed using Equation 3.

5 Discussion

The transactional CA we proposed is an original solution to translate reactive MAS into CA. We have shown the interest of such an approach on a diffusion-limited aggregation model allowing to find the three steps required to synchronously run this MAS in a CA framework. Then we proposed the first step of a generalization and an automation of this methodology. This work is based on the use of the classical influence-reaction model that is naturally related to the three-step approach of transactional CA. Possible extensions could rely on novel formulations of this model, such as the one proposed in [14].

Other solutions have already been proposed. For example, specific kinds of CA have been designed to model the movement of particles, like the *dimer cellular automata* that develops an asynchronous point of view of the dynamics, or the *lattice gas cellular automata* initially developed for simulating fluids [2]. The question of coding moving objects in CA also led authors to consider a two-step CA that prevents collisions [15] or to extend the neighborhood [16]. This approach is quite similar to a transactional CA. These solutions focus on the displacement of objects and have not been used yet in a more general context.

By contrast, transactional CA are developed in order to consider any kind of action. For example, we are currently applying the methodology presented in section 4, to generate a CA corresponding to a model of gathering virtual amoebae [17], an illustrative MAS that involves an active environment together with a weak exclusion principle. As shown on Figure 3, this result is promising and encourages us to use our approach on a broader range of problems.

References

1. Chopard, B., Droz, M.: Cellular Automata Modeling of Physical Systems. Collection Alea-Saclay: Monographs and Texts in Statistical Physics. Cambridge University Press, Cambridge (2005)
2. Deutsch, A., Dormann, S.: Cellular Automaton Modeling of Biological Pattern Formation Characterization, Applications, and Analysis. Modeling and Simulation in Science, Engineering and Technology, vol. 26. Birkhäuser, Basel (2005)
3. Adamatzky, A.: Computing in Nonlinear Media and Automata Collectives. Institute of Physics Publishing (2001)
4. Ferber, J.: Multi-Agent Systems, an Introduction to Distributed Artificial Intelligence. Addison-Wesley, Reading (1999)
5. Resnick, M.: Turtles, termites, and traffic jams: explorations in massively parallel microworlds. MIT Press, Cambridge (1994)
6. Regelous, S.: MASSIVE: Multiple agent simulation system in virtual environnement (2004), http://www.massivesoftware.com/
7. Simonin, O., Grunder, O.: A cooperative multi-robot architecture for moving a paralyzed robot. International Journal Mechatronics, The Science of Intelligent Machines (2009)
8. Halbach, M., Hoffmann, R.: Implementing cellular automata in FPGA logic. In: Parallel and Distributed Processing Symposium, International (2004)
9. Witten, T., Sanders, L.: Diffusion-limited aggregation, a kinetic critical phenomena. Phys. Rev. Lett. 47(19), 1400–1403 (1981)
10. Brooks, R.: A robust layered control system for a mobile robot. IEEE Journal of Robotics and Automation 2(1), 14–23 (1986)
11. Michel, F., Beurier, G., Gouaîch, A., Ferber, J.: The turtlekit platform: Application to multilevel emergence. In: ABS 4 Agent-Based Simulation 4 (2003)
12. Fatès, N.: Fiatlux CA simulator in Java (2008), http://webloria.loria.fr/~fates/fiatlux.html
13. Ferber, J., Muller, J.P.: Influences and reaction: a model of situated multiagent systems. In: Proc. of the 2nd Int. Conf. on MAS, pp. 72–79 (1996)
14. Chevrier, V., Fatès, N.: Multi-agent systems as discrete dynamical systems: Influences and reactions as a modelling principle. Technical report, INRIA - LORIA (2009), http://hal.inria.fr/inria-00345954
15. Hochberger, C., Hoffman, R., Waldschmidt, S.: CDL++ for the description of moving objects in cellular automata. In: Malyshkin, V.E. (ed.) PaCT 1999. LNCS, vol. 1662. Springer, Heidelberg (1999)
16. Nishidate, K., Baba, M., Gaylord, R.J.: Cellular automaton model for random walkers. Phys. Rev. Lett. 77(9), 1675–1678 (1996)
17. Fatès, N.: Decentralised gathering on a discrete field. Technical report, INRIA (2008), http://hal.inria.fr/inria-00132266/

Using Values to Turn Agents into Characters

Rossana Damiano and Vincenzo Lombardo

Dipartimento di Informatica
Centro Interdipartimentale per la Multimedialità e l'Audiovisivo
Università di Torino, Italy
{rossana,vincenzo}@di.unito.it

Abstract. Systems for interactive storytelling and drama rely on agent theories to model characters, and adopt various techniques to cope with non-determinism at the story level, such as story models, combining them according to sophisticate architectural designs. However, a consolidated approach has not emerged yet, that fully reconciles these two dimensions. In this paper, we propose a unifying framework to accommodate the tension between story control and character behavior. By using a model of agent to analyze a classical example, we show that this tension cannot be solved by discharging the distinguishing properties of agency. Rather, we claim that the accurate modeling of agency is a prerequisite to the success of any attempt to solve this tension, together with the possibility for the author to state the story direction in terms of explicitly declared values.

1   Introduction

In the last decade, a number of systems for entertainment and communication have appeared that – notwithstanding different design goals and conceptions – share a number of common features, including the use of artificial characters and the adoption of storytelling in the interaction with the user. These systems rely on multiple modalities for communicating with the user, such as natural language, graphics or virtual reality, and support different styles of interaction, such as dialogue, direct manipulation or embodiment.

For example, the Façade entertainment system involves the user as an active character in a dramatic situation in which a couple, whose marriage is falling apart, invites her/him for a drink, with wife and husband trying to bring the user on their respective sides. In the interactive storytelling system by [1], the user is immersed in a virtual reality environment, where he/she plays the part of one of the characters of the French novel "Madame Bovary", interacting with the other characters and affecting their feelings and behavior. In the Dramatour guide application for cultural heritage [2], a virtual character, the spider Carletto, accompanies the visitor in a historical location from a portable device, engaging in a dramatic monologue that exposes his psychological conflict between the roles of 'guide' and 'storyteller'. In the FearNot! edutainment system [3], the dynamics underlying bullying incidents in schools is dramatized in a cartoon-like environment, in which the child user intervenes as an empathic observer to give advice to the victim.

Given this heterogeneity of goals and instruments, a first, broad distinction has been established in the literature between story-based and character-based systems [4,1,5,6].

J. Filipe, A. Fred, and B. Sharp (Eds.): ICAART 2009, CCIS 67, pp. 283–296, 2010.

Story-based systems are characterized by centralized architectures, in which the behavior displayed by the system, possibly through characters' mediation, is driven by the unifying principle of a story. The story to be conveyed is usually underspecified in some way, so as to provide some support to non-determinism and interactivity. Character-based systems rely on the autonomous behaviour of characters to create situations, which are then interpreted as emergent narrative structures [7].

Whatever the chosen approach, it is widely acknowledged that the author's control over the story is related with communicative effectiveness; however, it must be traded off against the autonomy and the believability of the characters. For some specific forms of communication and entertainment, clear design strategies have emerged: for example, in video games, the quality of playability, anchored in carefully shaped and strongly constrained stories, is preferred over the definition of psychologically believable, autonomous characters. On the contrary, AI systems generally envisage interactivity as a main objective, sustained by a rich literature on interactive storytelling and drama [8,9,10]. However, a consolidated approach has not emerged yet that fully reconciles the two conflicting dimensions of story and characters.

In this paper, we sketch a formal system that systematizes the functions of story and characters in a unified theoretical framework. By using this framework, we accommodate the components of a variety of practical systems, pointing out the relevance of the notion of agency as a prerequisite to reconciling story and characters. Finally, we propose the notion of value to fill the gap between the language of agents and the author's control on drama.

2   A Formal Framework for Drama

The theoretical framework presented in this section lays out the 'language of drama', independently of the form and the media through which specific dramas are realized. Based on previous work by [11], revised here with a particular concern towards the contribution of AI and agents, the framework encompasses the major features of the practical systems mentioned above.

Following drama studies [12,13], for 'drama' we intend a form of narrative that describes the story via characters' action in present time and has a carefully crafted premise, i.e. an authorial *direction* that shapes the dramatic climax until its solution. Given this definition, the framework consists of two levels, the *directional level*, that encodes the specific traits of drama, and the *actional level*, that connects such traits with the notion of agency. The directional level of the system (Fig. 1, top) is centered on the notion of *drama unit*. A drama unit (*du*) is any segment of the drama that contributes to the story advancement, that is, it achieves a direction corresponding to a change in the emotional or belief state of one or more characters [13]. Originally expressed by Aristotle as "unity of action", a drama unit provides an effective direction for the story advancement operated by the segment of the story to which it is associated.

A drama unit can be recursively expanded into a number of children drama units, forming the *plot tree* (Fig. 1, top). The plot tree is licensed by the formal rules of drama, which pose constraints on the expansion relationship of *du*s (dominance relation in formal grammar terms). In particular, the direction of the parent *du* must include all the

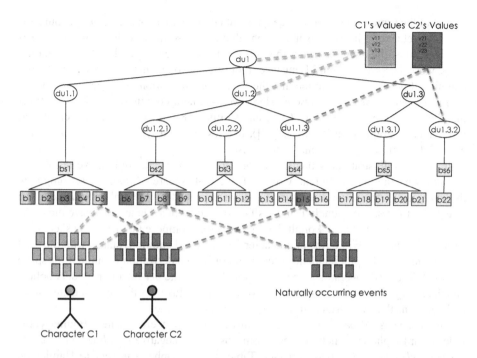

Fig. 1. A graphical representation of the formal system of drama. At the top, the directional level of drama, i.e. the way it achieves its direction through a sequence of elements situated at different levels of abstraction (*du*s), which manipulate characters' values; below, the actional level, represents the anchoring of the direction into the characters' actions contained in beats and rooted in a rational and emotional model of agency.

directions of children *du*s, with consistent initial and final states in case of subsequent changes in the belief or emotional states of the characters. The expansion of drama units into smaller drama units stops when, at the basic recursion, drama units are expanded into a sequence of *beats* (*b*s in Fig. 1, center), that will be exposed to the audience. Beats are the minimal units for the story advancement and consist of incident pairs (represented as pairs of adjacent boxes in the central part of Fig. 1), where incidents are characters' actions – executed as part of their plans – or unintentional events. The prescription that the direction is achieved through *conflict* is captured by the constraint that, for each beat, the intended outcome of the second incident should be incompatible with the intended outcome of the first incident.

The incidents that form beats are retrieved from the *actional* level of drama (Fig. 1, bottom), that includes a full-fledged model of agency, such as the belief–desire–intention (BDI) model [14]. This model provides the glue that links the behavior of characters into a coherent causal chain to sustain their believability. BDI agent architectures have proven an effective basis for the design and implementations of character-based storytelling systems [15,16,3,17]. The reason for their effectiveness in modeling characters lies in the fact that the BDI model provides a library of actions that encompass the actual behavior of characters in drama, including the actions that are executed as part of the typical agent loop: once an agent *commits* to a goal and *plans* a course of action to achieve it, it *executes* the actions that are prescribed by that plan. After executing actions, the agent *monitors* their effects through sensing and updates its model of the world accordingly, modifying its commitment through meta–deliberation and re–planning if necessary. [17] argue that the BDI characterization of agency is necessary also to model characters' emotions. Here, we assume a cognitive model of emotions like the one proposed by Ortony, Clore and Collins [18]; this model, largely employed in interactive drama (starting from [19]), defines emotion types based on an agent's appraisal of self and other agents' intentions.

In order to illustrate how this framework applies to specific drama, we resort to a well-known example, the 'nunnery scene' from Shakespeare's *Hamlet*, describing how the characters' rational and emotional states are plotted by the author to achieve the direction of the scene, namely Hamlet's failure to convince Ophelia to leave the court and his loss of trust in her.[1] For all the segments that compose the scene, we sketch the structure of the scene in terms of characters' beliefs and emotions states, with the goal of showing how the plot incidents affect them (a formal derivation is omitted for space reasons). The $feel$ operator models emotional states; the bel operator models beliefs and the has_goal operator models intentions. Table 2 illustrates the beliefs, goals, plans and emotions that we attribute to the characters in this scene.

The scene has a tripartite structure; the direction of the overall scene is Hamlet's attitude toward Ophelia, which changes from positive to negative ($feel(Hamlet, love\_for(Ophelia))$). In the first part (see Table 2, 1.1), Ophelia is sent to Hamlet by Claudius to induce him to reveal his inner feelings, $has\_goal(Ophelia, confess(Hamlet, inner\_feelings)))$, as a way to confirm the hypothesis that his madness is caused by his rejected love for Ophelia. In order to do so, she has devised a simple plan that consists of meeting Hamlet, starting an interaction with him, and asking him to talk about his inner feelings ($has\_plan(Ophelia, (meet(Ophelia, Hamlet), start\_interaction(Ophelia, Hamlet), ask(Ophelia, confess(Hamlet, inner\_feelings))))$). Hamlet does not want to meet Ophelia, so he tries to leave her, but is forced to answer by Ophelia's insistence, as she makes subsequent attempts to start an interaction with him. This segment of the scene establishes Claudius' belief that Hamlet is mad because of rejected love $bel(Claudius, mad(Hamlet))$; Ophelia's emotional state includes *fear* about Hamlet's madness, $feel(Ophelia, fear\_for(Hamlet\_madness))$ and *hope* (related with the fulfillment of her task to convince Hamlet to reveal his inner feelings.

At the beginning of the second part (Table 2, 1.2), Hamlet reactively forms the goal of saving Ophelia from the corruption of Elsinore court, $has\_goal(Hamlet, save(Hamlet, Ophelia))$, by inducing her to go to a nunnery ($has\_goal(Hamlet, has\_goal(Ophelia, go(Ophelia, nunnery))))$. In order to do so, he resorts to a rhetorical plan to convince her that moral and affective values do not hold anymore, and that everybody, including himself, is corrupted, with the intention that she would spontaneously decide to go to the nunnery ($bel(Hamlet, (bel(Ophelia, corrupted(court)) \rightarrow has\_goal(Ophelia, go(Ophelia, nunnery)))))$). Hamlet's emotional state is now set to *hope* about the effect of his rhetorical plan to convince Ophelia, $feel(Hamlet, hope$

---

[1] See [20] for a detailed analysis of the scene.

SCENE	CHARACTER	BELIEFS	GOAL	PLAN	EMOTIONS
1.1	Ophelia	has_goal (Claudius (know_if (Claudius, mad (Hamlet)))	confess (Hamlet, inner_feelings)	meet (Ophelia, Hamlet), start_ interaction (Ophelia, Hamlet), ask (Ophelia, confess (Hamlet, inner_feelings))	fear_for (mad (Hamlet)) love_for (Hamlet) hope (done (confess (Hamlet, inner_feelings)))
1.2	Hamlet	corrupted (court) in_danger (Ophelia) bel(Ophelia, corrupted (court)) → has goal (Ophelia, go (Ophelia, nunnery))	has_goal (Ophelia, go (Ophelia, nunnery))	convince (Hamlet, Ophelia, corrupted (court))	love_for (Ophelia) hope (has_goal (Ophelia, go (Ophelia, nunnery)))
	Ophelia				self_satisfaction (done (tell (Hamlet, inner_feelings)))
1.3	Hamlet	overhearing (Claudius)	know_if (bel (Ophelia, corrupted (court))) know_if (sincere (Ophelia))	ask (Hamlet, Ophelia, tell (Ophelia, where_is (Claudius)))	disappointment (not (done (convince (Hamlet, Ophelia, corrupted (court))) anger (not (sincere(Ophelia)))
	Ophelia	mad (Hamlet)			fear_confirmed (mad (Hamlet))
	Claudius	mad (Hamlet)			

Fig. 2. A representation of the rational and emotional state of the characters of Ophelia, Hamlet and Claudius along the "nunnery scene"

$(has\_goal(Ophelia, go\ (Ophelia, nunnery))))$, since he does not know if Ophelia's belief about the corruption of the court $(bel(Ophelia, corrupted(court)))$ has been established.

Finally, in the third part of the scene (Table 2, 1.3), Hamlet verifies whether he has convinced Ophelia that the court is corrupted. Prompted by the fact – a contingent event – that he has noticed Claudius and Polonius hidden behind the curtains, $bel(Hamlet, (overhearing(Claudius)))$, he realizes that she maybe be lying and asks her where her father is. For Hamlet, this is a way to verify if Ophelia is still subdued to her father's authority. As she answers that her father is at home, her lie, for Hamlet, counts as a confirmation that his plan has failed So, Hamlet is upset and angry at Ophelia's blameworthy behavior and starts feigning madness again, thus confirming Claudius' beliefs about his madness, $bel(Claudius, mad(Hamlet))$. The scene ends with Ophelia complaining about her tragic destiny. Hamlet's emotional attitude towards Ophelia has changed $(feel(Hamlet, love\_for(Ophelia))$ does not hold anymore), while his hope to convince Ophelia to go to the nunnery has turned into *disappointment* $(feel(Hamlet, hope\ has\_goal(Ophelia, (go(Ophelia, nunnery))))$ does not hold anymore as well), and *anger*, $feel(Hamlet, anger(not(sincere(Ophelia))))$. At the same time, Ophelia experiences the emotional state of *fear confirmed* about Hamlet's madness, $feel(Ophelia, fear\_confirmed(Hamlet\_madness))$, not mitigated by the fact of having accomplished her initial task $feel(Ophelia, self\_satisfaction$ $(done(confess$ $(Hamlet, inner\_feelings)))$.

As this short excerpt illustrates, at the actional level, even sophisticated characters like Shakespeare's show the typical behavior of rational agents: they recognizably have goals, they form plans to achieve them and monitor the execution of those plans, dropping the goals that have proven to be unachievable. For example, when Hamlet realizes that the integrity of Ophelia is threatened by Claudius' maneuvers, he reactively forms the goal to save her from the corruption of the court and devises a rhetorical plan to convince her to abandon the court. Later on, in the second part of the scene, he formulates the goal of knowing whether Ophelia is lying or not, as a way to monitor the outcome of his rhetorical plan. Finally, in the third part of the scene, when confronted with his failure to convince Ophelia to leave the court, he gives up, disappointed.

3   Applying the Framework to Practical Architectures

With respect to the dichotomy between story-based and character-based approaches, the adoption of a unifying formal framework brings the advantage of situating different systems against a background in which the two dimensions of story and characters are explicitly put in relation, thus providing a common evaluation grid. Moreover, the way each system copes with the tension between story design and characters' autonomy has relevant consequences for the role of the procedural author [8] who will input the knowledge in the system. In practical systems, setting the focus on the directional level corresponds to giving the author immediate control on the direction of the story and constraining, by converse, the autonomy of characters. Since the directional level manipulates facts that relate, more or less directly, with the properties of characters, the internal structure of characters – a sum of emotions and beliefs – must be represented in the system in some implicit way.

In general, story-based systems tend to operate at the directional level and incorporate sophisticate story models (object-level knowledge about the semantics of drama according to the drama framework) to account for the structural aspects of narration and drama, ranging from semiotic structuralism [21,22,23] to cognitive models of story understanding [24]. The knowledge about story generation has often been encoded in the form of logical rules, like in the DEFACTO [25] and the IDtension [21] systems. These systems closely resemble expert systems, mixing the empirical and theoretical knowledge of the author in a set of rules that the system uses to generate the story; the effectiveness of these systems seems to be directly connected to their ability to integrate actional operators and emotional aspects. A different approach resorts to AI planning for the generation of the story. In these systems, the planner may replace the author in devising the plot: planning operators are represented by a set of possible plot incidents [6], which can be combined in a consistent sequence to achieve a transition from an author-defined initial and final states, with explicit constraints on the ordering of incidents and their causal relations (where intentionality can represent a weak form of causation). Or, the planner may be in charge of solving a planning problem from the perspective of the story characters, delegating the control over the direction to the author's capability of encoding full-fledged drama units into joint planning operators [26].

The control over the story direction is largely preferred to the manipulation of characters when it comes to designing authoring tools, as exemplified by a recent generation of systems [27,28,29]. In particular, [29] propose a hybrid, layered language for drama representation in authoring tools, aimed at allowing the author to design the story at the directional level, as a partially ordered set of plot points, while leaving the story generation system the task of mapping this representation onto a planning formalism. Similarly, the Automated Story Director [30] confines the autonomy of characters to the generation of low-level behaviors based on action libraries, giving a 'drama manager' [4] the responsibility for monitoring the story advancement to preserve, during the interaction with the user, the story goals encoded by the author.

As for the techniques employed, they vary within each family of systems (rule-based, planning-based and hybrid). The Mimesis [6] storytelling system, for example, explicitly aims at constraining the interaction with the user to the realization of a specific direction. The story is generated offline by a partial order planner; the planner assembles actions executed by a set of characters in order to construct an overall plan that fulfills the initial and final states given by the author. The plan is analyzed to detect the potential impact of user's input, and converted into a conditional plan in which inputs are accounted for in advance. This powerful technique, called 'narrative mediation', allows the author to predict, in full detail, the possible forms of the plot. At the same time, the system sees to it that the intentions of the characters remain clearly recognizable to the user along the plan, and that all actions in the plan are part of the motivational structure of the characters. The core of this system clearly operates at the directional level; however, the representation of the characters as agents is acknowledged, though indirectly, by the use of "intention frames", motivational accounts with which the actions included in the plot are annotated. A drawback of this strategy is that the mental state of the characters are only accounted for indirectly, blurring the conceptual distinction between the goals that are individually pursued by the characters and the drama goal.

The Façade system [26] is designed to conduct an interactive drama to a clearly stated set of outcomes, in which the couple of protagonists either split or remain together (see Section 1), with the user being neutral or sympathizing for one of the two sides. In Façade, the richness of the user experience resides in becoming a story protagonist, triggering (but not controlling) the evolution of the plot towards one of the available directions. The generation of the plot is obtained through a hierarchical plan language (ABL), that encodes multi-agent, joint plans; the plan language also accounts for the role of the user in the joint action. The inclusion of the joint plans in the interactive story is then affected by the interaction with the user and guided by a measure of the plot emotional value. In Façade, the characters tend to be shaped by the actions they perform and the things they say in the space of the interpersonal relations, instead of being stated by an a priori definition of their mental state from which intentions can subsequently be derived. For this reason, even if this system operates at the directional level, it is situated mid-way between the directional level and the actional level.

Character-based systems are situated at the actional level of the drama framework. They tend to an improvisational approach to drama, close to the "comedy of the art" tradition, first translated in a computational architecture by the work of Hayes-Roth [31]. According to this paradigm, drama emerges from the interaction of a set of characters, constrained to specific roles. This approach – whose realization has been encouraged by the availability of conceptual and practical tools that implement the characters' deliberation – contrasts with the achievement of a specific direction. This situation corresponds to having a strongly underdetermined direction, subsuming a large variety of plots, or not having a specified direction at all. It is worth noting that practical systems tend to reduce the notion of character to mere deliberation and resort to planning techniques to implement it. The representation of the characters' mind is delegated to the use of truth maintaining techniques, while plan monitoring is delegated to the execution environments.

As an example of this approach, consider the "Friends" system described in [32], where the behavior of each character is generated by a hierarchical planner. Characters are committed to specific goals (like seducing another character), and reactively form intentions (i.e., partially instantiated plans) to achieve them. The user interacts with them as a 'deus ex machina' in a cooperative (i.g. giving suggestions) or conflictual way (i.g. preventing actions); so, user actions affect the behavior of the characters, either influencing their future deliberation (characters' high level plans are detailed out only when execution approximates, leaving room for the user's advice) or forcing them to replan. The characters' behaviors are executed concurrently, so that conflicts may emerge from their interaction. From the author's point of view, the use of the hierarchical task network (HTN) planning paradigm allows for a more direct control of the development of the plot, since it directly connects the characters' behavior with the specification of their goal; as a drawback, it reduces the responsiveness of the system, that must employ replanning techniques to display a flexible behavior. In practice, stating the drama direction in terms of the characters' goals releases the control over story; from a theoretical perspective proposed here, it collapses the distinction between the directional and the actional level, since the direction should abstract from how actions are represented in the system.

In the "Madame Bovary" based system by [1], the focus is on the responsiveness of characters' emotional attitudes. The behavior of each character is generated by a heuristic-search planner, and planning is limited to the selection of the next action, to cope with asynchronous user intervention without replanning. The planning operators represent how 'feelings' manipulate the characters' mental state. The author's control over the system is confined to the actional level, and consists of defining the initial mental state of the characters and a set of planning operators for each character. The resulting initial situation is open to opposite endings, depending on the user's input and the mental state of the characters. The notion of conflict is mostly internal to the characters' mental states, which can evolve towards different final outcomes. Within the conceptual dimension of the actional level of drama, this system privileges the accuracy of the emotional modeling of the characters; it does not contain an explicit notion of direction to provide a stronger control on the story development; the characters, in order to appear responsive and gain the user's engagement, exhibit an emotionally-based behavior rather than being driven by explicitly coded intentions, encouraging the user to explore, rather than 'direct', the advancement of the story. From the author's point of view, the possibility for the user to affect the beliefs and feelings of the characters at any time and the use of heuristic–search planning (HSP) lead to hypothesize a methodology consisting of iterative testing to tune the behavior of the system to the author's direction, which is not stated explicitly.

4   The Role of Values in Drama

The dichotomy between actional level and directional level systems, that is, between characters and story direction, can be reconciled through the notion of 'value': by extending the actional level definition of agent to include the values belonging to the directional level, agents can be leveraged to become characters.

The notion of value belongs to the realm of ethics and economics [33]. A value is an assignment of importance to some type of abstract or physical object; recognizably subjective, values are related with the regulation of behavior, but they retain a more abstract, symbolic meaning and do not exhibit a direct correspondence within the theories of rationality [34]; at the same time, they are very relevant for the establishment of the story direction. The notion of character's value has emerged in scriptwriting theory as a major propulsive force in story advancement. First stated in Egri's definition of drama premise [12], the notion of value underpins most of the subsequent work conducted in scriptwriting [35], until the recent formulation by McKee [13]. The progression of the story follows a cyclic pattern: a character follows line of action suitable to preserve its balance of values, when some event (typically, a twist of fate or an antagonist's action) occurs, that makes the character's line of action inadequate by putting some character's relevant value at stake. This constrains the character to abandon or suspend its current line of action and to devise yet another line of action to restore them.

Stories put characters' values at stake either by violating their values (e.g. fate of mankind in Bond movies) or by offering them the possibility of achieving the compliance with them (e.g, self-realization for Nora in Ibsen's A Doll's house). Values at stake possibly conflict with some other values (the hero's love for a beautiful girl in Bond movies, family values in Ibsen's drama). Notice that, in agent systems, strong limitations are posed to opportunistic behaviors, since they conflict with the requirement of behavior stability, and pose complexity issues. On the contrary, in fictional worlds, once the behavior of the character is consistently believable along the story, stability is not a desirable feature, because a story advances by changes rather than stability. The underlying assumption of inserting values in a computational framework of drama is that, if characters respond to explicitly declared values, the control of the author over the story direction can be improved, thus amending, at least partially, the problem of authorial control over storytelling systems, as it emerges from the survey conducted in the previous section. The use of values is motivated by the aim of lifting the authorial control from the need to deal with the inner mechanisms of characters that, in practical systems, are implemented through agent architectures developed without any specific concerns about storytelling.

So, we model characters as a 4-tuple $\{B, D, I, V\}$, where V represent character's values. Note that, here, we consider only the subjective dimension of values, and do not consider their relation with an external social system.

- Each character has a *set of values V*.
- Each value v is polarized, with a negative or positive *polarity p*, and is associated with a *condition c*. The negative polarity of a value means that, when the value condition holds in a certain state of the world, the value is violated; the positive polarity corresponds to the value being in force. In order to let characters arbitrate among their values, values are ranked according to their priority.

- A *value* v is defined by the construct $v(p, c, r)$ where p is a negative or positive polarity, c is a ground formula α and r is a priority (a real number).
- If the condition c of value v holds in a state of the world represented by the character's beliefs, that value is *at stake*.
- A character's *record of the values at stake*, VaS, is a subset of the constructs $v(p, c, r)$ such that c holds in the current state of the world or is expected to hold in the future, according to the character's beliefs.

The character's record of the values at stake (VaS) is a dynamic structure, updated along the progression of the story. When a character monitors its values, it considers not only the current state of the world, but also the world states that are to be generated by other characters' actions (the character's expectations). To model opportunism, we assume that the character monitors the possibility of bringing about its values, by verifying if their conditions are reachable from the current world state (self expectations). If the character is able to find a course of action that results in a state of the world in which the condition of the value holds, this state is added to the character's expectations.

When a character realizes that some value is at stake (either in the present or in its expectations), it reacts accordingly, by modifying its commitment in a way that restores the value. The way goals arise and are affected by values can be grasped effectively by the abstract goal architecture in [36], which provides a unifying account of goal types and describes how a goal state is transformed after the modifications of an agent's beliefs (Fig. 3). According to the goal framework in [36], *adopted* goals remain *suspended* until they are ready for execution; then, high priority goals become *active*, while active goals with lower priority are suspended. Goals can eventually be *dropped* if certain conditions hold [37].

The automaton in Fig 3 schematizes the framework in [36] and introduces the role of values at stake in goal adoption.

1. Given a construct $v(p, c, r)$ in VaS (the record of the values at stake), the character formulates and *adopts* a set of goals according to the following rules:
   (a) If the condition c holds in the character's *expectations* and the associated polarity p is *positive*, then the character forms the goal to achieve that state of affairs $g = c$ (*achievement goal*). Expectations model the opportunity of achieving the compliance with one's own values.

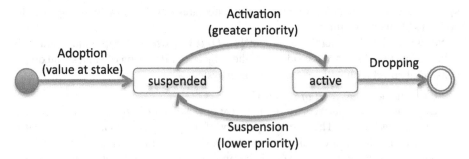

Fig. 3. A graphical representation of the goal states and transitions (inspired by [36])

(b) If the condition c holds in the *present* or in the character's *expectations* and the associated polarity p is *negative*, then the character forms the goal to achieve a state of the world s in which that condition does not hold (any s such that $c \notin s$). This includes the case in which the character has to contrast the behavior of an antagonist whose behavior is expected to bring about just a state of affairs in which the condition c holds.

2. Goals that are adopted because of some value v become *suspended* immediately after the adoption. The activation of goals depends on the priority of the associated value v.

(a) If the priority of the value at stake $v(p, c, r)$ is higher than the priority of the current values at stake, the goal becomes *active*; the character can *generate a plan* to achieve it and start to *execute* it (if the active goal was active in the past, the character may resume a previously suspended plan, if still valid).

(b) Else, the new goal does not become active immediately (but can activated later depending on its priority).

3. A goal is *dropped* if the character realizes that it is impossible to find a plan to achieve it, it has already been achieved, or it is not relevant anymore, independently of its priority.

The "nunnery scene", analyzed in Section 2 in terms of characters' goals and emotions, is better accounted for in a framework that explicitly includes values. The value pursued by Claudius is the "control over the court", put at stake by behavior of Hamlet, who has been playing the fool since his arrival at Elsinore. From the point view of Claudius, Hamlet's behavior is associated with the violation of this value, since the control over the court presupposes, among other conditions, that all the members of the court obey the social rules of the court itself. So, the value-driven mechanism of goal formation establishes Claudius' goal of making Hamlet inoffensive (by rule $1.a$): finding out the cause for Hamlet's strange behavior is the first step of Claudius' plan to make him inoffensive.

Ophelia, although she is in love with Hamlet, is subdued by Claudius' authority. When she is ordered to investigate the cause for Hamlet's madness, coherently with her values, she reacts by forming the goal of making Hamlet confess his inner feelings ($has\_goal(Ophelia, confess(Hamlet, inner\_feelings))$) and formulates a plan to achieve it. Later on, when she is required by Hamlet to tell where her father is, the value of "obedience" conflicts with the value of "moral integrity", since the compliance with the value of "moral integrity", that would be enforced by the fact of telling the truth (by rule $1.b$), becomes incompatible with the value of obedience (expectedly violated by the revelation of Claudius eavesdropping their conversation – rule $1.a$). When Ophelia gives priority to the value of 'obedience' (by rule $2.a$), as a result, she exposes her scale of values to the audience, an event that is extremely meaningful in dramatic terms and constitutes the culmination of the climax of the scene.

Hamlet's behavior is driven by the value of "justice", put at stake by her father's murder by Claudius that constitutes, unpunished, a blatant violation of the value, prompting Hamlet's goal to take revenge over Claudius (by rule $1.a$). In the nunnery scene, however, this value is temporarily suspended in favor of the value of "moral integrity", triggered by Hamlet realization that Ophelia's integrity is currently threatened by her submission to Claudius and Polonius, but can be restored if she abandons the court. So, by rule 1.b, he forms the goal of saving Ophelia from the corruption of the court ($has\_goal(Hamlet, save(Hamlet, Ophelia))$) by inducing her to go to a nunnery ($has\_goal(Hamlet, has\_goal(Ophelia, go(Ophelia, nunnery)))$). Hamlet's effort to comply with the value of "moral integrity", however, is frustrated by Ophelia's behavior, dictated by her obedience to her father.

In summary, the direction states the frustration of a character's (Hamlet) attempt to preserve the value of "moral integrity", a frustration that is determined by another characters' (Ophelia) choice to comply with the value of "obedience". The conflict between "obedience" and "moral integrity" is solved in favor of the value of "obedience". The value-driven analysis of this scene also reveals the key role of emotions in establishing the direction: the conflict between "obedience" and "moral integrity" is exacerbated by the characters' emotions, whose polarity is changed as a result of the solution of this conflict: Hamlet's love for Ophelia switches from positive to negative, while Ophelia's feeling of hope is turned into despair by her belief that Hamlet is eventually fallen into madness.

The integration of values in the rational model of characters posits a computational system for storytelling in the position to account for the basic patterns of drama. When a character suspends the current goal (and the related plan) in favor of a new goal, associated with a value at stake of higher priority, the new goal becomes the active goal. The pattern is repeated until the story reaches its climax. At that point, the suspended goals, if any, are normally resumed in the inverse order. Some plans may not be relevant anymore, since their goals may have been achieved or become unachievable or irrelevant in the meantime (fortuitous events, expired deadlines, other characters' activities).

5   Discussion and Conclusions

Although there is a general agreement about the role played by characters and story in practical systems, the complex relations between the two have received much less attention. This situation that can be partly attributed to the fact that the notions of character and story, taken in isolation, rely on well established models (autonomous agents, planning and graph theories). However, the importance of the story direction has been amply recognized because of the difficulty encountered by the interactive systems to reconcile the authorial control and the interaction with the user.

In this paper we have sketched a formal framework to represent the interdependencies of the two levels of drama, i.e., direction and characters. The analysis of a classical example through this framework points out that a detailed model of agency is necessary to grasp the behavior of characters in drama. For example, agency is required to model the monitoring of actions by the characters, the reasoning about one's own intentions, and the meta-deliberation capabilities.

We have proposed to extend agency with the notion of "characters' values", providing a definition of how they are accounted for in characters' deliberation and a method for the formation of new goals based on values. The language through which the author expresses its message takes advantage from the introduction of explicit values, since it allows the author to control the advancement of the story by abstracting from the functioning of characters as agents at the actional level of drama. Values can be explicitly manipulated by the author in the creation of interactive stories, mediating between the author's control of the story and the freedom of the user in the interaction with the characters.

References

1. Pizzi, D., Charles, F., Lugrin, J., Cavazza, M.: Interactive storytelling with literary feelings. In: Paiva, A.C.R., Prada, R., Picard, R.W. (eds.) ACII 2007. LNCS, vol. 4738, pp. 630–641. Springer, Heidelberg (2007)
2. Damiano, R., Lombardo, V., Nunnari, F., Pizzo, A.: Dramatization meets information presentation. In: Proceedings of ECAI 2006, Riva del Garda, Italy (2006)
3. Aylett, R., Vala, M., Sequeira, P., Paiva, A.: Fearnot!–an emergent narrative approach to virtual dramas for anti-bullying education. In: Cavazza, M., Donikian, S. (eds.) ICVS-VirtStory 2007. LNCS, vol. 4871, pp. 202–205. Springer, Heidelberg (2007)
4. Mateas, M., Stern, A.: Integrating plot, character and natural language processing in the interactive drama Façade. In: TIDSE 2003 (2003)
5. Theune, M., Faas, S., Heylen, D., Nijholt, A.: The virtual storyteller: Story creation by intelligent agents. In: Proceedings TIDSE 2003, pp. 204–215. Fraunhofer IRB Verlag (2003)
6. Riedl, M., Young, M.: From linear story generation to branching story graphs. IEEE Journal of Computer Graphics and Applications, 23–31 (2006)
7. Spierling, U.: Adding Aspects of "Implicit Creation" to the Authoring Process in Interactive Storytelling. In: Cavazza, M., Donikian, S. (eds.) ICVS-VirtStory 2007. LNCS, vol. 4871, pp. 13–25. Springer, Heidelberg (2007)
8. Murray, J.: Hamlet on the Holodeck. The Future of Narrative in Cyberspace. The MIT Press, Cambridge (1998)
9. Ryan, M.: Avatars of Story. University of Minnesota Press (2006)
10. Wardrip-Fruin, N., Harrigan, P.: First Person: New Media As Story, Performance, and Game. MIT Press, Cambridge (2004)
11. Damiano, R., Lombardo, V., Pizzo, A.: Formal encoding of drama ontology. In: Subsol, G. (ed.) ICVS-VirtStory 2005. LNCS, vol. 3805, pp. 95–104. Springer, Heidelberg (2005)
12. Egri, L.: The Art of Dramatic Writing. Simon and Schuster, New York (1946)
13. McKee, R.: Story. Harper Collins, New York (1997)
14. Bratman, M., Israel, D.J., Pollack, M.E.: Plans and resource-bounded practical reasoning. Computational Intelligence 4, 349–355 (1988)
15. Norling, E., Sonenberg, L.: Creating Interactive Characters with BDI Agents. In: Proceedings of the Australian Workshop on Interactive Entertainment IE 2004 (2004)
16. Rank, S., Petta, P.: Appraisal for a character-based story-world. In: Panayiotopoulos, T., Gratch, J., Aylett, R.S., Ballin, D., Olivier, P., Rist, T. (eds.) IVA 2005. LNCS (LNAI), vol. 3661, pp. 495–496. Springer, Heidelberg (2005)
17. Peinado, F., Cavazza, M., Pizzi, D.: Revisiting Character-based Affective Storytelling under a Narrative BDI Framework. In: Proc. of ICIDIS 2008, Erfurt, Germany (2008)
18. Ortony, A., Clore, G., Collins, A.: The Cognitive Stucture of Emotions. Cambrigde University Press (1988)
19. Bates, J., Loyall, A., Reilly, W.: An architecture for action, emotion, and social behaviour. In:Artificial Social Systems. LNCS (LNAI), vol. 830. Springer, Heidelberg (1994)
20. Damiano, R., Pizzo, A.: Emotions in drama characters and virtual agents. In: AAAI Spring Symposium on Emotion, Personality, and Social Behavior (2008)

21. Szilas, N.: Idtension: a narrative engine for interactive drama. In: Proc. 1st International Conference on Technologies for Interactive Digital Storytelling and Entertainment (TIDSE 2003), Darmstadt, Germany (2003)

22. Peinado, F., Gervás, P.: Transferring Game Mastering Laws to Interactive Digital Storytelling. In: Göbel, S., Spierling, U., Hoffmann, A., Iurgel, I., Schneider, O., Dechau, J., Feix, A. (eds.) TIDSE 2004. LNCS, vol. 3105, pp. 48–54. Springer, Heidelberg (2004)

23. Hartmann, K., Hartmann, S., Feustel, M.: Motif definition and classification to structure non-linear plots and to control the narrative flow in interactive dramas. In: Subsol, G. (ed.) ICVS-VirtStory 2005. LNCS, vol. 3805, pp. 158–167. Springer, Heidelberg (2005)

24. Swartjes, I., Theune, M.: A fabula model for emergent narrative. In: Göbel, S., Malkewitz, R., Iurgel, I. (eds.) TIDSE 2006. LNCS, vol. 4326, pp. 49–60. Springer, Heidelberg (2006)

25. Sgouros, N.: Dynamic generation, management and resolution of interactive plots. Artificial Intelligence 107(1), 29–62 (1999)

26. Mateas, M., Stern, A.: Structuring content in the façade interactive drama architecture. In: Proceedings of Artificial Intelligence and Interactive Digital Entertainment (2005)

27. Weiss, S., Müller, W., Spierling, U., Steimle, F.: Scenejo–an interactive storytelling platform. In: Subsol, G. (ed.) ICVS-VirtStory 2005. LNCS, vol. 3805, pp. 77–80. Springer, Heidelberg (2005)

28. Iurgel, I.: Cyranus–an authoring tool for interactive edutainment applications. In: Pan, Z., Aylett, R.S., Diener, H., Jin, X., Göbel, S., Li, L. (eds.) Edutainment 2006. LNCS, vol. 3942, pp. 577–580. Springer, Heidelberg (2006)

29. Medler, B., Magerko, B.: Scribe: A tool for authoring event driven interactive drama. In: Göbel, S., Malkewitz, R., Iurgel, I. (eds.) TIDSE 2006. LNCS, vol. 4326, pp. 139–150. Springer, Heidelberg (2006)

30. Riedl, M., Stern, A.: Believable Agents and Intelligent Story Adaptation for Interactive Storytelling. In: Göbel, S., Malkewitz, R., Iurgel, I. (eds.) TIDSE 2006. LNCS, vol. 4326, pp. 1–12. Springer, Heidelberg (2006)

31. Hayes-Roth, B., Brownston, L., van Gent, R.: Multi-agent collaboration in directed improvisation. In: First International Conference on Multi-Agent Systems, San Francisco (1995)

32. Cavazza, M., Charles, F., Mead, S.: Interacting with virtual characters in interactive storytelling. In: Proc. of the First Int. Joint Conf. on Autonomous Agents and Multiagent Systems (2002)

33. Anderson, E.: Value in Ethics and Economics. Harvard University Press, Cambridge (1993)

34. van Fraassen, B.: Values and the heart's command. Journal of Philosophy 70(1), 5–19 (1973)

35. Campbell, J.: The Hero with a Thousand Faces. Princeton University Press, Princeton (1949)

36. van Riemsdijk, M., Dastani, M., Winikoff, M.: Goals in Agent Systems: A Unifying Framework. In: Proceedings of AAMAS 2008 (2008)

37. Cohen, P.R., Levesque, H.J.: Intention is choice with commitment. Artificial Intelligence 42, 213–261 (1990)

Author Index